Impact Mechanics

Impact mechanics is concerned with the reaction forces that develop during a collision and the dynamic response of structures to these reaction forces. The subject has a wide range of engineering applications, from designing sports equipment to improving the crashworthiness of automobiles.

This book develops a range of different methodologies that are used to analyze collisions between various types of structures. These range from rigid body theory for structures that are stiff and compact, to vibration and wave analysis for flexible structures. The emphasis is on low-speed impact where any damage is local to the small region of contact between the colliding bodies. The analytical methods combine mechanics of contact between elastic–plastic or viscoplastic bodies with dynamics of structural response. These methods include representations of the source of contact forces – the forces that cause sudden changes in velocity – consequently the analytical methods are firmly based on physical interactions and less dependent on *ad hoc* assumptions than have been achieved hitherto.

Intended primarily as a text for advanced undergradute and graduate students, *Impact Mechanics* builds upon foundation courses in dynamics and strength of materials. It includes numerous industrially relevant examples and end-of-chapter homework problems drawn from industry and from sports such as golf, baseball, and billiards. Practicing engineers will also find the methods presented in this book very useful in calculating the response of mechanical systems to impact.

Bill Stronge is a recognized expert on impact mechanics and his research has had a major influence on current understanding of collisions that involve friction. He conducts research on impact response of plastically deforming solids aimed at applications in the design of light, crashworthy structures and energy-absorbing collision barriers. Dr. Stronge is the Reader of Applied Mechanics in the Department of Engineering at the University of Cambridge and a Fellow of Jesus College.

Impact Mechanics

W. J. STRONGE

University of Cambridge

CAMBRIDGE
UNIVERSITY PRESS

PUBLISHED BY THE PRESS SYNDICATE OF THE UNIVERSITY OF CAMBRIDGE
The Pitt Building, Trumpington Street, Cambridge, United Kingdom

CAMBRIDGE UNIVERSITY PRESS
The Edinburgh Building, Cambridge CB2 2RU, UK
40 West 20th Street, New York NY 10011–4211, USA
477 Williamstown Road, Port Melbourne, VIC 3207, Australia
Ruiz de Alarcón 13, 28014 Madrid, Spain
Dock House, The Waterfront, Cape Town 8001, South Africa

http://www.cambridge.org

First published 2000
First paperback edition 2004

Typeface Times Roman 10.25/12.5 pt.   *System* LaTeX2$_\varepsilon$ [TB]

A catalogue record for this book is available from the British Library

Library of Congress Cataloguing-in-Publication Data

Stronge, W. J. (William James), 1937–
  Impact mechanics / W.J. Stronge.
    p.   cm.
  ISBN 0 521 63286 2 hardback
  1. Impact.   I. Title.
 TA354 .S77   2000
  620.1′125 – dc21
                                99–044947

ISBN 0 521 63286 2 hardback
ISBN 0 521 60289 0 paperback

To Katerina and Jaime

Contents

List of Symbols

Man is not a circle with a single center; he is an ellipse with two foci. Facts are one, ideas are the other.

Victor Hugo, *Les Miserables*

a	radius of cylinder or sphere; radius of contact area
$\bar{a}$	$= a_c/a_Y$, nondimensional maximum contact radius
b	width, thickness
c	dashpot force coefficient
c_0	longitudinal wave speed, uniaxial stress (thin bars)
c_{cr}	critical dashpot force coefficient
c_g	$= d\omega/dk$, group velocity of propagating waves
c_p	$= \omega/k$, phase velocity of propagating waves
e, e_0, e_*	kinematic, kinetic, energetic coefficient of restitution
f, g, h	functions
h_O	moment of momentum about point O
g	$= 9.81$ m s^{-2}, gravitational constant
i	$= \sqrt{-1}$ imaginary unit; typical number in series
k	$= 2\pi/\lambda$, wave number
k_r	area radius of gyration for cross-section of bar about centroid
$\hat{k}_r$	mass radius of gyration of body B for center of mass
m	$= (M^{-1} + M'^{-1})^{-1}$, effective mass
m_{ij}	inertia matrix for contact point C
m	generalized inertia matrix ($r \times r$, where r = number of generalized coordinates)
n	number of particles in system
p	$= p_3$, normal component of reaction impulse at point of contact
p_c	normal impulse at transition from period of compression to restitution
p_f	normal impulse at termination of restitution period
p_s	normal impulse at termination of initial period of sliding
q	transverse force per unit length
q_r	generalized coordinate
r	radial coordinate
r_i, r_i'	position vectors from centers of mass G and G' to point of contact C
s	$= \sqrt{v_1^2 + v_2^2}$, sliding speed at any normal impulse p

$\hat{s}$	$= \mathrm{sgn}(v_1)$, direction of sliding (planar changes of velocity)
t	time
t_1	time of transition from initial stick to sliding
t_2	time of transition from sliding to stick
u_i	components of displacement
$\dot{u}_I$	particle velocity in wave incoming to interface
$\dot{u}_R$	particle velocity in wave reflected from interface
$\dot{u}_T$	particle velocity in wave transmitted through interface
v	$= v_3$, normal component of relative velocity of coincident contact points
v_0	normal component of relative velocity of contact points at incidence
v_1, v_2	tangential components of relative velocity of contact points
v_f	normal component of relative velocity at termination of restitution
x	axial coordinate
y	transverse coordinate, nondimensional indentation
z	depth coordinate
A	area of cross-section
A_i	constant
D_i	dissipation of energy from work of ith component of force
E, E'	Young's moduli of material in bodies B, B$'$
E_*	$= EE'/(E + E')$, effective Young's modulus at contact
F	$= F_3$, normal force at contact point
F_i	components of contact force
G	shear modulus of material
I	moment of inertia for cross-section of beam
I_{ij}, I'_{ij}	moments, products of inertia for bodies B, B$'$ about respective centers of mass
L	length
$\mathcal{M}$	bending moment at section of beam
M, M'	masses of rigid bodies B, B$'$, respectively
P, P'	normal components of impulse acting on bodies B, B$'$, respectively
R	radius of cylinder, sphere
R_*	$RR'/(R + R')$, effective radius of contact curvature
$\bar{R}_*$	effective radius of contact curvature after plastic deformation
S	shear force at section of beam
T	kinetic energy of system of colliding bodies
T_0	incident kinetic energy of system
T_f	final kinetic energy of system at termination of period of restitution
U	potential energy (e.g. gravitational potential)
V_i, V'_i	components of velocity at contact points C, C$'$
$\hat{V}_i, \hat{V}'_i$	components of velocity at centers of mass of bodies B, B$'$
W_n, W_3	work of normal component of reaction force at C
W_1, W_2	work of tangential components of reaction force at C
W_c	$= W_3(p_c)$, work of normal force during compression
W_f	$= W_3(p_f)$, final work of normal force
$\bar{W}_f$	$= W_1(p_f) + W_2(p_f) + W_3(p_f)$, final total work of contact reaction
X	nondimensional displacement
Y	yield stress

Z	$= dX/d(\omega t)$, nondimensional velocity
α	$= M/\rho AL$, mass ratio
$\beta_1, \beta_2, \beta_3$	inertia coefficients (planar changes in velocity)
γ	$= EI/\rho A$
γ	$= \beta_2^2/\beta_1\beta_2$, inertia parameter
γ_R	$= (\Gamma - 1)/(\Gamma + 1)$, reflection coefficient
γ_T	$= 2(A_1\Gamma/A_2)/(\Gamma + 1)$, transmission coefficient
γ_1	$= [p(t_f) - p(t_1)]/p_c$, ratio of impulse during final slip to p_c
γ_0	shear warping at neutral axis
$\bar{\gamma}$	$= \Xi\gamma_0$, shear rotation of cross-section
δ	relative indentation at contact point
ε_{ijk}	permutation tensor
ε_{ij}	components of strain
ζ	$= c/c_{\mathrm{cr}}$, damping ratio
η	local coordinate
η	$= x - c_0 t$, Galilean coordinate
η	square root of ratio of tangential to normal compliance
θ	$= dw/dx$, rotation of section; inclination of body
ϑ	ratio of kinetic energy of toppling group to that of leading element
ϑ_Y	ratio of mean fully plastic indentation pressure to uniaxial yield stress
κ	stiffness coefficient of spring element
λ	wavelength of propagating disturbance
μ_0	dashpot force coefficient
μ	Amontons–Coulomb coefficient of limiting friction (dry friction)
$\bar{\mu}$	coefficient for stick
ν	Poisson's ratio
ξ	local coordinate
ξ	$= x + ct$, Galilean coordinate
$\bar{\xi}$	$= 2d/a$, characteristic depth for plane strain deformation field
ρ	mass density
σ_{ij}	components of stress
τ	nondimensional time; characteristic time
φ	$= \omega_{i+1}(-)/\omega_{i+1}(+)$, ratio angular speeds before and after impact
ϕ	$= \tan^{-1}(v_2/v_1)$, sliding direction in tangent plane
$\hat{\phi}$	isoclinic direction of slip
$\hat{\phi}_*$	separatrix direction of slip
χ	stiffness ratio
ψ_0	angle of incidence for relative velocity at contact point
ψ_f	angle of rebound for relative velocity at contact point
ω	tangential resonant frequency
ω_0	initial angular velocity
ω_0	$= \kappa/m$, characteristic frequency of oscillation
ω_c	cutoff frequency for propagation
ω_d	damped resonant frequency
ω_i, ω_i'	angular velocity vectors for bodies B, B$'$
ω_i	angular speed for contact at point i
Γ	$= A_2\rho_2 c_2/A_1\rho_1 c_1$, impedance ratio

$\overline{\Gamma}$	$= \mu_0/\rho c_0 A$, dashpot force ratio
Ξ	Timoshenko beam coefficient
Σ_0	$= F_0/A$, negative pressure
Φ	matrix
Ψ	geometry of polygonal solid
Ω	normal resonant frequency

Vectors & Dyadics

$\mathbf{e}$	unit vector parallel to common tangent plane
$\mathbf{h}_O$	moment of momentum about point O
$\hat{\mathbf{h}}$	moment of momentum about center of mass
$\mathbf{n}$	$\equiv \mathbf{n}_3$ unit vector normal to common tangent plane
$\mathbf{r}_i$	position vector of ith particle relative to center of mass
$\hat{\mathbf{s}}$	$\equiv \mathbf{v}_e/|\mathbf{v}_e|$, direction of sliding (3D)
$\mathbf{v}$	$= \mathbf{V} - \mathbf{V}'$ relative velocity across contact point
$\mathbf{F}_i$	force on ith particle
$\hat{\mathbf{I}}$	moment of inertia for center of mass
$\mathbf{P}_i$	impulse on ith particle
$\mathbf{V}_i$	velocity of ith particle
$\hat{\mathbf{V}}$	velocity of center of mass
$\boldsymbol{\rho}_i$	position vector of ith particle relative to inertial reference frame
$\boldsymbol{\omega}$	angular velocity of rigid body

Preface

Caminante, no hay camino.
Se hace camino al ander.
"Traveller, there is no path
Paths are made by walking."

> A. Machado, popular song from Latin America

When bodies collide, they come together with some relative velocity at an initial point of contact. If it were not for the contact force that develops between them, the normal component of relative velocity would result in overlap or interference near the contact point and this interference would increase with time. This reaction force deforms the bodies into a compatible configuration in a common contact surface that envelopes the initial point of contact. Ordinarily it is quite difficult and laborious to calculate deformations that are geometrically compatible, that satisfy equations of motion and that give equal but opposite reaction forces on the colliding bodies. To avoid this detail, several different approximations have been developed for analyzing impact: rigid body impact theory, Hertz contact theory, elastic wave theory, etc. This book presents a spectrum of different theories for collision and describes where each is applicable. The question of applicability largely depends on the materials of which the bodies are composed (their hardness in the contact region and whether or not they are rate-dependent), the geometric configuration of the bodies and the incident relative velocity of the collision. These factors affect the relative magnitude of deformations in the contact region in comparison with global deformations.

A collision between hard bodies occurs in a very brief period of time. The duration of contact between a ball and bat, a hammer and nail or an automobile and lamppost is no more than a few milliseconds. This brief period has been used to justify *rigid body impact theory* in which bodies instantaneously change velocity when they collide. As a consequence of the instantaneous period of contact, the bodies have negligible displacement during the collision. For any analysis of changes in momentum occurring during impact, the approximation that displacements are negligibly small greatly simplifies the analysis. With this approximation, the changes in velocity can be calculated without integrating accelerations over the contact period. Along with this simplification, however, there is a hazard associated with loss of information about the contact forces that cause these changes in velocity – without forces the changes in velocity cannot be directly associated with deformability of the bodies.

   In order to solve more complex problems, particularly those involving friction, we develop a method that spreads out the changes in velocity by considering that they are a continuous function of impulse rather than time. With this approach, the approximation of negligible displacement during a very brief period of contact results in an equation of motion with constant coefficients; this equation is trivially integrable to obtain changes in momentum of each body as a function of the impulse of the contact force. This permits the analyst to follow the evolution of contact and variation in relative velocity across the contact patch as a function of impulse.

   For rigid body impact theory the equations of dynamics are not sufficient to solve for the changes in velocity – an additional relation is required. Commonly this relation is provided by the *coefficient of restitution*. Most books on mechanics treat the coefficient of restitution as an impact law; i.e., for the contact points of colliding bodies, they consider the coefficient of restitution to be an empirical relationship for the normal component of relative velocity at incidence and separation. This has been satisfactory for collisions between smooth bodies, but for bodies with rough surfaces where friction opposes sliding during impact, the usual kinematic (or Newton) coefficient of restitution has a serious deficiency. In the technical and scientific literature the topic of rigid body impact was re-opened in 1984 largely as a consequence of some problems where calculations employing the kinematic coefficient gave solutions which were patently unrealistic – for collisions in which friction opposes small initial slip, such calculations predicted an increase in kinetic energy as a consequence of the collision. In order to rectify this problem and clearly separate dissipation due to friction from that due to irreversible internal deformations near the contact point of colliding bodies, a different definition of the coefficient of restitution (termed the energetic coefficient of restitution) was proposed and is used throughout this book. In those problems where friction is negligible or where slip is unidirectional during contact, the energetic and kinematic coefficients are equal; if the direction of slip changes during contact, however, these coefficients are distinct.

   These methods will be illustrated by analyses of practical examples. Many of these are taken from sport; e.g. the bounce of a hockey puck, the spin (and consequent hook or slice) resulting from mis-hitting a golf shot, and batting for maximum range.

   While rigid body impact theory is effective for analyzing the response of hard bodies, more complex analytical descriptions are required if a colliding body is soft or deformable, i.e. if the collision generates significant structural deformations far from the contact region. This occurs if the impact occurs near a slender section of a colliding body or if the body is hollow, as in the case of an inflated ball. To calculate the response of deformable bodies a time-dependent analysis is required, since the contact force depends on local deformation of the body. In this case the response depends on the compliance of the contact regions in addition to the inertia properties and initial relative velocities that determine the outcome for rigid body impacts. The compensation for the additional complexity of time-dependent impact analysis for deformable bodies is that an empirically determined coefficient of restitution is no longer required to relate the final and initial states of the system. This relationship can be calculated for any given material and structural properties. For colliding bodies that are compact in shape and composed of hard materials, the contact stresses rapidly diffuse, so that substantial deformations occur only near the point of initial contact; in this case the change of state resulting from impact can be calculated on the basis of quasistatic continuum mechanics. On the other hand, for impacts that are transverse to some slender member, the collision generates vibratory motion far from the site of impact so that the calculation must be based on structural dynamics of beams, plates or shells. Examples are provided for impact between elastic–plastic solids and for collisions against slender elastic plates and beams.

This textbook evolved from lecture notes prepared for an upper division course presented originally at the University of California–Davis. A later version of the course was tested on students at the National University of Singapore. Those notes were expanded with additional material developed subsequently by myself, my students and my colleagues. I was aided in improving the presentation by helpful criticisms from Mont Hubbard, Chwee Teck Lim, Victor Shim and Jim Woodhouse. Our interest has been in developing more physically based analytical models in order to improve the accuracy of calculations of impact response and to increase the range of applicability for any measurement of collision properties of a system. In these respects this book is complementary to the neoclassical treatise *Impact, the Theory and Physical Behavior of Colliding Solids*, by W. Goldsmith – a monograph which provides a wealth of experimental data on collision behavior of metals, glass and natural materials. The present text has stepped off from this base to incorporate the physically based knowledge of mechanics of collision that has been developed in the last 40 years. In order to appreciate the analytical methods described here, the background required is an undergraduate engineering course including dynamics, strength of materials and vibrations.

CHAPTER 1

Introduction to Analysis of Low Speed Impact

Philosophy is written in this grand book – I mean the universe –
which stands continuously open to our gaze, but cannot be understood unless one first learns to comprehend the language in which
it is written. It is written in the language of mathematics and its
characters are triangles, circles and other geometric figures, without
which it is humanly impossible to understand a single word of it;
without these one is wandering about in a dark labyrinth.

Galileo Galilei, *Two New Sciences*, 1632

When a bat strikes a ball or a hammer hits a nail, the surfaces of two bodies come together with some relative velocity at an initial instant termed *incidence*. After incidence there would be interference or interpenetration of the bodies were it not for the interface pressure that arises in a small *area of contact* between the two bodies. At each instant during the contact period, the pressure in the contact area results in local deformation and consequent indentation; this indentation equals the interference that would exist if the bodies were not deformed.

At each instant during impact the interface or contact pressure has a resultant force of action or reaction that acts in opposite directions on the two colliding bodies and thereby resists interpenetration. Initially the force increases with increasing indentation and it reduces the speed at which the bodies are approaching each other. At some instant during impact the work done by the contact force is sufficient to bring the speed of approach of the two bodies to zero; subsequently, the energy stored during compression drives the two bodies apart until finally they separate with some relative velocity. For impact between solid bodies, the contact force that acts during collision is a result of the local deformations that are required for the surfaces of the two bodies to conform in the contact area.

The local deformations that arise during impact vary according to the incident relative velocity at the point of initial contact as well as the hardness of the colliding bodies. Low speed collisions result in contact pressures that cause small deformations only; these are significant solely in a small region adjacent to the contact area. At higher speeds there are large deformations (i.e. strains) near the contact area which result from plastic flow; these large localized deformations are easily recognizable, since they have gross manifestations such as cratering or penetration. In each case the deformations are consistent with the contact force that causes velocity changes in the colliding bodies. The normal impact speed required to cause large plastic deformation is between $10^2 \times V_Y$ and $10^3 \times V_Y$ where V_Y is the minimum relative speed required to initiate plastic yield in the softer body (for metals the normal incident speed at yield V_Y is of the order of 0.1 m s^{-1}). This text explains how the dynamics of low speed collisions are related to both local and global deformations in the colliding bodies.

1.1   Terminology of Two Body Impact

1.1.1   Configuration of Colliding Bodies

As two colliding bodies approach each other there is an instant of time, termed *incidence*, when a single *contact point* C on the surface of the first body B initially comes into contact with point C' on the surface of the second body B'. This time $t = 0$ is the initial instant of impact. Ordinarily the surface of at least one of the bodies has a continuous gradient at either C or C' (i.e., at least one body has a topologically smooth surface) so that there is a unique *common tangent plane* that passes through the coincident contact points C and C'. The orientation of this plane is defined by the direction of the normal vector **n**, a unit vector which is perpendicular to the common tangent plane.

Central or Collinear Impact Configuration:

If each colliding body has a center of mass G or G' that is on the common normal line passing through C, the impact configuration is *collinear*, or central. This requires that the position vector $\mathbf{r}_C$ from G to C, and the vector $\mathbf{r}'_C$ from G' to C', both be parallel to the common normal line as shown in Fig. 1.1a:

$$\mathbf{r}_C \times \mathbf{n} = \mathbf{r}'_C \times \mathbf{n} = \mathbf{0}.$$

Collinear impact configurations result in equations of motion for normal and tangential directions that can be decoupled. If the configuration is not collinear, the configuration is *eccentric*.

Eccentric Impact Configuration:

The impact configuration is eccentric if at least one body has a center of mass that is off the line of the common normal passing through C as shown in Fig. 1.1b. This occurs if either

$$\mathbf{r}_C \times \mathbf{n} \neq \mathbf{0} \quad \text{or} \quad \mathbf{r}'_C \times \mathbf{n} \neq \mathbf{0}.$$

If the configuration is eccentric and the bodies are rough (i.e., there is a tangential force

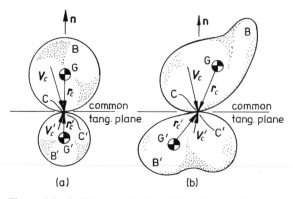

Figure 1.1. Colliding bodies B and B' with (a) collinear and (b) noncollinear impact configurations. In both cases the angle of incidence is oblique; i.e. $\phi_0 \neq 0$.

of friction that opposes sliding), the equations of motion each involve both normal and tangential forces (and impulses). Thus eccentric impact between rough bodies involves effects of friction and normal forces that are not separable.

1.1.2 Relative Velocity at Contact Point

At the instant when colliding bodies first interact, the coincident contact points C and C′ have an initial or *incident relative velocity* $\mathbf{v}_0 \equiv \mathbf{v}(0) = \mathbf{V}_C(0) - \mathbf{V}'_C(0)$. The initial relative velocity at C has a component $\mathbf{v}_0 \cdot \mathbf{n}$ normal to the tangent plane and a component $(\mathbf{n} \times \mathbf{v}_0) \times \mathbf{n}$ parallel to the tangent plane; the latter component is termed *sliding*. The *angle of obliquity at incidence*, ψ_0, is the angle between the initial relative velocity vector $\mathbf{v}_0$ and the unit vector $\mathbf{n}$ normal to the common tangent plane,

$$\psi_0 \equiv \tan^{-1}\left(\frac{(\mathbf{n} \times \mathbf{v}_0) \times \mathbf{n}}{\mathbf{v}_0 \cdot \mathbf{n}}\right).$$

Direct impact occurs when in each body the velocity field is uniform and parallel to the normal direction. Direct impact requires that the angle of obliquity at incidence equals zero ($\psi_0 = 0$); on the other hand, *oblique impact* occurs when the angle of obliquity at incidence is nonzero ($\psi_0 \neq 0$).

1.1.3 Interaction Force

An interaction force and the impulse that it generates can be resolved into components normal and parallel to the common tangent plane. For particle impact the impulse is considered to be normal to the contact surface and due to short range interatomic repulsion. For solid bodies, however, contact forces arise from local deformation of the colliding bodies; these forces and their associated deformations ensure compatibility of displacements in the contact area and thereby prevent interpenetration (overlap) of the bodies. In addition a tangential force, *friction*, can arise if the bodies are *rough* and there is sliding in the contact area. Dry friction is negligible if the bodies are *smooth*.

Conservative forces are functions solely of the relative displacement of the interacting bodies. In an *elastic collision* the forces associated with attraction or repulsion are conservative (i.e. reversible); it is not necessary however for friction (a nonconservative force) to be negligible. In an *inelastic collision* the interaction forces (other than friction) are nonconservative, so that there is a loss of kinetic energy as a result of the cycle of compression (loading) and restitution (unloading) that occurs in the contact region. The energy loss can be due to irreversible elastic–plastic material behavior, rate-dependent material behavior, elastic waves trapped in the separating bodies, etc.

1.2 Classification of Methods for Analyzing Impact

In order to classify collisions into specific types which require distinct methods of analysis, we need to think about the deformations that develop during collision, the distribution of these deformations in each of the colliding bodies, and *how these deformations affect the period of contact*. In general there are four types of analysis for low speed collisions, associated with particle impact, rigid body impact, transverse impact on flexible bodies (i.e. transverse wave propagation or vibrations) and axial impact

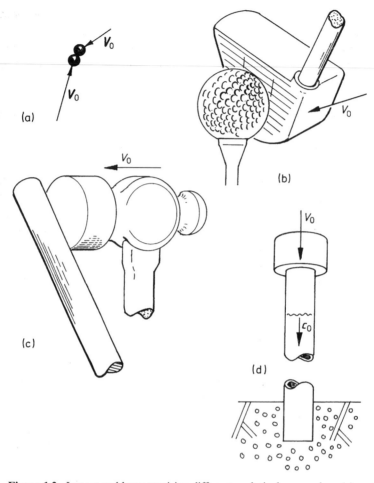

Figure 1.2. Impact problems requiring different analytical approaches: (a) particle impact (stereo-mechanical), (b) rigid body impact, (c) transverse deformations of flexible bodies and (d) axial deformation of flexible bodies.

on flexible bodies (i.e. longitudinal wave propagation). A typical example where each method applies is illustrated in Fig. 1.2.

(a) ***Particle impact*** is an analytical approximation that considers a normal component of interaction impulse only. By definition, particles are smooth and spherical. The source of the interaction force is unspecified, but presumably it is strong and the force has a very short range, so that the period of interaction is a negligibly small instant of time.

(b) ***Rigid body impact*** occurs between compact bodies where the contact area remains small in comparison with all section dimensions. Stresses generated in the contact area decrease rapidly with increasing radial distance from the contact region, so the internal energy of deformation is concentrated in a small region surrounding the interface. This small deforming region has large stiffness and acts much like a short but very stiff spring separating the colliding bodies at the

contact point. The period of contact depends on the normal compliance of the
contact region and an effective mass of the colliding bodies.

    (c) ***Transverse impact on flexible bodies*** occurs when at least one of the bodies
suffers bending as a result of the interface pressures in the contact area; bending
is significant at points far from the contact area if the depth of the body in the
direction normal to the common tangent plane is small in comparison with dimensions parallel to this plane. This bending reduces the interface pressure and
prolongs the period of contact. Bending is a source of energy dissipation during
collision in addition to the energy loss due to local deformation that arises from
the vicinity of contact.

    (d) ***Axial impact on flexible bodies*** generates longitudinal waves which affect the
dynamic analysis of the bodies only if there is a boundary at some distance from
the impact point which reflects the radiating wave back to the source; it reflects
the outgoing wave as a coherent stress pulse that travels back to its source essentially undiminished in amplitude. In this case the time of contact for an impact
depends on the transit time for a wave travelling between the impact surface and
the distal surface. This time can be less than that for rigid body impact between
hard bodies with convex surfaces.

1.2.1 Description of Rigid Body Impact

For bodies that are hard (i.e. with small compliance), only very small deformations are required to generate very large contact pressures; if the surfaces are initially
nonconforming, the small deformations imply that the contact area remains small throughout the contact period. The interface pressure in this small contact area causes the initially
nonconforming contact surfaces to deform until they conform or touch at most if not all
points in a small contact area. Although the contact area remains small in comparison
with cross-sectional dimensions of either body, the contact pressure is large, and it gives
a large stress resultant, or *contact force*. The contact force is large enough to rapidly
change the normal component of relative velocity across the small deforming region that
surrounds the contact patch. The large contact force rapidly accelerates the bodies and
thereby limits interference which would otherwise develop after incidence if the bodies
did not deform.

Hence in a small region surrounding the contact area the colliding bodies are subjected
to large stresses and corresponding strains that can exceed the yield strain of the material. At quite modest impact velocities (of the order of 0.1 m s^{-1} for structural metals)
irreversible plastic deformation begins to dissipate some energy during the collision; consequently there is some loss of kinetic energy of relative motion in all but the most benign
collisions. Although the stresses are large in the contact region, they decay rapidly with
increasing distance from the contact surface. In an elastic body with a spherical coordinate
system centered at the initial contact point, the radial component of stress, σ_r, decreases
very rapidly with increasing radial distance r from the contact region (in an elastic solid σ_r
decreases as r^{-2} in a 3D deformation field). For a hard body the corresponding rapid decrease in strain means that significant deformations occur only in a small region around
the point of initial contact; consequently the deflection or indentation of the contact area
remains very small.

Since the region of significant strain is not very deep or extensive, hard bodies have very small compliance (i.e., a large force generates only a small deflection). The small region of significant deformation is like a short stiff spring which is compressed between the two bodies during the period of contact. This spring has a large spring constant and gives a very brief period of contact. For example, a hard-thrown baseball or cricket ball striking a bat is in contact for a period of roughly 2 ms, while a steel hammer striking a nail is in contact for a period of about 0.2 ms. The contact duration for the hammer and nail is less because these colliding bodies are composed of harder materials than the ball and bat. Both collisions generate a maximum force on the order of 10 kN (i.e. roughly one ton).

From an analytical point of view, the most important consequence of the small compliance of hard bodies is that very little movement occurs during the very brief period of contact; i.e., despite large contact forces, there is insufficient time for the bodies to displace significantly during impact. This observation forms a fundamental hypothesis of *rigid body impact theory*, namely, that for hard bodies, analyses of impact can consider the period of contact to be vanishingly small. Consequently any changes in velocity occur instantaneously (i.e. in the initial or incident configuration). The system configuration at incidence is termed the *impact configuration*. This theory assumes there is no movement during the contact period.

Underlying Premises of Rigid Body Impact Theory
   (a) In each of the colliding bodies the contact area remains small in comparison with both the cross-sectional dimensions and the depth of the body in the normal direction.
   (b) The contact period is sufficiently brief that during contact the displacements are negligible and hence there are no changes in the system configuration; i.e., the contact period can be considered to be instantaneous.

If these conditions are approximately satisfied, rigid body impact theory can be applicable. In general this requires that the bodies are hard and that they suffer only small local deformation in collision. For a solid composed of material that is rate-independent, a small contact area results in significant strains only in a small region around the initial contact point. If the body is hard, the very limited region of significant deformations causes the compliance to be small and consequently the contact period to be very brief. This results in two major simplifications:

   (a) Equations of planar motion are trivially integrable to obtain algebraic relations between velocity changes and the reaction impulse.[1]
   (b) Finite active forces (e.g. gravitational or magnetic attraction) which act during the period of contact can be considered to be negligible, since these forces do no work during the collision.

During the contact period the only significant active forces are reactions at points of contact with other bodies; these reactions are induced by displacement constraints.

Figure 1.3 shows a collision where application of rigid body impact theory is appropriate. This series of high speed photographs shows development of a small area of contact when an initially stationary field hockey ball is struck by a hockey stick at an incident

---

[1] Because velocity changes can be obtained from algebraic relations, rigid body impact was one of the most important topics in dynamics before the development of calculus in the late seventeenth century.

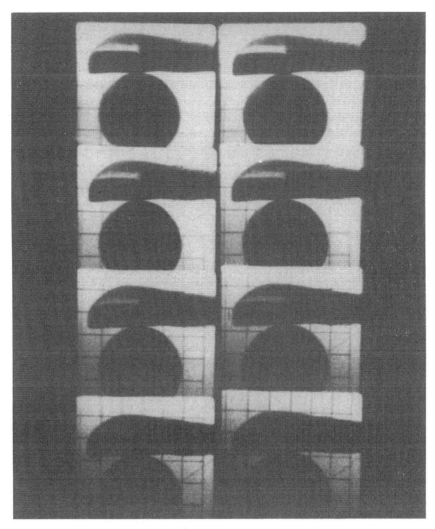

Figure 1.3. High speed photographs of hockey stick striking at 18 m s^{-1} (40 mph) against a stationary field hockey ball (dia. $D' = 74$ mm, mass $M' = 130$ g). Interframe period $\tau = 0.0002$ s, contact duration $t_f \approx 0.0015$ s, and maximum normal force $F_c \approx 3900$ N.

speed of 18 m s^{-1}. During collision the contact area increases to a maximum radius a_c that remains small in comparison with the ball radius R'; in Fig. 1.3, $a_c/R' < 0.3$. The relatively small contact area is a consequence of the small normal compliance (or large elastic modulus) of both colliding bodies and the initial lack of conformation of the surfaces near the point of first contact.

A useful means of postulating rigid body impact theory is to suppose that two colliding bodies are separated by an infinitesimal deformable particle.[2] The deformable particle is

---

[2] The physical construct of a deformable particle separating contact points on colliding rigid bodies is mathematically equivalent to Keller's (1986) asymptotic method of integrating with respect to time the equations for relative acceleration of deformable bodies and then taking the limit as compliance (or contact period) becomes vanishingly small.

located between the point of initial contact on one body and that on the other, although these points are coincident. The physical construct of an infinitesimal compliant element separating two bodies at a point of contact allows variations in velocity during impact to be resolved as a function of the normal component of impulse. This normal component of impulse is equivalent to the integral of the normal contact force over the period of time after incidence. Since collisions between bodies with nonadhesive contact surfaces involve only compression of the deformable particle – never extension – the normal component of impulse is a monotonously increasing function of time after incidence. Thus variations in velocity during an instantaneous collision are resolved by choosing as an independent variable the normal component of impulse rather than time. This gives velocity changes which are a continuous (smooth) function of impulse.

There are three notable classes of impact problems where rigid body impact theory is not applicable if the impact parameters representing energy dissipation are to have any range of applicability.

(a) The first involves impulsive couples applied at the contact point. Since the contact area between rigid bodies is negligibly small, impulsive couples are inconsistent with rigid body impact theory. To relate a couple acting during impulse to physical processes, one must consider the distribution of deformation in the contact region. Then the couple due to a distribution of tangential force can be obtained from the law of friction and the first moment of tractions in a finite contact area about the common normal through the contact point.

(b) A second class of problems where rigid body impact theory does not apply is axial impact of collinear rods with plane ends. These are problems of one dimensional wave propagation where the contact area and cross-sectional area are equal because the contacting surfaces are conforming; in this case the contact area may not be small. For problems of wave propagation deformations and particle velocities far from the contact region are not insignificant. As a consequence, for one dimensional waves in long bars, the contact period is dependent on material properties and depth of the bars in a direction normal to the contact plane rather than on the compliance of local deformation near a point of initial contact.

(c) The third class of problems where rigid body theory is insufficient are transverse impacts on beams or plates where vibration energy is significant.

Collisions with Compliant Contact Region Between Otherwise Rigid Bodies
While most of our attention will be directed towards rigid body impact, there are cases where distribution of stress is significant in the region surrounding the contact area. These problems require consideration of details of local deformation of the colliding bodies near the point of initial contact; they are analyzed in Chapters 6 and 8. The most important example may be collisions against multibody systems where the contact points between bodies transmit the action from one body to the next; in general, this case requires consideration of the compliance at each contact. Considerations of local compliance may be represented by discrete elements such as springs and dashpots or they can be obtained from continuum theory.

For collisions between systems of hard bodies, it is necessary to consider *local displacement* in each contact region although *global displacements* are negligibly small; i.e. different scales of displacement are significant for different analytical purposes. The relatively small displacements that generate large contact forces are required to analyze interactions between spatially discrete points of contact. If the bodies are hard however, these same displacements may be sufficiently small so that they have negligible effect on the inertia properties; i.e. during collision any changes in the inertia properties are insignificant despite the small local deformations.

1.2.2   Description of Transverse Impact on Flexible Bodies

Transverse impact on plates, shells or slender bars results in significant flexural deformations of the colliding members both during and following the contact period. In these cases the stiffness of the contact region depends on flexural rigidity of the bodies in addition to continuum properties of the region immediately adjacent to the contact area; i.e., it is no longer sufficient to suppose that a small deforming region is surrounded by a rigid body. Rather, flexural rigidity is usually the more important factor for contact stiffness when impact occurs on a surface of a plate or shell structural component.

1.2.3   Description of Axial Impact on Flexible Bodies

Elastic or elastic–plastic waves radiating from the impact site are present in every impact between deformable bodies – in a deformable body it is these radiating waves that transmit variations in velocity and stress from the contact region to the remainder of the body. Waves are an important consideration for obtaining a description of the dynamic response of the bodies, however, only if the period of collision is determined by wave effects. This is the case for axial impact acting uniformly over one end of a slender bar if the far, or *distal*, end of the bar imposes a reflective boundary condition. Similarly, for radial impact at the tip of a cone, elastic waves are important if the cone is truncated by a spherical surface with a center of curvature at the apex. In these cases where the impact point is also a focal point for some reflective distal surface, the wave radiating from the impact point is reflected from the distal surface and then travels back to the source, where it affects the contact pressure. On the other hand, if different parts of the outgoing stress wave encounter boundaries at various times and the surfaces are not normal to the direction of propagation, the wave will be reflected in directions that are not towards the impact point; while the outgoing wave changes the momentum of the body, this wave is diffused rather than returning to the source as a coherent wave that can change the contact pressure and thereby affect the contact duration.

1.2.4   Applicability of Theories for Low Speed Impact

This text presents several different methods for analyzing changes in velocity (and contact forces) resulting from low speed impact, i.e. impact slow enough that the bodies are deformed imperceptibly only. These theories are listed in Table 1.1 with descriptions of the differences and an indication of the range of applicability for each.

The stereomechanical theory is a relationship between incident and final conditions; it results in discontinuous changes in velocity at impact. In this book a more sophisticated *rigid body* theory is developed – a theory in which the changes in velocity are a continuous function of the normal component of the impulse p at the contact point. This theory

Table 1.1. *Applicability of Theories for Oblique, Low Speed Impact*

Impact Theory	Independent Variable	Coeff. of Restitution[a]	Angle of Incidence at Impact Point,[b] ψ_0	Spatial Gradient of Contact Compliance,[c] χ^{-1}	(Impact Point Compliance)/ (Structural Compliance)[d]	Computational Effort	Illustration
Stereomechanical[e]	None	e, e_0	$> \tan^{-1}(\mu\beta_1/\beta_3)$	>1	$\gg 1$	Low	
Rigid body[f]	Impulse p	e_*	$> \tan^{-1}(\mu\beta_1/\beta_3)$[g]	>1 (sequential) $\ll 1$(simultaneous)	$\gg 1$	Low	
Compliant contact[f]	Time t	e_*	All	All	$\gg 1$	Moderate	
Continuum[f]	Time t	None	All	All	All	High	

[a] $e, e_0, e_* =$ kinematic, kinetic, energetic coefficients of restitution.
[b] $\mu =$ Amontons–Coulomb coefficient of limiting friction; $\beta_1, \beta_3 =$ inertia coefficients.
[c] Distributed points of contact.
[d] Flexible bodies.
[e] Nonsmooth dynamics.
[f] Smooth dynamics.
[g] Or negligible tangential compliance.

results from considering that the coincident points of contact on two colliding bodies are separated by an infinitesimal deformable particle – a particle that represents local deformation around the small area of contact. With this artifice, the analysis can follow the process of slip and/or slip–stick between coincident contact points if the contact region has negligible tangential compliance. Rigid body theories are useful for analyzing two body impact between compact bodies composed of stiff materials; however, they have limited applicability for multibody impact problems.

When applied to multibody problems, rigid body theories can give accurate results only if the normal compliance of the point of external impact is very small or large in comparison with the compliance of any connections with adjacent bodies. If compliance of the point of external impact is much smaller than that of all connections to adjacent bodies, at the connections the maximum reaction force occurs well after the termination of contact at the external impact point, so that the reactions essentially occur sequentially. Small impact compliance results in a wave of reaction that travels away from the point of external impact at a speed that depends on the inertia of the system and the local compliance at each connecting joint or contact point. On the other hand, if the normal compliance of the point of external impact is very large in comparison with compliance of any connections to adjacent bodies, the reactions at the connections occur simultaneously with the external impact force. Only in these limiting cases can the dynamic interaction between connected bodies be accurately represented with an assumption of either sequential or simultaneous reactions. Generally the reaction forces at points of contact arise from infinitesimal relative displacements that develop during impact; these reaction forces are coupled, since sometimes they overlap.

If however other points of contact or cross-sections of the body have compliance of the same order of magnitude as that at any point of external impact, then the effect of these flexibilities must be incorporated into the dynamic model of the system. If the compliant elements are local to joints or other small regions of the system, an analytical model with local compliance may be satisfactory; e.g. see Chapter 8. On the other hand, if the body is slender, so that significant structural deformations develop during impact, either a wave propagation or a structural vibration analysis may be required; see Chapter 7 or 9. Whether the distributed compliance is local to joints or continuously distributed throughout a flexible structure, these theories require a time-dependent analysis to obtain reaction forces that develop during contact and the changes in velocity caused by these forces.

Hence the selection of an appropriate theory depends on structural details and the degree of refinement required to obtain the desired information.

1.3   Principles of Dynamics

1.3.1   Particle Kinetics

The fundamental form of most principles of dynamics is in terms of the dynamics of a particle. A *particle* is a body of negligible or infinitesimal size, i.e. a point mass. The particle is the building block that will be used to develop the dynamics of impact for either rigid or deformable solids. A particle of mass M moving with velocity $\mathbf{V}$ has *momentum* $M\mathbf{V}$. If a resultant force $\mathbf{F}$ acts on the particle, this causes a change in momentum in accord with Newton's second law of motion.

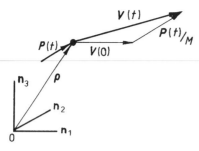

Figure 1.4. Change in velocity of particle with mass M resulting from impulse $\mathbf{P}(t)$.

Law II: The momentum $M\mathbf{V}$ of a particle has a rate of change with respect to time that is proportional to and in the direction of any resultant force $\mathbf{F}(t)$ acting on the particle[3]:

$$d(M\mathbf{V})/dt = \mathbf{F} \tag{1.1}$$

Usually the particle mass is constant, so that Eq. (1.1) can be integrated to obtain the changes in velocity as a continuous function of the *impulse* $\mathbf{P}(t)$:

$$\mathbf{V}(t) - \mathbf{V}(0) = M^{-1}\int_0^t \mathbf{F}(t')\,dt' \equiv M^{-1}\mathbf{P}(t) \tag{1.2}$$

This vector expression is illustrated in Fig. 1.4.

The interaction of two particles B and B′ that collide at time $t = 0$ generates active forces $\mathbf{F}(t)$ and $\mathbf{F}'(t)$ that act on each particle respectively, during the period of interaction, $0 < t < t_f$ – these forces of interaction act to prevent interpenetration. The particular nature of interaction forces depends on their source: whether they are due to contact forces between solid bodies that cannot interpenetrate, or are interatomic forces acting between atomic particles. In any case the force on each particle acts solely in the radial direction. These interaction forces are related by Newton's third law of motion.

Law III: Two interacting bodies have forces of action and reaction that are equal in magnitude, opposite in direction and collinear:

$$\mathbf{F}' = -\mathbf{F} \tag{1.3}$$

Laws II and III are the basis for impulse–momentum methods of analyzing impact. Let particle B have mass M, and particle B′ have mass M'. Integration of (1.3) gives equal but opposite impulses $-\mathbf{P}'(t) = \mathbf{P}(t)$, so that equations of motion for the *relative velocity* $\mathbf{v} \equiv \mathbf{V} - \mathbf{V}'$ can be obtained as

$$\mathbf{v}(t) = \mathbf{v}(0) + m^{-1}\mathbf{P}(t), \qquad m^{-1} = M^{-1} + M'^{-1} \tag{1.4}$$

---

[3] Newton's second law is valid only in an inertial reference frame or a frame translating at constant speed relative to an inertial reference frame. In practice a reference frame is usually considered to be fixed relative to a body, such as the earth, which may be moving. Whether or not such a reference frame can be considered to be inertial depends on the magnitude of the acceleration being calculated in comparison with the acceleration of the reference body, i.e. whether or not the acceleration of the reference frame is negligible.

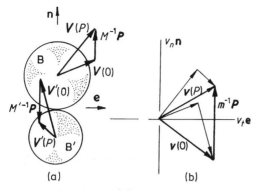

Figure 1.5. (a) Equal but opposite normal impulses **P** on a pair of colliding bodies with masses M and M' result in velocity changes $M^{-1}\mathbf{P}$ and $-M'^{-1}\mathbf{P}$ respectively. (b) The light lines are the initial and the final velocity for each body, while the heavy lines are the initial relative velocity $\mathbf{v}(0)$, the final relative velocity $\mathbf{v}(P)$ and the change $m^{-1}\mathbf{P}$ in relative velocity.

where m is the *effective mass*. The change of variables from velocity $\mathbf{V}(t)$ in an inertial reference frame to relative velocity $\mathbf{v}(t)$ is illustrated in Fig. 1.5. Equation (1.4) is an equation of relative motion that is applicable in the limit as the period of contact approaches zero $(t_f \to 0)$; this equation is the basis of smooth dynamics of collision for particles and rigid bodies.

Example 1.1   A golf ball has mass $M = 61$ g. When hit by a heavy club the ball acquires a speed of 44.6 m s^{-1} (100 mph) during a contact duration $t_f = 0.4$ ms. Assume that the force–deflection relation is linear, and calculate an estimate of the maximum force F_{max} acting on the ball.

Solution

Effective mass $m = 0.061$ kg.
Initial relative velocity $v(0) = v_0 = -44.6$ m s^{-1}.

(a) Linear spring $\Rightarrow$ simple harmonic motion for relative displacement δ at frequency ω where $\omega t_f = \pi$.
(b) Change in momentum of relative motion $=$ impulse, Eq. (1.4):

$$m v_0 = \int_0^{t_f} F(t)\,dt = \int_0^{t_f} F_{\text{max}} \sin(\omega t)\,dt$$

$$\Rightarrow \quad F_{\text{max}} = 21.4 \text{ kN} (\approx 2 \text{ tons})$$

1.3.2   Kinetics for a Set of Particles

For a set of n particles where the ith particle has mass M_i and velocity $\mathbf{V}_i$ the equations of translational motion can be expressed as

$$\frac{d}{dt}\sum_{i=1}^{n} M_i \mathbf{V}_i = \sum_{i=1}^{n} \mathbf{F}_i + \sum_{i=1}^{n}\sum_{k=1}^{n} \mathbf{F}'_{ik}, \qquad k \neq i$$

where $\mathbf{F}_i$ is an external force acting on particle i and $\mathbf{F}'_{ik}$ is an internal interaction force of particle k on particle i. Since the internal forces are equal but opposite ($\mathbf{F}'_{ik} = -\mathbf{F}'_{ki}$), the sum of these forces over all particles vanishes; hence

$$\frac{d}{dt} \sum_{i=1}^{n} M_i \mathbf{V}_i = \sum_{i=1}^{n} \mathbf{F}_i. \tag{1.5}$$

The *moment of momentum* $\mathbf{h}_O$ of particle i about point O is defined as $\mathbf{h}_O \equiv \rho_i \times M_i \mathbf{V}_i$, where ρ_i is the position vector of the particle from O and M_i is the mass of the particle. Thus the set of particles has a moment of momentum about O,

$$\mathbf{h}_O = \sum_{i=1}^{n} \rho_i \times M_i \mathbf{V}_i.$$

For a set of n particles the rate of change of moment of momentum about O is related to the moment about O of the external forces acting on the system:

$$\frac{d\mathbf{h}_O}{dt} = \frac{d}{dt} \sum_{i=1}^{n} \rho_i \times M_i \mathbf{V}_i = \sum_{i=1}^{n} \rho_i \times \mathbf{F}_i. \tag{1.6}$$

If the configuration of the system does not change during the period of time t, integration of (1.6) with respect to time gives

$$\mathbf{h}_O(t) - \mathbf{h}_O(0) = \sum_{i=1}^{n} \rho_i \times \int_0^t \mathbf{F}_i(t')\,dt' = \sum_{i=1}^{n} \rho_i \times \mathbf{P}_i(t). \tag{1.7}$$

1.3.3   Kinetic Equations for a Rigid Body

A rigid body can be represented as a set of particles separated by fixed distances. When the body is moving, the only relative velocity between different points on the body is due to *angular velocity* $\boldsymbol{\omega}$ of the body and the distance between the points.

Translational Momentum and Moment of Momentum

Suppose a body of mass M is composed of n particles with individual masses M_i, $i = 1, \ldots, n$ where the position vector of each particle ρ_i can be expressed as a set of coordinates in an inertial reference frame with origin at O. The location $\hat{\rho}$ of the center of mass of this set of particles is given by

$$\hat{\rho} = M^{-1} \sum_{i=1}^{n} M_i \rho_i, \qquad \text{where} \quad M = \sum_{i=1}^{n} M_i$$

as illustrated in Fig. 1.6. The center of mass has velocity $\hat{\mathbf{V}} \equiv d\hat{\rho}/dt = M^{-1} \sum_{i=1}^{n} M_i \mathbf{V}_i$. Hence for the set of particles that constitute this body, the translational equation of motion (1.5) can be expressed as

$$\frac{d}{dt}(M\hat{\mathbf{V}}) = \sum_{i=1}^{n} \mathbf{F}_i. \tag{1.8}$$

For a rigid body with mass M, this integrated form of Newton's law states that the temporal rate of change of translational momentum $M\hat{\mathbf{V}}$ is equal to the resultant of external forces acting on the body, $\sum_{i=1}^{n} \mathbf{F}_i$.

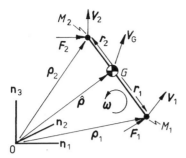

Figure 1.6. Elementary rigid body consisting of particles B_1 and B_2 that are linked by a light rigid bar. The particles have masses M_1 and M_2 respectively; the particles are subject to forces F_1 and F_2.

The moment of momentum about O for the body can be expressed in terms of the moment of the translational momentum of the body $M\hat{\mathbf{V}}$ plus the couple from the rotational momentum of the body. First note that the position ρ_i of a particle can be decomposed into the position vector $\hat{\rho}$ of the center of mass plus the position vector $\mathbf{r}_i$ of the ith particle relative to the center of mass; i.e., $\rho_i = \hat{\rho} + \mathbf{r}_i$. Thus the moment of momentum about O becomes

$$\mathbf{h}_O = \hat{\mathbf{h}} + \hat{\rho} \times M\hat{\mathbf{V}}$$

where $\hat{\mathbf{h}}$ is the moment of momentum for the system about the center of mass G,

$$\hat{\mathbf{h}} = \sum_{i=1}^{n} \mathbf{r}_i \times M_i \mathbf{V}_i = \sum_{i=1}^{n} M_i \mathbf{r}_i \times (\omega \times \mathbf{r}_i) \equiv \hat{\mathbf{I}} \cdot \omega$$

and $\hat{\mathbf{I}}$ is the *inertia dyadic* for the center of mass G.[4]

Thus the equation of motion (1.6) can be expressed as

$$\frac{d\mathbf{h}_O}{dt} = \frac{d}{dt}(\hat{\rho} \times M\hat{\mathbf{V}}) + \frac{d}{dt}(\hat{\mathbf{I}} \cdot \omega) = \sum_{i=1}^{n}(\hat{\rho} + \mathbf{r}_i) \times \mathbf{F}_i. \tag{1.9}$$

This decomposition of the particle position vector $\rho_i = \hat{\rho} + \mathbf{r}_i$ has separated the equation for the rate of change of moment of momentum about O into a term for the moment of translational momentum of the body acting at the center of mass G and a second term for the moment of momentum relative to G.

Noting that the differential of an applied impulse $d\mathbf{P}_i = \mathbf{F}_i\, dt$ we obtain from (1.8) and (1.9) that for a rigid body there are three independent equations of motion in terms of the applied impulse $d\mathbf{P} = \sum_{i=1}^{n} d\mathbf{P}_i$,

$$d(M\hat{\mathbf{V}}) = d\mathbf{P} \tag{1.10a}$$

$$d(\hat{\rho} \times M\hat{\mathbf{V}}) = \hat{\rho} \times d\mathbf{P} \tag{1.10b}$$

$$d(\hat{\mathbf{I}} \cdot \omega) = \sum_{i=1}^{n} \mathbf{r}_i \times d\mathbf{P}_i \tag{1.10c}$$

---

[4] For body-fixed Cartesian coordinates aligned with mutually perpendicular unit vectors $\mathbf{n}_j$, $j = 1, 2, 3$, and origin at G, the inertia dyadic is expressed as $\hat{\mathbf{I}} \equiv \hat{I}_{jk}\mathbf{n}_j\mathbf{n}_k$, where subscripts denote Cartesian coordinate directions. This rotational inertia of the body has coefficients obtained by integrating over the mass of the body, $\hat{I}_{jj} = \int(\mathbf{r} \cdot \mathbf{r} - r_j r_j)\,dM$ and $\hat{I}_{jk} = -\int r_j r_k\,dM$, $j \neq k$, where r_j is the jth component of the position vector $\mathbf{r}$ in a Cartesian coordinate system and no summation is implied by repeated subscripts.

Equation (1.10b) is the differential of the moment about O of translational momentum of the body.

Kinetic Energy

For some problems it is preferable to use a scalar measure of activity of a body; e.g. the kinetic energy T rather than the vectorial representations (1.10). Consider a body composed of n particles so that the kinetic energy T is expressed as

$$T = \frac{1}{2} \sum_{i=1}^{n} M_i \mathbf{V}_i \cdot \mathbf{V}_i$$

The kinetic energy of a rigid body T can be resolved into translational kinetic energy T_v and rotational kinetic energy T_ω

$$T_v \equiv \tfrac{1}{2} M \hat{\mathbf{V}} \cdot \hat{\mathbf{V}} = \tfrac{1}{2} M (\hat{V}_1 \cdot \hat{V}_1 + \hat{V}_2 \cdot \hat{V}_2 + \hat{V}_3 \cdot \hat{V}_3) \tag{1.11a}$$

$$\begin{aligned} T_\omega &\equiv \tfrac{1}{2} \omega \cdot \hat{\mathbf{I}} \cdot \omega \\ &= \tfrac{1}{2} \big(\omega_1^2 \hat{I}_{11} + \omega_2^2 \hat{I}_{22} + \omega_3^2 \hat{I}_{33} + 2\omega_1\omega_2 \hat{I}_{12} + 2\omega_2\omega_3 \hat{I}_{23} + 2\omega_3\omega_1 \hat{I}_{31} \big) \end{aligned} \tag{1.11b}$$

where $\hat{\mathbf{V}}$ is the velocity of the center of mass, ω is the angular velocity of the body and $\hat{\mathbf{I}}$ is the inertia dyadic for the center of mass.

Taking the scalar product of (1.10a) with $\hat{\mathbf{V}}$ and (1.10c) with ω we obtain the equation of motion,

$$dT_v = \hat{\mathbf{V}} \cdot d\mathbf{P}, \qquad dT_\omega = \omega \cdot \sum_{i=1}^{n} \mathbf{r}_i \times d\mathbf{P}_i. \tag{1.12}$$

These equations of motion have a right-hand side that is the differential of the rate-of-work of applied impulses and the differential of the rate-of-work of applied torques about the center of mass, respectively. Expressions (1.12) are scalar equations for the change of state of a rigid body subject to a number n of active impulses $\mathbf{P}_i$.

Example 1.2 A cube resting on a flat level plane has mass M and sides of length $2a$. The cube has slightly convex sides so only the edges touch the plane. One edge of the cube is raised slightly and then released so that when the opposite edge C strikes the surface, the cube is rotating with angular velocity ω_-. The impact is perfectly plastic so there is no bounce of edge C. Find the angular velocity ω_+ immediately after impact at C and calculate the part of the initial kinetic energy that is lost at impact, $(T_- - T_+)/T_-$.

Solution

Moment of momentum about C:

$$\mathbf{h}_C(t) = \hat{I}_{33}\omega(t) + \mathbf{r}_{G/C} \times M\hat{\mathbf{V}}(t).$$

Polar moment of inertia for G:

$$\hat{I}_{33} = 2Ma^2/3$$

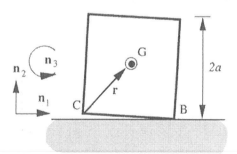

Velocity of center of mass:

$$\hat{\mathbf{V}}_- = \mathbf{V}_{B-} + \omega_- \times \mathbf{r}_{G/B} = a\omega_-(-\mathbf{n}_1 - \mathbf{n}_2)$$
$$\hat{\mathbf{V}}_+ = \mathbf{V}_{C+} + \omega_+ \times \mathbf{r}_{G/C} = a\omega_+(-\mathbf{n}_1 + \mathbf{n}_2).$$

Moment of momentum about impact point C:

$$\mathbf{h}_{C-} = \frac{2Ma^2}{3}\omega_- \mathbf{n}_3 + a(\mathbf{n}_1 + \mathbf{n}_2) \times Ma\omega_-(-\mathbf{n}_1 - \mathbf{n}_2) = \frac{2Ma^2\omega_-}{3}\mathbf{n}_3$$

$$\mathbf{h}_{C+} = \frac{2Ma^2}{3}\omega_+ \mathbf{n}_3 + a(\mathbf{n}_1 + \mathbf{n}_2) \times Ma\omega_+(-\mathbf{n}_1 + \mathbf{n}_2) = \frac{8Ma^2\omega_+}{3}\mathbf{n}_3.$$

Impulsive moment about impact point C equals zero during impact [Eq. (1.6)]:

$$\mathbf{h}_{C+} = \mathbf{h}_{C-} \quad \Rightarrow \quad \omega_+/\omega_- = 1/4.$$

Kinetic energy (planar motion):

$$T = \tfrac{1}{2}M\hat{\mathbf{V}} \cdot \hat{\mathbf{V}} + \tfrac{1}{2}\hat{I}_{33}\boldsymbol{\omega} \cdot \boldsymbol{\omega}$$

$$T_- = Ma^2\omega_-^2 + \tfrac{1}{3}Ma^2\omega_-^2, \qquad T_+ = Ma^2\omega_+^2 + \tfrac{1}{3}Ma^2\omega_+^2.$$

Part of initial kinetic energy absorbed in impact:

$$\frac{T_- - T_+}{T_-} = \frac{15}{16}.$$

Equation (1.6) relates the moment of forces about the origin to the rate of change of moment of momentum about the origin. Is there a similar relation for moment of forces about a point that is moving?

1.3.4 Rate of Change for Moment of Momentum of a System about a Point Moving Steadily Relative to an Inertial Reference Frame

Let point A be coincident with point O but moving steadily at velocity $\mathbf{V}_A$ relative to an inertial reference frame, and denote the position vector of the ith particle relative to A by ρ_i. If $\mathbf{h}_A$ denotes the moment of momentum of the system of n particles with respect to A, then Eq. (1.6) gives the following theorem.

For a system of particles, the rate of change of moment of momentum with respect to point A is equal to the moment of external forces about A if and only if

(i) $\mathbf{V}_A = \mathbf{0}$, so that point A is fixed in an inertial reference frame; or
(ii) $\mathbf{V}_A$ and $\hat{\mathbf{V}}$ are parallel.

In either case,

$$\frac{d\mathbf{h}_A}{dt} = \sum_{i=1}^{n} \rho_i \times \mathbf{F}_i. \qquad (1.13)$$

During a period t in which the configuration of the system does not vary, Eq. (1.13) can be integrated with respect to time to give

$$\mathbf{h}_A(t) - \mathbf{h}_A(0) = \sum_{i=1}^{n} \rho_i \times \mathbf{P}_i(t). \qquad (1.14)$$

Example 1.3 A system consists of two particles A and B that are connected by a light inextensible string. Particle A with mass $2M$ is located at $\rho_A = (1, 1, 0)$ and has velocity $\mathbf{V}_A = (1, 0, 0)$ in a Cartesian reference frame, while particle B with mass M is located at $\rho_B = (3, 2, 0)$ and has velocity $\mathbf{V}_B = (\frac{1}{2}, \dot{y}, 0)$. Find (i) the component of velocity $\dot{y}$ and the out-of-plane component ω of the angular velocity of the string; and (ii) the moment of momentum of the system about the origin O.

Solution
Position of B relative to A,

$$\rho_{B/A} \equiv \rho_B - \rho_A = (2, 1, 0).$$

Velocity of B relative to A,

$$\mathbf{V}_{B/A} \equiv \mathbf{V}_B - \mathbf{V}_A = \left(-\tfrac{1}{2}, \dot{y}, 0\right).$$

Inextensible string requires

$$0 = \mathbf{V}_{B/A} \cdot \rho_{B/A} = -1 + \dot{y}.$$

Thus (i) velocity of B is

$$\mathbf{V}_B = (\tfrac{1}{2}, 1, 0).$$

Angular velocity ω causes

$$\begin{aligned}\mathbf{V}_{B/A} = \omega \times \rho_{B/A} &= (0, 0, \omega) \times (2, 1, 0)\\ &= \omega(-1, 2, 0)\end{aligned}$$

$$\therefore \quad \omega = \tfrac{1}{2}.$$

Thus (ii) the moment of momentum about O is

$$\mathbf{h}_O = \sum_{i=1}^{2} \rho_i \times M_i \mathbf{V}_i$$

$$= 2M(0, 0, -1) + M(0, 0, 2) = 0.$$

1.4 Decomposition of a Vector

Any vector $\mathbf{v}$ can be decomposed into a component $v_n \mathbf{n}$ in direction $\mathbf{n}$ and a component $v_t \mathbf{e}$ in a direction perpendicular to $\mathbf{n}$:

$$\mathbf{v} = \mathbf{v} \cdot \mathbf{nn} + (\mathbf{n} \times \mathbf{v}) \times \mathbf{n} = v_n \mathbf{n} + v_t \mathbf{e}$$

where

$$v_n = \mathbf{n} \cdot \mathbf{v}, \qquad v_t = |\mathbf{n} \times \mathbf{v}|, \qquad \mathbf{e} = v_t^{-1}(\mathbf{n} \times \mathbf{v}) \times \mathbf{n}.$$

This relation is particularly helpful for analyzing oblique impact of rigid bodies; it is used to resolve the relative velocity and impulse at the contact point into components normal and tangential to the surfaces of the colliding bodies.

It is worth noting that the vector $\mathbf{v}$ has a component perpendicular to $\mathbf{n}$ that can be expressed alternatively as

$$v_t \mathbf{e} = (\mathbf{n} \times \mathbf{v}) \times \mathbf{n} = \mathbf{v} - \mathbf{v} \cdot \mathbf{nn}. \tag{1.15}$$

1.5 Vectorial and Indicial Notation

Frequently there is a need to express a vector variable such as the velocity $\mathbf{V}$ of a point in terms of the components of velocity in some reference frame. Let $\mathbf{n}_i$ be a set of mutually perpendicular unit vectors fixed in reference frame $\mathfrak{R}$. The components of velocity in this reference frame have magnitudes that can be expressed as $V_i \equiv \mathbf{V} \cdot \mathbf{n}_i$, $i = 1, 2, 3$. In vectorial notation the vector is expressed in terms of its components $\mathbf{V} = \sum_{i=1}^{3} \mathbf{V} \cdot \mathbf{n}_i \mathbf{n}_i \equiv V_i \mathbf{n}_i$, where the last equality follows from repeated subscripts implying summation over the range of spatial coordinates. On the other hand, in indicial or shorthand notation $\mathbf{V} \equiv V_i$, i.e., unit vectors are implied rather than written explicitly. Similarly, if $\mathbf{r}$ denotes the position of C relative to a point O fixed in $\mathfrak{R}$, then in vectorial notation $\mathbf{r} \equiv r_i \mathbf{n}_i$ and in indicial notation $\mathbf{r} \equiv r_i$.

As a consequence of these definitions, the *inner* or *dot product* can be expressed in alternative forms,

$$\mathbf{r} \cdot \mathbf{V} = r_i V_i \equiv \sum_{i=1}^{3} r_i V_i \tag{1.16}$$

where a repeated subscript (e.g. i) implies summation over the range of spatial coordinates. Likewise the *vector* or *cross product* can be expressed as

$$\mathbf{r} \times \mathbf{V} = \varepsilon_{ijk} r_j V_k \equiv \sum_{j=1}^{3} \sum_{k=1}^{3} \varepsilon_{ijk} r_j V_k \tag{1.17}$$

where ε_{ijk} is the permutation tensor which has values $\varepsilon_{ijk} = +1$ if indices are in cyclic order, $\varepsilon_{ijk} = -1$ if indices are in anticyclic order and $\varepsilon_{ijk} = 0$ if an index is repeated.

PROBLEMS

1.1 A wood block of mass M is struck at the center of one side by a bullet of mass M' that strikes at a normal incident speed V_0. The bullet is captured by the block. Find the final speed of the block and the fraction of the bullet's initial kinetic energy T_0 that is transformed to heat.

1.2 Direct impact of a body with mass M (g) against a rigid wall at an incident velocity V_0 (m s^{-1}) results in a contact duration t_f (ms). For collision conditions (M, V_0, t_f), calculate an estimate of the maximum force $F_{\max}$ (N) during a collision of each of the following bodies: tennis ball (45, 44, 4.0); golf ball (61, 44, 0.6); Ping Pong ball (2.5, 44, 0.5), steel hammer $(10^3, 1, 0.2)$.

1.3 For the two particle system in Ex. 1.3, find the center of mass $\hat{\rho}$ of the system, the velocity $\mathbf{V}_G$ of the center of mass, and the inertia dyadic $\hat{\mathbf{I}}$ and moment of momentum $\hat{\mathbf{h}}$ relative to the center of mass. Verify that these relations satisfy $\hat{\mathbf{h}} = \hat{\mathbf{I}} \cdot \boldsymbol{\omega}$.

1.4 A rigid uniform bar of mass M and length $3L$ lies across two parallel rails B and B' that are separated by width L. The bar is transverse to the rails and centered. The end closest to B' is raised a small distance and then released so that the bar is rotating with angular speed ω_- when it strikes B'.
   (a) Find the ratio ω_+/ω_-, where ω_+ is the angular speed of the bar at the instant immediately after impact, if the impact with rail B' is perfectly plastic.
   (b) Find ω_+/ω_- if the impact with rail B' is elastic.

(c) Obtain the fraction of the incident kinetic energy T_- that finally is lost as a result of collision (a).

1.5 A prismatic cylinder with polygonal cross-section has n equal sides, where $n \geq 4$. Between adjacent sides a regular polygon has an included angle 2α, where $\alpha = \pi/n$. Let the prismatic cylinder of mass M have a radius a from the center O to each vertex C_i. Find that this prismatic cylinder has a polar moment of inertia $\hat{I} = Ma^2(2 + \cos 2\alpha)/6$ about the center of mass O.

If the cylinder is rolling on a flat and level surface and before vertex C_i impacts with the surface it has angular speed ω_-, obtain an expression for the angular speed ω_+ the instant after C_i strikes the surface. Assume that after impact, C_i remains in contact with the surface.

1.6 A wheel of radius a and radius of gyration $\hat{k}_r = (\hat{I}/M)^{1/2}$ about the center of mass rolls upon a rough horizontal surface which may be idealized as a series of uniform serrations of pitch $2b$. The wheel rolls without slip or elastic rebound on the tips of the serrations under the action of a constant horizontal force F applied at its center.

(a) Show that at each impact the ratio of the angular speed at separation to that at incidence satisfies

$$\frac{\omega_+}{\omega_-} = \frac{\hat{k}_r^2 + a^2 - 2b^2}{\hat{k}_r^2 + a^2}.$$

(b) Find that when steady conditions are reached the wheel moves with a fluctuating forward velocity and

$$F = \frac{Mbv_1^2}{a^2 - b^2} \frac{\hat{k}_r^2 + a^2 - b^2}{a^2 + \hat{k}_r^2}.$$

where v_1 is the maximum forward component of velocity in each cycle.

1.7 For vectors $\mathbf{r} = \{r_1, r_2, r_3\}^T$ and $\mathbf{V} = \{V_1, V_2, V_3\}^T$ verify that $\mathbf{r} \times \mathbf{V} = \varepsilon_{ijk} r_j V_k$.

CHAPTER 2

Rigid Body Theory for Collinear Impact

The value of a formalism lies not only in the range of problems to
which it can be successfully applied but equally in the degree to
which it encourages physical intuition in guessing the solution of
intractable problems.

Sir Alfred Pippard, *Physics Bulletin* **20**, 455, 1969.

Two bodies, labeled B and B', collide when they come together with an initial
difference in velocity. Ordinarily they first touch at a point that will be termed the *contact
point* C. During a very brief period of contact, the point C on the surface of body B is
coincident with point C' on the surface of body B'. If at least one of the bodies, B or
B', has a surface that is topologically smooth at the contact point (i.e., the surface has
continuous curvature), there is a plane tangent to this surface at C; the coincident contact
points C and C' lie in this *tangent plane*. If both bodies are convex and the surfaces have
continuous curvature near the contact point, then this tangent plane is tangential to both
surfaces that touch at C; i.e., the surfaces of the colliding bodies have a *common tangent
plane*. The direction of the normal to the tangent plane is specified by a unit vector **n**;
this direction is termed the *common normal direction*. The contact force and changes
in relative velocity at the contact point C will be resolved into components normal and
tangential to the common tangent plane.

2.1   Equation of Relative Motion for Direct Impact

Consider two colliding bodies B and B' that have masses M and M' and time-
dependent velocities $V(t)$ and $V'(t)$ in the direction parallel to **n**. In a direct collision
these bodies are not rotating when they collide, so that the velocity is uniform in each
body (i.e. the same at every point). During contact there are equal but opposite compressive reaction forces which develop at contact points C and C'; these forces oppose
interference or overlap of the contact surfaces. In the case of *direct impact* between
collinear bodies the relative velocity between the contact points C and C' remains parallel to the common normal direction throughout the contact period. A reaction force
develops at the contact point as a consequence of compression of the local contact
region; this force opposes relative motion during contact. In a direct collision the reaction force acts in the normal direction; i.e. parallel to the velocities, as illustrated in
Fig. 2.1. If the colliding bodies are hard, the contact force is very large in comparison with
any body force; consequently, in rigid body impact theory any body or applied contact

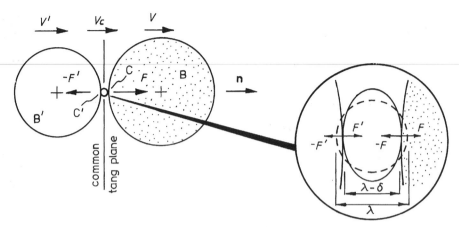

Figure 2.1. Collinear impact of two rigid bodies with contact points separated by an infinitesimal deformable particle. The particle represents small local deformation of the contact region.

forces of finite magnitude are negligibly small in comparison with the reaction at the contact point C. The finite body forces are ignorable because they do no work during the vanishingly small displacements that develop during an almost instantaneous collision. This is why a body force such as gravity does not affect the changes in velocity occurring in a collision. During impact between hard bodies the only active forces are reactions at points of contact. These reaction forces are indefinitely large; nevertheless, they produce a finite impulse that continuously changes the relative velocity during the instant of contact.

The assumption that the deformation of colliding bodies can be lumped in an infinitesimal deformable particle between the contact points is a key to obtaining changes in velocity as a function of *impulse* during the infinitesimal contact period. At the contact point, bodies B and B' are subjected to contact forces $\mathbf{F}(t)$ and $\mathbf{F}'(t)$ which have normal components of force $F(t) \equiv \mathbf{F} \cdot \mathbf{n}$ and $F'(t) \equiv \mathbf{F}' \cdot \mathbf{n}$ respectively. These reactions generate normal components of impulse $P(t)$ and $P'(t)$ where

$$dP = F\,dt \quad \text{and} \quad dP' = F'\,dt.$$

The translational equation of motion for each body in direction $\mathbf{n}$ can be expressed as

$$M\,dV = dP \quad \text{and} \quad M'\,dV' = dP'.$$

Let the normal component of *relative velocity* across the deformable particle at the contact point be $v = V - V'$.

This modeling recognizes that because the deformable particle has negligible mass, the reaction impulses acting on either side of this particle are equal in magnitude but opposite in direction. Thus the same reaction impulse acts on each colliding body but the directions of these impulses are opposed,

$$dP = -dP'.$$

Defining the impulse on body B as positive, $dp \equiv dP$, and substituting the translational equations of motion into the definition of relative velocity between the contact points gives the differential equation for changes in the normal component of relative velocity,

$$dv = m^{-1} dp \qquad (2.1)$$

where the effective mass m is defined as

$$m \equiv (M^{-1} + M'^{-1})^{-1} = \frac{MM'}{M + M'} \qquad (2.2)$$

After integration of Eq. (2.1) and application of the initial condition $v_0 \equiv v(0) = V(0) - V'(0)$, the normal relative velocity $v(p)$ is obtained as a function of normal impulse p,

$$v = v_0 + m^{-1}p, \qquad \text{where} \quad v_0 < 0 \qquad (2.3)$$

Thus during collision the normal component of relative velocity is a linear function of the normal impulse.

The key to calculating changes in velocity during impact is to find a means of evaluating the terminal impulse p_f at separation. The theory of rigid body impact will be more useful if the terminal impulse can be based on physical considerations – more useful in the sense that if experimentally obtained values of physical parameters are representative of the underlying sources of dissipation, these impact parameters will be applicable within a range of incident velocities. In Sect. 2.5 the terminal impulse is related to the *energetic coefficient of restitution*; this coefficient represents dissipation of (kinetic) energy due to inelastic deformation in the region surrounding the contact point.

2.2   Compression and Restitution Phases of Collision

After the colliding bodies first touch, the contact force $F(t)$ rises as the deformable particle is compressed. Let δ be the indentation or compression of the deformable particle. (The particle represents the compliance of the small part of the total mass which has significant deformation, i.e. the region of the bodies that surrounds the contact point C.) Without detailed information about the compliance of the colliding bodies there is no way of obtaining δ directly. Nevertheless, if compliance is rate-independent, the maximum indentation and maximum force occur simultaneously when the normal component of relative velocity vanishes.[1] Figure 2.2a illustrates the normal contact force as a function of indentation δ, while Fig. 2.2b shows this force as a function of time. The latter graph shows the separation of the contact period into an initial phase of approach or *compression* and a subsequent phase of *restitution*. During compression, *kinetic energy of relative motion is transformed into internal energy of deformation* by the contact force – the contact force does work that reduces the initial normal relative velocity of the colliding bodies while simultaneously an equal but opposite contact force does work that increases the internal deformation energy of the deformable particle. The compression

---

[1] Chapter 5 considers impact of bodies composed of viscoelastic or rate-dependent materials. For impacts where the contact region is viscoelastic the maximum force does not necessarily occur simultaneously with vanishing of the normal relative velocity.

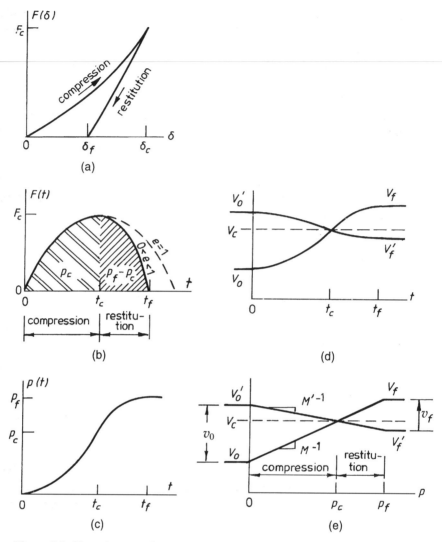

Figure 2.2. Normal contact force F as a function of (a) relative displacement δ and (b) time t; (c) normal impulse $p(t)$ as function of time t; and changes of normal velocities V, V' of colliding bodies as functions of (d) time t and (e) normal impulse p.

phase terminates and restitution begins when the normal relative velocity of the contact points vanishes. During the subsequent phase of restitution, the elastic part of the internal energy is released. Elastic strain energy stored during compression generates the force that drives the bodies apart during restitution – the work done by this force restores part of the initial kinetic energy of relative motion. The compliance of the deforming region during restitution is smaller than that during compression, so when contact terminates there is finally some residual compression δ_f of the deformable particle.

At any time t after incidence, the normal component of contact force F has an impulse p which equals the area under the curve of force shown in Fig. 2.2b. Since normal force is always compressive, the normal component of impulse increases monotonically as illustrated in Fig. 2.2c. Thus the normal impulse p can replace time t as an independent variable. During compression the increasing impulse slows body B′ and increases the speed of B as illustrated in Fig. 2.2d. Let the instant when indentation changes from compression to restitution be t_c. The colliding bodies have a relative velocity between contact points that vanishes at the end of compression: $v(t_c) = 0$; i.e., compression terminates when the contact points have the same speed V_c in the normal direction. Figure 2.2e illustrates that the contact point of each body experiences a change in velocity that is directly proportional to the normal reaction impulse at the contact point C as expressed by Eq. (2.3). The reaction impulse $p_c = \int_0^{t_c} F(t) \, dt$ which brings the two bodies to a common speed is termed the *normal impulse for compression*; this impulse is a characteristic which is useful for analyzing collision processes. The normal impulse for compression is obtained from Eq. (2.3) and the condition that at the end of compression the normal component of relative velocity vanishes [$v(p_c) = 0$]. Hence the normal impulse for compression is the product of the effective mass and the initial relative velocity at C,

$$p_c = -mv_0, \quad \text{where} \quad v_0 < 0. \tag{2.4}$$

2.3   Kinetic Energy of Normal Relative Motion

During collision each body undergoes changes in velocity that can be expressed as a function of the normal component of impulse, p:

$$V = V_0 + M^{-1}p$$
$$V' = V_0' - M'^{-1}p. \tag{2.5}$$

These equations give changes in the normal component of relative velocity $v(p)$:

$$v = v_0 + m^{-1}p, \quad \text{where} \quad v = V - V'$$

Compression terminates when both bodies have a common normal component of velocity $V_c \equiv V(p_c) = V'(p_c)$; this occurs when the normal impulse for compression $p_c = -mv_0$. At this normal reaction impulse the bodies have normal velocities that can be expressed as

$$V(p_c) = V_0 - \frac{m}{M}v_0 \quad \text{and} \quad V'(p_c) = V_0' + \frac{m}{M'}v_0. \tag{2.6}$$

If the system has an initial kinetic energy T_0, where

$$T_0 = \tfrac{1}{2}MV_0^2 + \tfrac{1}{2}M'V_0'^2$$

then at the transition from compression to restitution the system kinetic energy $T_c \equiv T(p_c) = (M + M')V_c^2/2$ will have been reduced to

$$T_c = T_0 - \tfrac{1}{2}mv_0^2. \tag{2.7}$$

The kinetic energy lost during compression is referred to as the *kinetic energy of normal relative motion at* C. For collinear impact between bodies B and B′, this equals that part of the initial kinetic energy of relative motion which is due to the normal component of relative velocity. During compression contact forces do work on colliding bodies that transforms this kinetic energy of normal relative motion into internal energy of deformation. For elastic deformations, this internal energy that is stored in the bodies is known as *strain energy*. The elastic strain energy is the source of the normal component of force that drives the contact points apart during the restitution phase of collision.

2.4   Work of Normal Contact Force

Work done by the normal contact force during separate phases of compression and restitution gives a relationship between impulse applied during compression, p_c, and the final or terminal impulse at separation, p_f. During compression the normal contact force does work on the deformable particle (in fact, on the small deforming region in each body surrounding the initial point of contact); this work deforms the particle and raises its internal energy. Of course the counterpart to the force that compresses the particle is the equal but opposite force that reduces the kinetic energy of normal relative motion during compression. Part of the energy absorbed during compression of the particle is recoverable during restitution; the recoverable part is known as elastic strain energy.

The work W_n done on the compressible particle by the normal component of force F can be calculated by recognizing that the force is related to the differential of impulse, $dp = F \, dt$, so that

$$W_n = \int_0^t Fv \, dt' = \int_0^p v \, dp'. \tag{2.8}$$

2.5   Coefficient of Restitution and Kinetic Energy Absorbed in Collision

Unless the impact speed is extremely small, there is energy dissipated in a collision. This can be due to plastic deformation, elastic vibrations and also rate-dependent processes such as viscoplasticity. Whatever the cause, it results in an inelastic or irreversible relation between the normal component F of the contact force and the compression δ. This dissipation results in smaller compliance during unloading (restitution) than was present during loading (compression); i.e., the force–deflection curve given in

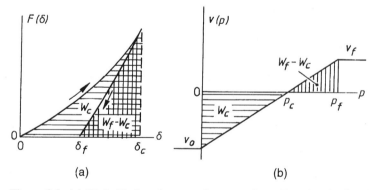

Figure 2.3. (a) Work W_c done by normal contact force F against bodies during period of compression, and work $W_f - W_c$ recovered during restitution, as functions of normal relative displacement δ at contact point. (b) Work of normal contact force related to changes in normal relative velocity during periods of compression ($p < p_c$) and restitution ($p_c < p < p_f$).

Fig. 2.3a exhibits hysteresis. The kinetic energy of relative motion that is transformed to internal energy of deformation during loading equals the area under the loading curve in Fig. 2.3a; this area is denoted by $W_c \equiv W_n(p_c)$. On the other hand, the area under the unloading curve equals the elastic strain energy released from the deforming region during restitution; in Fig. 2.3a this is denoted by $W_f - W_c \equiv W_n(p_f) - W_n(p_c)$. In the restitution phase the contact force generated by elastic unloading increases the kinetic energy of relative motion. These transformations of energy are due to work done by the contact force. This work done by the reaction force can easily be calculated for the separate phases of compression and restitution if changes in relative velocity are obtained as a function of normal impulse as illustrated in Fig. 2.3b; after initiation of contact the work done during these separate phases is proportional to the area between the horizontal axis and the line describing the normal relative velocity at any impulse.

In a direct collision the reaction force is normal to the common tangent plane; during compression the impulse of the normal contact force does work $W_n(p_c)$ on the rigid bodies that surround the small deforming region – this work equals the internal energy of deformation absorbed in compressing the deformable region. An expression for this work is obtained by integrating (2.3) and recalling that $p_c = -mv_0$:

$$W_n(p_c) = \int_0^{p_c} v(p)\, dp = v_0 p_c + \frac{1}{2} m^{-1} p_c^2 = -\frac{1}{2} m v_0^2. \tag{2.9}$$

This is just the kinetic energy of normal relative motion that is lost during compression. During the succeeding phase of restitution the rigid bodies regain some of this kinetic energy of normal relative motion due to the work $W_n(p_f) - W_n(p_c)$ done by contact forces,

$$W_n(p_f) - W_n(p_c) = \int_{p_c}^{p_f} (v_0 + m^{-1} p)\, dp = \frac{m v_0^2}{2} \left(1 - \frac{p_f}{p_c}\right)^2, \qquad v_0 < 0. \tag{2.10}$$

This work comes from and is equal to the elastic strain energy released during restitution.

These expressions for work done by the contact force during separate parts of the collision (period) are used to express the part of the initial kinetic energy of normal relative motion that is lost due to hysteresis of contact force. Expressions (2.9) and (2.10) give the part of this transformed energy that is irreversible, and this can be used to define an *energetic coefficient of restitution*, e_*.

Definition. The square of the coefficient of restitution, e_*^2, is the negative of the ratio of the elastic strain energy released during restitution to the internal energy of deformation absorbed during compression,

$$e_*^2 = -\frac{W_n(p_f) - W_n(p_c)}{W_n(p_c)}. \tag{2.11}$$

This coefficient has values in the range $0 \leq e_* \leq 1$, where 0 implies a perfectly plastic collision (i.e., no final separation, so that none of the initial kinetic energy of normal relative motion is recovered), while 1 implies a perfectly elastic collision (i.e., no loss of kinetic energy of normal relative motion). For direct impact the final impulse p_f and the total loss of kinetic energy are both directly related to the coefficient of restitution e_* by expressions (2.9) and (2.10). Using the negative of the square root of Eq. (2.11), the impulse at separation is obtained as

$$p_f = -mv_0(1 + e_*) = p_c(1 + e_*).\qquad(2.12)$$

For direct collinear impact the normal component of relative velocity at separation is given by

$$v_f = v_0 + m^{-1}p_f.\qquad(2.13)$$

Hence the ratio of final to initial relative velocity and the ratio of normal impulse during restitution to that during compression are also directly related to e_*:

$$e_* = -\frac{v_f}{v_0} \quad \text{and} \quad e_* = \frac{p_f - p_c}{p_c}$$

The first of the expressions above is the same as the definition of the *kinematic coefficient of restitution*, $e \equiv -v_f/v_0$. Newton first defined this impact parameter on the basis of his measurements of energy loss in collisions between identical balls. Newton presumed (incorrectly) that the coefficient of restitution is a material property. In his *Principia* (1686) he gave values for spheres made of steel ($e = 5/9$), glass ($e = 15/16$), cork ($e = 5/9$) and compressed wool ($e = 5/9$). The relationship between the *normal impulse* for restitution and that for compression is termed the *kinetic coefficient of restitution*, $e_o \equiv (p_f - p_c)/p_c$. This coefficient was defined by Poisson (1811), who recognized that it is equivalent to the kinematic coefficient of restitution for direct impact between rough bodies if the direction of slip is constant.[2] Subsequently, in Chapter 3, it will be shown that *the kinematic, kinetic and energetic definitions of the coefficient of restitution are equivalent unless the bodies are rough, the configuration is eccentric and the direction of slip varies during collision.*

2.6   Velocities of Contact Points at Separation

With terminal impulse p_f, at each contact point C or C' the normal component of velocity can be obtained from Eq. (2.6) and (2.12). At separation each contact point has a velocity $V_f \equiv V(p_f)$ or $V'_f \equiv V'(p_f)$ that is given by

$$V_f = V_0 + M^{-1}p_f = V_0 - \frac{mv_0}{M}(1 + e_*)$$

$$V'_f = V'_0 - M'^{-1}p_f = V'_0 + \frac{mv_0}{M'}(1 + e_*).\qquad(2.14)$$

At separation Eq. (2.13) gives a normal relative velocity $v_f \equiv v(p_f)$ as

$$v_f = v_0 + m^{-1}p_f = -e_*v_0.\qquad(2.15)$$

This is the same as that given by the kinematic definition of the coefficient of restitution.

Alternatively the velocity changes during restitution can be obtained from the changes in kinetic energy during this phase of impact. The final velocities are expressed in a form somewhat like that of (2.6),

$$V_f = V_c + (m/M)v_f \quad \text{and} \quad V'_f = V_c - (m/M')v_f.\qquad(2.16)$$

---

[2] Rough bodies have a tangential force – friction – that opposes relative tangential motion, or *slip*, at the contact point C.

This gives a terminal kinetic energy

$$T_f = T_c + m v_f^2/2$$

where $T_c = (M + M') V_c^2/2$. Hence for a frictionless collinear impact the terminal relative velocity v_f is related to changes in kinetic energy by

$$-\frac{v_f}{v_0} = \left(\frac{T_f - T_c}{T_0 - T_c}\right)^{1/2} = e_*. \tag{2.17}$$

If irreversibility of the contact force is the only source of dissipation during collision, then at separation the system suffers a final loss of kinetic energy,

$$T_0 - T_f = (1 - e_*^2) m v_0^2/2$$

This loss of energy depends directly on the *kinetic energy of normal relative motion at the contact point* C.

Example 2.1 A stationary sphere of mass M is struck at a normal angle of obliquity by an identical ball moving at 10 m s^{-1}. (i) Calculate the work done by the contact force during compression W_c and the part of the initial kinetic energy T_0 that is transformed during the period of compression. (ii) Calculate the work done during restitution, $W_f - W_c$. (iii) Obtain an expression for the terminal impulse p_f from the energetic coefficient of restitution and the ratio of work done during restitution to the negative of that done during compression. (iv) Calculate the relative velocity between the spheres at separation and the final velocity of each sphere as functions of the coefficient of restitution e_*.

Solution
Effective mass:

$$m = M/2.$$

Initial relative velocity:

$$v_0 = 0 - V'(0).$$

Initial momentum:

$$M V(0) + M' V'(0) = 0 - 2m v_0.$$

Initial kinetic energy:

$$T_0 = 0 + \frac{M' V'(0)^2}{2} = 2\frac{m v_0^2}{2}.$$

Equation of relative motion:

$$v(p) = v_0 + m^{-1} p.$$

Impulse terminating compression:

$$p_c = -m v_0, \qquad v(p_c) = 0.$$

(i)   Work during compression:

$$W_c = \int_0^{p_c} v(p)\, dp = -p_c^2/2m.$$

Kinetic energy at end of compression:

$$T_c = T_0 + W_c = mv_0^2/2.$$

Part of initial kinetic energy T_0 transformed during compression:

$$\frac{T_0 - T_c}{T_0} = \frac{1}{2}.$$

(ii) Work during restitution:

$$W_f - W_c = \int_{p_c}^{p_f} v(p)\,dp$$

$$= (p_f - p_c)^2/2m.$$

(iii) Definition of coefficient of restitution,

$$e_*^2 = -\frac{W_f - W_c}{W_c} = \frac{(p_f - p_c)^2}{p_c^2}.$$

gives impulse at separation

$$p_f = -mv_0(1 + e_*), \qquad v_0 < 0.$$

(iv) Relative velocity at separation:

$$v_f = v_0 + m^{-1}p_f = -e_*v_0.$$

Velocity of center of inertia:

$$V_c = V(p_c) = V'(0)/2.$$

Velocity at separation:

$$V_f = V(p_f) = V_c + \frac{m}{M}v_f = V'(0).$$

$$V_f' = V'(p_f) = V_c - \frac{m}{M'}v_f = 0.$$

2.7   Partition of Loss of Kinetic Energy

For a collision between dissimilar bodies, the loss of kinetic energy during the period of compression can be divided into a kinetic energy loss for each body by considering separately the work done on each body by contact forces. In order to achieve this partition the change in velocity of each body relative to the *center of inertia* must be considered separately. The center of inertia is a reference frame moving in the common normal direction with constant speed V_c throughout the period of contact. This speed is obtained from the principle of conservation of translational momentum and recognition that at the termination of the compression phase, the two bodies have a common velocity

$$V_c = \frac{MV(0) + M'V'(0)}{M + M'}. \tag{2.18}$$

During a collision between bodies B and B', let the normal component of velocities relative to the center of inertia be defined as $v_1 \equiv V - V_c$ and $v_2 \equiv V_c - V'$. After substituting into (2.14), a relation between the momenta of relative motion is obtained, $Mv_1 = M'v_2$. Hence the kinetic energy $T(p)$ can be expressed as

$$T = 0.5M(V_c + v_1)^2 + 0.5M'(V_c - v_2)^2$$
$$= 0.5\left[(M + M')V_c^2 + Mv_1^2 + M'v_2^2\right]$$

i.e., the kinetic energy separates into a part associated with the velocity of the center of inertia (which does not change during impact) and parts due to the relative motion of each body separately. After using the momentum relation, the final loss of kinetic energy for the impact pair will be

$$T_0 - T_f = 0.5\left(1 + \frac{M}{M'}\right)M\left[v_1^2(0) - v_1^2(p_f)\right]$$
$$= 0.5\left(1 + \frac{M'}{M}\right)M'\left[v_2^2(0) - v_2^2(p_f)\right]. \tag{2.19}$$

Each body dissipates some kinetic energy of relative motion (relative to the velocity of the steadily moving center of inertia). At the termination of compression there is a normal impulse p_c and the loss of kinetic energy for each body can be expressed as

$$T_0 - T_c = 0.5\left(1 + \frac{M}{M'}\right)Mv_1^2(0) = 0.5\left(1 + \frac{M'}{M}\right)M'v_2^2(0).$$

Hence the final loss of energy for the system in comparison with the energy absorbed during compression is obtained as

$$\frac{T_0 - T_f}{T_0 - T_c} = 1 - \frac{v_1^2(p_f)}{v_1^2(0)} = 1 - \frac{v_2^2(p_f)}{v_2^2(0)} \tag{2.20}$$

implying that $v_1(p_f)/v_1(0) = v_2(p_f)/v_2(0)$. Consequently the final change in relative velocity for each body is in proportion to the incident velocity relative to the velocity of the center of inertia. This partition of change of relative velocities is independent of any difference in compliance of the bodies.

Example 2.2   Smooth billiard balls B and B′ with equal masses M are rolling on a level table before they collide; at incidence the centers of mass have initial velocities $\hat{\mathbf{V}}_0$ and $\hat{\mathbf{V}}'_0$, respectively, while the point where each ball touches the table is stationary. Before impact the centers of mass are moving in directions at angles θ and θ', respectively, relative to the common normal direction, as shown in Fig. 2.4. Find at the termination of impact the velocity of the point on each ball which touches the table. These velocities are the initial post-impact sliding velocity of each ball on the table.

Solution   During impact the normal component of relative velocity between contact points C and C′ is $v = \hat{\mathbf{V}} \cdot \mathbf{n} - \hat{\mathbf{V}}' \cdot \mathbf{n}$. The balls have a center of mass O or O′, respectively, and the initial velocities of the contact points are given by

$$\hat{\mathbf{V}}_0 = \hat{V}_0 \sin\theta\, \mathbf{e}_1 + \left(\hat{V}_0 \cos\theta + \frac{p}{M}\right)\mathbf{n}$$

$$\hat{\mathbf{V}}'_0 = \hat{V}'_0 \sin\theta'\, \mathbf{e}_1 + \left(\hat{V}'_0 \cos\theta' - \frac{p}{M'}\right)\mathbf{n}.$$

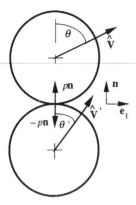

Figure 2.4. Plan view of initial velocities for centers of two colliding spheres.

Since the balls are spheres, the centers of mass and the contact points C and C' have the same normal component of relative velocity v. During impact this is a function of the impulse,

$$v = v_0 + p/m$$

$$v_0 = \hat{V}_0 \cos \theta - \hat{V}'_0 \cos \theta'.$$

At the end of compression the normal impulse is $p_c = -mv_0$. Thus at separation the centers of mass have velocities

$$\hat{V}_f = \hat{V}_0 \sin \theta\, \mathbf{e}_1 + \left(\hat{V}_0 \cos \theta - (1 + e_*)\frac{mv_0}{M} \right)\mathbf{n}$$

$$\hat{V}'_f = \hat{V}'_0 \sin \theta'\, \mathbf{e}_1 + \left(\hat{V}'_0 \cos \theta' + (1 + e_*)\frac{mv_0}{M'} \right)\mathbf{n}$$

and at the contact point there is a normal relative velocity $v_f = (\hat{V}_f - \hat{V}'_f) \cdot \mathbf{n} = -e_* v_0$. Meanwhile the points of contact between the spheres and the table that initially were stationary, now slide at velocities

$$V_f = -(1 + e_*)\frac{mv_0}{M}\mathbf{n}$$

$$V'_f = (1 + e_*)\frac{mv_0}{M}\mathbf{n}.$$

PROBLEMS

Normal Incidence

2.1 A railway boxcar with mass 50,000 kg is rolling at 2 m s^{-1} when it is struck by an overtaking boxcar of mass 100,000 kg rolling in the same direction at 3 m s^{-1}. For a coefficient of restitution $e_* = 0.5$, find the velocity of each boxcar at separation and calculate the loss of kinetic energy in the collision.

2.2 Regulations require that a baseball has mass $M = 150$ g and a coefficient of restitution $e_* = 0.546 \pm 0.03$ at an impact speed of 26 m s^{-1}. Suppose a baseball is thrown at 40 m s^{-1} and is hit by a bat with mass $M' = 800$ g; the ball is hit at the center of mass of the bat which is traveling at a speed of 25 m s^{-1} in a direction opposite to that of the ball. Find the speed of bat and ball at the instant of separation. Also show that the part of the incident kinetic energy that is dissipated in the collision equals $1 - e_*^2$. (Assume the angle of incidence is normal, the rotational kinetic energy of the bat is negligible and the coefficient of restitution is independent of the impact speed.)

2.3 Three elastic spherical balls B_1, B_2 and B_3 have their centers aligned. Balls B_2 and B_3 are stationary and slightly separated before ball B_1, traveling at speed V_0, collides against B_2. Prove that the final speed $V_3(p_f)$ of ball B_3 is a maximum if ball B_2 has mass M_2 equal to the geometric mean of the masses of B_1 and B_3; i.e. $M_2 = (M_1 M_3)^{1/2}$. Find the maximum value of the ratio $V_3(p_f)/V_0$ by letting the mass ratio $M_3/M_1 = \alpha^2$ (proof originally by Huyghens).

2.4 In Problem (2.3) let the number of balls be an arbitrary number n rather than 3. To obtain a maximum ratio of the final to the initial speed $V_n(p_f)/V_0$, prove that the mass of each ball is related to that of its neighbors by $M_i = (M_{i-1} M_{i+1})^{1/2}$, where $1 < i < n$. Find the maximum ratio $V_n(p_f)/V_0$ as a function of n.

Oblique Incidence

2.5 An incident particle B$'$ scatters through angle θ' when it strikes a stationary particle B. After this elastic impact the velocity of particle B makes an angle θ with the initial direction of the incident particle. Find the mass ratio of the two particles M'/M as a function of angles θ and θ'. (Hint: first obtain that for each body the change in momentum in the common normal direction equals $2m V_0' \cos \theta$.)

2.6 A deuteron is the nucleus of an atom of heavy hydrogen in a molecule of heavy water; it is composed of a proton and a neutron and has about double the mass of a neutron. If a neutron traveling at 100 m s^{-1} in a nuclear reactor suffers an elastic collision with a deuteron that is initially at rest and the neutron is scattered through an angle of $30°$, find the scattering angle of the deuteron. Also find the final speed of each particle.

2.7 A smooth sphere traveling at an initial speed V_0 collides against an initially stationary identical sphere with angle θ between the common normal direction and the path of the traveling sphere. Show that the course of the initially moving sphere will be deflected through an angle $\pi/2 - \theta - \tan^{-1}[(1 - e_*)(2 \tan \theta)^{-1}]$. For the second (initially stationary) sphere, find the final direction of travel as an angle measured from the initial course of the first sphere.

2.8 A compound pendulum with mass M' and radius of gyration k_r rotates freely about its pivot with angular speed ω_0. A stationary sphere of mass M has a direct collision with the pendulum at a distance λ from the pivot. If the collision is elastic and at separation the pendulum is stationary while the sphere has speed $\lambda \omega_0$, show that the impact occurs at distance $\lambda = k_r$ from the pivot. If instead the collision is inelastic with coefficient of restitution e_*, find the angular speed of the pendulum and the speed of the ball at separation if the point of impact is located at $\lambda = k_r$.

2.9 A uniform rod of mass M' and length $2L$ represents a baseball bat. The bat has a radius of gyration $\hat{k}_r$ about the center of mass, $\hat{k}_r = L/\sqrt{3}$. The rod pivots about a frictionless pin at one end. Assume that the initially stationary bat is driven about the pin by a couple C that is constant. After rotation through an angle $\theta = \pi/2$ the bat strikes a stationary ball of mass M. (The rotational speed of the bat when it strikes the ball is given by $\dot{\theta}_- = \sqrt{\pi C/M'(L^2 + \hat{k}_r^2)}$.)

(a) Find the impact point on the bat where a direct collision produces no reaction at the pin. Show that this point, the *center of percussion*, is located a distance $\xi + L$ away from the pivot where $\xi L = \hat{k}_r^2$.

(b) For a bat that strikes the ball at the center of percussion, find that a maximum impulse is imparted to the ball if the mass ratio of bat to ball, $M'/M = 4$. For this optimal bat striking a baseball, find an expression for the maximum speed of the batted ball if the collision is elastic.

2.10 Your local Little League team has asked you to provide advice on the optimal length of baseball bats in order to hit the ball as far as possible. Let the bat have a uniform density per unit length ρ and length L, and suppose that the kinetic energy of the bat at impact is a constant T_0. The ball with mass M has speed u_0 when it is struck by the bat. Assume that the batter's hands can be considered to be a pinned joint that cannot restrain the changes in velocity during impact. Furthermore, assume that there is direct impact between ball and bat and that the coefficient of restitution is e_*.

(a) Explain why optimal range is achieved if the ball is struck near the end of the bat.

(b) Find the length L of the bat for maximum range.

CHAPTER 3

Rigid Body Theory for Planar or 2D Collisions

Nature uses only the longest threads to weave her patterns, so each
small piece of her fabric reveals the organization of the entire tapestry.

R.P. Feynman, *The Character of Physical Law*, 1965

Eccentric impact configurations have the center of mass of at least one of
the colliding bodies off the common normal line which passes through the contact
point. A consequence of impact in an eccentric configuration is that the impulse acting at the contact point gives an impulsive moment about the center of mass. This
causes changes of the angular as well as the translational velocities during the period of
contact.

3.1 Equations of Relative Motion at Contact Point

Let bodies B and B' with masses M and M' collide at point C. Let the rotational
inertia for the center of mass of each body about the axis normal to the plane be specified by
radii of gyration[1] $\hat{k}_r$ and $\hat{k}'_r$ and let the impact configuration be defined by position vectors
$r_i = (r_1, r_3)$ and $r'_i = (r'_1, r'_3)$ from the center of mass of each body to the contact point
C. The bodies have angular velocities ω_i and ω'_i and velocities $\hat{V}_i$ and $\hat{V}'_i$ at the centers
of mass, respectively. These position and velocity vectors are defined in a Cartesian
coordinate system n_1, n_3, where $n_3 \equiv n$ is normal to the common tangent plane through
the contact point C and n_1 is tangential to this plane, as depicted in Fig. 3.1. At the contact
point of each body, a reaction force F_i or F'_i develops that opposes interpenetration of
the bodies during impact. These forces give differentials of impulse dP_i and dP'_i at C and
C', respectively:

$$dP_i = F_i \, dt \quad \text{and} \quad dP'_i = F'_i \, dt, \qquad i = 1, 3.$$

With these terms defined, Newton's second law gives equations of motion for translation
of the centers of mass,

$$
\begin{aligned}
d\hat{V}_i &= M^{-1} \, dP_i \\
d\hat{V}'_i &= M'^{-1} \, dP'_i.
\end{aligned}
\tag{3.1}
$$

---

[1] Here the diacritical mark $^\wedge$ over a symbol expresses that this variable is for the center of mass or for
axes passing through the center of mass. Usually, for $\hat{k}_r$, the radius of gyration about the center of mass,
this diacritical mark will be suppressed.

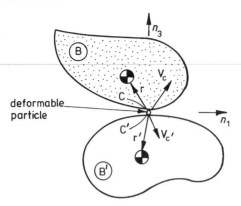

Figure 3.1. Colliding rigid bodies with contact points separated by deformable particle.

and for planar rotation of the bodies,

$$d\omega_i = \left(M\hat{k}_r^2\right)^{-1}\varepsilon_{ijk}r_j\,dP_k$$
$$d\omega_i' = \left(M'\hat{k}_r'^2\right)^{-1}\varepsilon_{ijk}r_j'\,dP_k'$$

(3.2)

where a repeated index indicates summation and the permutation tensor is given by $\varepsilon_{ijk} = +1$ if the indices are in cyclic order, $\varepsilon_{ijk} = -1$ if the indices are in anticyclic order and $\varepsilon_{ijk} = 0$ if there are repeated indices. (The transformation from vector to index notation is $\mathbf{r} \equiv r_i$ and $\mathbf{P} \equiv P_i$. In index notation the vector product is expressed as $\mathbf{r} \times d\mathbf{P} \equiv \varepsilon_{ijk}r_j\,dP_k$.) The normal (or polar) radius of gyration about the center of mass of some common shapes of bodies can be found in Table 3.1.

With the construct of an infinitesimal deformable particle between the points of contact, changes in relative velocity between the bodies at the contact point C are obtained as a function of impulse P_i during the contact. Since these are rigid bodies, the velocity of the contact point on each body is related to the velocity of the center of mass of the same body by

$$V_i = \hat{V}_i + \varepsilon_{ijk}\omega_j r_k, \qquad V_i' = \hat{V}_i' + \varepsilon_{ijk}\omega_j' r_k'.$$

(3.3)

The *relative velocity* at C is defined as the velocity difference

$$v_i \equiv V_i - V_i'.$$

After substituting Eqs (3.1)–(3.3) into an expression for the differential of relative velocity at the contact point $dv_i \equiv dV_i - dV_i'$, the differential equations for velocity changes are obtained in terms of the components of impulse. In addition, since the deformable particle has negligible mass, the differential impulses on either side of the particle are equal but opposite: $dp_i \equiv dP_i = -dP_i'$. Consequently, for planar velocity changes we obtain $dv_i = m_{ij}^{-1}\,dp_j$, or

$$\begin{Bmatrix} dv_1 \\ dv_3 \end{Bmatrix} = m^{-1}\begin{bmatrix} \beta_1 & -\beta_2 \\ -\beta_2 & \beta_3 \end{bmatrix}\begin{Bmatrix} dp_1 \\ dp_3 \end{Bmatrix}$$

(3.4)

Table 3.1. *Radius of Gyration $\hat{k}$ for Axes through Center of Mass*

Plane Sections

Shape	Area A	In-plane Radius of Gyration, $\hat{k}$	Polar Radius of Gyration, $\hat{k}_2 = \hat{k}_r$
Rectangle – height a, base b	ab	$a/\sqrt{12}$	$\sqrt{(a^2+b^2)/12}$
Thin circular disk – radius a	πa^2	$a/2$	$a/\sqrt{2}$
Thin circular ring – radius a, thickness b	$2\pi ab$	$a/2$	$a/\sqrt{2}$
Triangle – height a, base b	$ab/2$	$a/\sqrt{18}$	$\sqrt{(a^2+b^2)/18}$
Ellipse – axes a, b	πab	$a/2$ or $b/2$	$\sqrt{a^2+b^2}/2$

3D Bodies

Body	Volume V	Cross-section Radius of Gyration, $\hat{k}$	Axial Radius of Gyration, $\hat{k}_2$
Circular cylinder – radius a, length L	$\pi a^2 L$	$\sqrt{(a^2+b^2)/12}$	$a/\sqrt{2}$
Thin-walled circular cylinder – radius a, thickness b, length L	$2\pi abL$	$\sqrt{(6a^2+L^2)/12}$	a
Sphere – radius a	$4\pi a^3/3$	$a\sqrt{2/5}$	$a\sqrt{2/5}$
Thin-walled spherical shell – radius a, thickness b	$4\pi a^2 b$	$a\sqrt{2/3}$	$a\sqrt{2/3}$
Ellipsoid – axes $2a$, $2b$, $2c$	$4\pi abc/3$	$\sqrt{(b^2+c^2)/5}$	$\sqrt{(a^2+b^2)/5}$

where the elements β_i of the configuration matrix can be expressed as

$$\beta_1 = 1 + mr_3^2/M\hat{k}_r^2 + mr_3'^2/M'\hat{k}_r'^2$$

$$\beta_2 = mr_1 r_3/M\hat{k}_r^2 + mr_1' r_3'/M'\hat{k}_r'^2 \qquad (3.5)$$

$$\beta_3 = 1 + mr_1^2/M\hat{k}_r^2 + mr_1'^2/M'\hat{k}_r'^2.$$

To proceed, the tangential and normal components of reaction force need to be related in order to express the differential equations (3.4) in terms of a single, monotonously increasing independent variable – the normal impulse $p \equiv p_3$.

3.2   Impact of Smooth Bodies

Smooth surfaces have negligible friction or adhesion; in this case the only component of force is that which is normal to the tangent plane. In collisions between bodies with smooth contact surfaces, the tangential differential impulse vanishes:

$$dp_1 = dp_2 = 0 \qquad \text{for smooth contact surfaces.} \qquad (3.6)$$

Example 3.1 A smooth solid sphere B′ with mass M and radius R that is rolling on a rough level surface collides with an identical ball B that is initially stationary. Before the collision the center of mass G′ has speed $\hat{V}'(0)$. Suppose the collision is elastic ($e_* = 1$).

The contact between the balls is smooth, but for sliding between the ball and the table there is a coefficient of friction μ. Find the final velocity of each center of mass after the collision at a time when sliding ceases.

Solution To analyze the oblique impact shown in Fig. 3.2, establish unit vectors n_1, n_3 in the plane of the table with $n_3 \equiv n$ in the direction of the normal to the common tangent plane. Let θ be the angle of obliquity between the normal to the tangent plane and the initial velocity vector $\hat{V}_i'(0)$ for the moving center of mass, so that $\hat{V}_i'(0) = \hat{V}'(0)(-\sin\theta\, n_1 + \cos\theta\, n_3)$. Ball B' is initially rolling (no slip), so that the point on the ball which touches the table A' is stationary; hence, at incidence the ball has components of angular velocity

$$\omega_1'(0) = \frac{\hat{V}'(0)}{R}\cos\theta, \qquad \omega_3'(0) = \frac{\hat{V}'(0)}{R}\sin\theta.$$

When impact commences the contact point C' on ball B' has tangential and normal components of velocity

$$V_{iC'}' = \hat{V}_i' + e_{ijk}\omega_j'r_k'$$

where C' is located at $r_i' = G'C'$ relative to the center of mass G'. At incidence the contact point C' has components of relative velocity $v_i = V_{iC} - V_{iC'}'$ expressed as

$$v_1(0) = 0 + \hat{V}'(0)\sin\theta = V_0'\tan\theta$$
$$v_2(0) = 0 + \hat{V}'(0)\cos\theta = V_0'$$
$$v_3(0) = 0 - \hat{V}'(0)\cos\theta \equiv -V_0'$$

where V_0' is the normal component of incident relative velocity at C'. Notice that the coordinate system has been set up so that the initial value of the normal component of relative velocity $v_3(0) < 0$ while the in-plane component $v_1(0) > 0$.

For impact between two identical bodies with smooth (frictionless) contact surfaces the normal component of the equation of motion can be integrated over an initial part of the contact period to give

$$v_3(p) = v_3(0) + m^{-1}p, \qquad m^{-1} = 2M^{-1}.$$

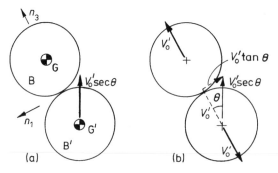

Figure 3.2. Oblique impact of a rolling sphere against a stationary sphere, where the normal component of velocity of G' at impact is $\hat{V}_0'$. Arrows indicate (a) incident velocity of center of mass, and (b) components of incident velocity (dashed) and change in velocity at G and G' during impact (solid).

At the end of compression the normal component of relative velocity vanishes $[v_3(p_c) = 0]$, giving a normal impulse for compression $p_c = MV_0'/2$ and, for elastic impact where $e_* = 1$, a normal impulse at separation $p_f = (1 + e_*)p_c = MV_0'$. Hence the colliding bodies suffer equal but opposite changes in velocity,

$$\Delta \hat{V}_i = MV_0' n_3, \qquad \Delta \hat{V}_i' = -MV_0' n_3.$$

In each ball these changes in velocity are uniform. Thus the initially stationary ball moves off in direction θ with a uniform velocity $MV_0' n_3$; i.e., at separation the struck ball is sliding on the table in direction θ and it has not begun to rotate.

At the instant of separation, the ball B' that was initially rolling has a stationary center of mass and a velocity $-V_0' n_3$ at point A' that is touching the table; i.e., this ball now slides in direction $-n_3$ with the same angular velocity it had before impact, $\omega_i(0) = (V_0'/R, 0, (V_0' \tan \theta)/R)$. After separation from impact, changes in velocity of the center of mass of ball B' occur in the n_1, n_3 plane due to friction at point A where the ball is sliding on the table. If friction satisfies the Amontons–Coulomb law with a limiting coefficient of friction μ and the collision occurs on a level surface where the gravitational constant is g, then for the period of sliding τ that immediately follows separation, integration of components of equations of translational and rotational motion of the ball give the following normal component of velocity of the center of mass and the component of angular velocity in direction n_1,

$$M\hat{V}_3' = 0 + \mu Mgt$$

$$\frac{2MR^2}{5}\omega_1' = \frac{2MR^2}{5}\left(\frac{V_0'}{R}\right) - \mu MgRt, \qquad t \leq \tau.$$

Sliding continues until the angular speed has slowed to the angular speed for rolling, $\omega_1' = \hat{V}_3'/R$. Together with the preceding equations, this gives a period of sliding $\tau_{B'} = 2V_0'/7\mu g$, and at the end of this period, a velocity $\hat{V}_i'(\tau_{B'}) = (-V_0' \tan \theta, 0, 2V_0'/7)$ for the center of mass. Figure 3.3a shows the velocity of the center of mass of each sphere at separation and the change of velocity that occurs as the sphere slides on the table. At time $\tau_{B'}$ after impact the ball once again begins to roll. While the ball is sliding the friction force is constant, consequently, ball B' travels a distance x_3 while sliding and

$$x_3 = 2\hat{V}_0'^2/49\mu g.$$

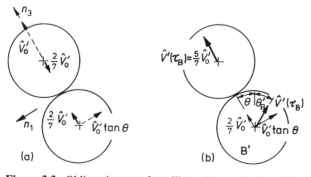

Figure 3.3. Oblique impact of a rolling sphere against a stationary sphere: (a) velocity of G and G' at separation (dashed arrows) and change in velocity during postimpact sliding (solid arrows); (b) terminal velocities of G and G' at termination of sliding.

This distance is not very far. For $V_0' = 1$ m s^{-1}, a coefficient of limiting friction $\mu = 0.1$ and $g = 9.81$ m s^{-2} we obtain a distance $x_3 = 40$ mm, which is less than the diameter of a billiard ball. As a consequence of the collision the path of ball B$'$ has been diverted through an angle $\theta_{B'} = \tan^{-1}[(\sin 2\theta)/(1.8 - \cos 2\theta)]$.

3.3   Friction from Sliding of Rough Bodies

3.3.1   Amontons–Coulomb Law for Dry Friction

Dry friction due to roughness of contacting surfaces can be represented by the Amontons–Coulomb law of sliding friction (Johnson, 1985). This law relates tangential and normal components of reaction force at the contact point by introducing a *coefficient of limiting friction* μ which acts if there were sliding. For negligible tangential compliance the sliding speed s is given by $s = \sqrt{v_1^2 + v_2^2}$. Denoting the normal component of a differential increment of impulse by $dp \equiv dP \equiv n \cdot dp_i$, the law of friction takes the form

$$\sqrt{(dp_1)^2 + (dp_2)^2} < \mu \, dp \qquad \text{if} \quad v_1^2 + v_2^2 = 0 \tag{3.7}$$

$$dp_1 = -\frac{\mu v_1}{\sqrt{v_1^2 + v_2^2}} \, dp, \quad dp_2 = -\frac{\mu v_2}{\sqrt{v_1^2 + v_2^2}} \, dp \qquad \text{if} \quad v_1^2 + v_2^2 > 0. \tag{3.8}$$

Equation (3.7) expresses an upper bound on the ratio of tangential to normal force for rolling contact; i.e., if a force ratio less than μ can satisfy the constraint of zero sliding, then if prior to separation the sliding speed vanishes ($s \equiv \sqrt{v_1^2 + v_2^2} = 0$), subsequently the contact *sticks*.

If sliding is present ($s > 0$), Eq. (3.8) represents a tangential increment of impulse or friction force which, at any impulse, acts in a direction directly opposed to sliding. The sliding direction can be defined by angle $\phi \equiv \tan^{-1}(v_2/v_1)$ which is the angle in the tangent plane measured from n_1. Thus in three dimensional problems the components of slip are

$$v_1 = s \cos\phi, \qquad v_2 = s \sin\phi$$

and (3.8) can be expressed as

$$dp_1 = -\mu \cos\phi \, dp, \quad dp_2 = -\mu \sin\phi \, dp, \qquad s > 0.$$

For planar impact $v_2 = 0$ and $\phi = 0$ or π.

For sliding contact the Amontons–Coulomb law expresses that the ratio of tangential to normal force is a constant; i.e., friction is independent of both sliding speed and normal pressure. During impacts of hard bodies with convex contact surfaces the normal pressure in the contact area is very large – certainly large enough to increase the true contact area due to plastic deformation of surface asperities (see Greenwood, 1996). It was Morin (1845) who, at the suggestion of Poisson, performed experiments which showed that the dynamic coefficient of friction for impact was equal to that measured in quasistatic tests at steady, relatively low speeds of sliding.

During sliding the tangential force is related to the normal force by the law of friction (3.7)–(3.8), hence the equations of motion (3.4) can be expressed in terms of a single independent variable – the normal reaction impulse. Since the normal contact force is always compressive, this impulse is a monotonously increasing scalar function during the contact period; i.e., the collision process can be resolved as a function of the independent variable p.

3.3.2 Equations of Planar Motion for Collision of Rough Bodies

The law of friction is the key that relates tangential to normal impulse if the contact is sliding. On the other hand, if friction is sufficient to prevent the development of sliding, this law provides the coefficient of limiting friction for stick – i.e., it represents a constraint on friction force. At any impulse during contact, a means of discriminating whether the next increment of impulse is sliding or sticking is obtained by assuming that there is no tangential component of relative velocity (or slip) at C and then comparing the ratio of the differentials of tangential constraint impulse to normal impulse for stick with the coefficient of friction μ.

Equations of Planar Motion for Stick
For an arbitrary normal impulse $p \equiv p_3$, if there is no tangential acceleration, then Eq. (3.4) gives

$$dv_1 = m^{-1}(\beta_1 \, dp_1 - \beta_2 \, dp) = 0. \tag{3.9}$$

Suppose any initial sliding vanishes $v_1(p) = 0$ at impulse p_s, where $0 \leq p_s < p_f$. In order to maintain stick during $p > p_s$, a specific ratio of tangential to normal reaction force is required; this is termed the *coefficient for stick*, $\bar{\mu}$. Thus

$$dp_1 = \bar{\mu} \, dp \;\Rightarrow\; dv_1 = 0 \qquad \text{during} \quad p > p_s$$

where for planar impact, $\bar{\mu}$ is defined as

$$\bar{\mu} \equiv \beta_2/\beta_1. \tag{3.10}$$

The coefficient for stick depends solely on the distribution of mass; it can be either positive or negative.

Stick occurs if the tangential relative velocity $v_1(p) = 0$; this implies that the coefficient of limiting friction $\mu \geq |\bar{\mu}|$ since otherwise the contact begins to slide.[2] If the contact sticks ($v_3 = 0$ and $\mu \geq |\bar{\mu}|$) during the period $p > p_s$ the ratio of reaction forces at C is $\bar{\mu}$ and the direction of the tangential force depends on sgn(β_2). While the contact point sticks, any changes in the relative velocity at C satisfy the following equations of motion:

$$\begin{aligned} dv_1/dp &= 0 \\ dv_3/dp &= m^{-1}(\beta_3 - \bar{\mu}\beta_2), \qquad p > p_s. \end{aligned} \tag{3.11}$$

Equations of Planar Motion for Sliding
On the other hand, if the contact is already sliding ($\bar{\mu}\mu$), then the law of friction relates the tangential and normal components of reaction force: $dp_1 = -\hat{s}\mu \, dp$, where

---

[2] Notice that $\beta_1 > 1$ but for collinear impact, $\beta_2 = 0$. Hence for a collinear collision, if initial sliding is brought to a halt before separation, subsequently the contact sticks.

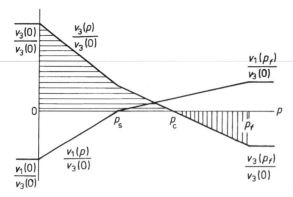

Figure 3.4. Changes in normal and tangential components of relative velocity at C as functions of normal impulse p for Coulomb friction and initial slip that is first brought to a halt at impulse p_s and then reverses in direction.

$\hat{s} \equiv \text{sgn}\,(v_1) \equiv v_1/|v_1|$. For Eq. (3.4) this gives

$$dv_1/dp = m^{-1}[-\beta_2 - \hat{s}\mu\beta_1] \tag{3.12a}$$

$$dv_3/dp = m^{-1}[\beta_3 + \hat{s}\mu\beta_2]. \tag{3.12b}$$

These equations can be integrated to give the relative velocity at any impulse during an initial period of unidirectional slip. For a normal component of relative velocity that is negative at incidence, $v_3(0) < 0$, this gives

$$v_1(p) = v_1(0) - m^{-1}[\beta_2 + \hat{s}\mu\beta_1]p$$
$$v_3(p) = v_3(0) + m^{-1}[\beta_3 + \hat{s}\mu\beta_2]p. \tag{3.13}$$

Henceforth in this chapter (and without loss of generality), *the coordinate system will be set up so that at incidence, the tangential component of relative velocity is positive*: $v_1(0) > 0$. This convention will eliminate the need to carry along in calculations the sign of the current direction of slip, $\hat{s}$.

With expressions (3.12)–(3.13), the equations of motion can be examined to identify the range of inertia and friction parameters for different contact processes. Figure 3.4 illustrates the changes in both normal and tangential components of relative velocity as a function of normal impulse p for an eccentric collision in which initial sliding $v_1(0)$ is brought to a halt at impulse p_s and then reverses. The rates of change of velocity with impulse are constant for ranges of impulse $0 < p < p_s$ and $p_s < p < p_f$, where p_f is the terminal impulse when contact points separate.

3.3.3   Contact Processes and Evolution of Sliding during Impact

For all impact configurations the definitions (3.5) give $\beta_1 > 0, \beta_3 > 0$ and $\beta_1\beta_3 > \beta_2^2$. Moreover, in order for normal force to oppose indentation, Eq. (3.12b) requires[3]

$$\mu\beta_2/\beta_3 > -1 \quad \text{to give} \quad dv_3/dp > 0. \tag{3.14}$$

[3] For eccentric impact configurations ($\beta_2 \neq 0$) inequality (3.14) provides an upper bound on the coefficient of friction in order to avoid jam (Stronge, 1990).

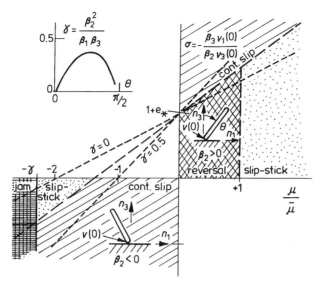

Figure 3.5. Regions of incident angle of obliquity and friction for different types of sliding processes at the contact point C.

The process of slip is described by (3.12a). For $\hat{s} > 0$ slip is retarded if $dv_1/dp < 0$, whereas for $\hat{s} < 0$ retardation requires $dv_1/dp > 0$. If initial retardation causes slip to vanish at impulse p_s before separation, reversal in direction of slip requires $dv_1/dp > 0$ during $p > p_s$. Conditions of this type give the range of parameters wherein it is possible to have a particular slip process; e.g., the direction of tangential acceleration is constant so if sliding vanishes before separation, thereafter either the contact sticks or slip reverses:

can stick if $\qquad 1 < |\mu/\bar{\mu}|$ $\qquad\qquad\qquad\qquad\qquad\qquad$ (3.15a)

can reverse if $\qquad 0 < \mu/\bar{\mu} < 1.$ $\qquad\qquad\qquad\qquad\qquad$ (3.15b)

Reversal can occur only if $\bar{\mu} > 0$ [assuming $\hat{s}(0) > 0$]. Figure 3.5 illustrates the range of these different types of contact processes as a function of a parameter $\mu/\bar{\mu}$ that combines friction and inertia properties. In this figure the vertical bands are given by inequalities (3.14) and (3.15).

Stick or slip reversal can occur only if the speed of slip vanishes before separation; otherwise sliding continues in the initial direction throughout the contact period. Integration of (3.12) indicates that sliding vanishes during compression if

$$-\frac{v_1(0)}{v_3(0)} < \frac{\beta_2(1 + \mu/\bar{\mu})}{\beta_3(1 + \gamma\mu/\bar{\mu})}, \qquad \gamma \equiv \frac{\beta_2^2}{\beta_1\beta_3} \qquad (3.16a)$$

while sliding vanishes during restitution if

$$\left[\frac{\beta_2(1 + \mu/\bar{\mu})}{\beta_3(1 + \gamma\mu/\bar{\mu})}\right] < -\frac{v_1(0)}{v_3(0)} < \left[\frac{\beta_2(1 + \mu/\bar{\mu})}{\beta_3(1 + \gamma\mu/\bar{\mu})}\right]\frac{p_f}{p_c} \qquad (3.16b)$$

where p_c is the normal impulse at transition from compression to restitution and p_f is the terminal normal impulse at separation. When sliding vanishes [$v_1(p_s) = 0$], the normal impulse p_s can be obtained from (3.13a),

$$p_s = (\beta_2 + \mu\beta_1)^{-1}mv_1(0) \tag{3.17}$$

and the normal component of relative velocity is $v_3(p_s) = v_3(0) + m^{-1}(\beta_3 + \mu\beta_2)p_s$. If sliding vanishes during compression, an additional impulse $p_c - p_s$ occurs before compression terminates. For parameters that give slip reversal during compression, the compression phase of collision has normal impulse

$$p_c = -\frac{mv_3(0)}{\beta_3 - \mu\beta_2}\left\{1 + \left(\frac{2\mu\beta_2}{\beta_2 + \mu\beta_1}\right)\frac{v_1(0)}{v_3(0)}\right\}, \qquad 0 < p_s < p_c \tag{3.18a}$$

whereas for unidirectional slip

$$p_c = -(\beta_3 + \mu\beta_2)^{-1}mv_3(0), \qquad p_s > p_c. \tag{3.18b}$$

In Fig. 3.5 the diagonal line represents the ratio of incident velocities $v_1(0)/v_3(0) = \tan\psi_0$, where the initial slip is brought to a halt just as the contact points separate [Eq. (3.16b)]. Angles of incidence that give initial sliding speeds outside this line, $|\psi_0| > \tan^{-1}[(1 + e_*) \times (1 + \mu/\bar{\mu})(1 + \gamma\mu/\bar{\mu})^{-1}]$, result in slip that will slow but not halt before separation.

3.4   Work of Reaction Impulse

3.4.1   Total Work Equals Change in Kinetic Energy

The total work done on the colliding bodies by contact forces $\bar{W}_f = \Sigma_{i=1}^3 W_i(p_f)$ can be calculated from the sum over all components of the partial work done by each separate component of impulse,

$$\bar{W}_f = \int_0^{t_f} F_i v_i \, dt = \int_0^{p_i(p_f)} v_i \, dp_i. \tag{3.19}$$

This work equals the change ΔT in the kinetic energy of the system. For a collision between bodies B and B' that occurs at contact point C, the change in kinetic energy and hence the work can be expressed as

$$\bar{W}_f = \Delta T = \frac{M}{2}[\hat{V}_i(t_f) - \hat{V}_i(0)][\hat{V}_i(t_f) + \hat{V}_i(0)]$$

$$+ \frac{\hat{I}_{ij}}{2}[\omega_i(t_f) - \omega_i(0)][\omega_j(t_f) + \omega_j(0)]$$

$$+ \frac{M'}{2}[\hat{V}_i'(t_f) - \hat{V}_i'(0)][\hat{V}_i'(t_f) + \hat{V}_i'(0)]$$

$$+ \frac{\hat{I}_{ij}'}{2}[\omega_i'(t_f) - \omega_i'(0)][\omega_j'(t_f) + \omega_j'(0)]$$

where matrices $\hat{I}_{ij}$ and $\hat{I}_{ij}'$ contain the moments (and products) of inertia of each body for axes through each respective center of mass. During the brief period t_f the reaction force at contact point C imparts an impulse to each body, $P_i(t_f) = M[\hat{V}_i(t_f) - \hat{V}_i(0)]$ and $P_i'(t_f) = M[\hat{V}_i'(t_f) - \hat{V}_i'(0)]$, and an impulsive torque about each center of mass, $e_{ijk}r_j P_k = \hat{I}_{ij}[\omega_j(t_f) - \omega_j(0)]$ and $e_{ijk}r_j'P_k' = \hat{I}_{ij}'[\omega_j'(t_f) - \omega_j'(0)]$, respectively. Thus the change in kinetic energy can be written as

$$\Delta T = \frac{P_i}{2}[\hat{V}_i(t_f) + \hat{V}_i(0)] + \frac{e_{ijk}r_j P_k}{2}[\omega_i(t_f) + \omega_i(0)]$$
$$+ \frac{P_i'}{2}[\hat{V}_i'(t_f) + \hat{V}_i'(0)] + \frac{e_{ijk}r_j' P_k'}{2}[\omega_i'(t_f) + \omega_i'(0)].$$

Together with the expression for equal but opposed contact forces $p_i \equiv P_i = -P_i'$ and the relation for relative velocity across the intermediate deformable particle, $[v_i(t) = [\hat{V}_i(t) - e_{ijk}r_j\omega_k(t)] - [\hat{V}_i'(t) - e_{ijk}r_j'\omega_k'(t)]$, the preceding equation gives the total work done by reaction forces,

$$\bar{W}_f = \frac{p_i}{2}[v_i(t_f) + v_i(0)]. \tag{3.20}$$

This relation was derived first by Thomson and Tait (1879) and subsequently appeared in Routh (1905, Art 346). The total work equals the sum of the work done by individual components of impulse $\bar{W}_f = \Sigma_{i=1}^{3} W_i(p_f)$; in general, however, no expression such as (3.20) applies to the partial work $W_i(p_f)$ done by the ith component of the reaction impulse.

3.4.2 Partial Work by Component of Impulse

The energetic coefficient of restitution depends on the work done by the normal component of contact force during the period of contact. A useful method of calculating this work is to use the following theorem for each separate period of slip (wherein the components of contact force are proportional) and then sum the results for the period of collision.

The partial work $W_{\tilde{n}}$ done on colliding bodies by the component of reaction impulse in direction $\tilde{n}_i$ during any period of unidirectional slip $\Delta t = t_2 - t_1$ equals the scalar product of this component of reaction impulse $\Delta p_{\tilde{n}}$ and half the sum of the components in direction $\tilde{n}_i$ of the initial and final relative velocities across the contact point:

$$W_{\tilde{n}} = \Delta p_{\tilde{n}}[v_{\tilde{n}}(t_2) + v_{\tilde{n}}(t_1)]/2 \qquad (\text{no summation on } \tilde{n}) \tag{3.21}$$

where $\Delta p_{\tilde{n}} = \tilde{n}_i[p_i(t_2) - p_i(t_1)]$, $v_{\tilde{n}}(t) = \tilde{n}_i v_i(t)$ and $\tilde{n}_i$ is a unit vector $\tilde{n}_i \tilde{n}_i = 1$.

PROOF Let Δp_i be the impulse acting on a body during a period $t_2 - t_1$. The aim is to calculate the partial work $W_{\tilde{n}}$ done by the component of impulse that acts in a direction parallel to a unit vector $\tilde{n}_i$.[4] At the contact point, changes in relative velocity are obtained from the second law of motion,

$$v_i(t) = v_i(t_1) + m_{ij}^{-1}\Delta p_j, \qquad t > t_1 \tag{3.22}$$

where m_{ij}^{-1} is the inverse of the inertia matrix for C that was given in Eq. (3.4). Here we assume that the contact period is brief so that during contact the bodies do not change in orientation. The impulse Δp_i has a component in direction $\tilde{n}_i$ that equals

---

[4] Components of vectors are defined in relation to a reference frame composed of a triad of unit vectors n_i.

$(\Delta p_{\tilde{n}})\tilde{n}_i = (\tilde{n}_j \, \Delta p_j)\tilde{n}_i \equiv \tilde{n}_i \Sigma_{j=1}^{3} \tilde{n}_j \, \Delta p_j$. If the components of Δp_j increase proportionally during $t > t_1$, then an expression similar to (3.19) can be integrated to give the partial work done by the component of impulse acting in direction $\tilde{n}_i$,

$$W_{\tilde{n}} = (\tilde{n}_j \, \Delta p_j)\{\tilde{n}_i [v_i(t_1) + m_{ik}^{-1} \, \Delta p_k/2]\}.$$

With (3.22) and after rearranging, this can be expressed as

$$W_{\tilde{n}} = \tilde{n}_i \, \Delta p_i \, \{\tilde{n}_j [v_j(t_2) + v_j(t_1)]\}/2 = \Delta p_{\tilde{n}} \, [v_{\tilde{n}}(t_2) + v_{\tilde{n}}(t_1)]/2. \qquad \text{q.e.d.}$$

Theorem (3.21) is especially useful for impulsive forces, i.e. in the limit as $t_f \to 0$. For a single point of collision between two smooth (frictionless) bodies there is only a normal component of impulse, so (3.20) and (3.21) are equivalent. With friction, however, tangential sliding during contact results in tangential impulses in addition to the normal impulse. For friction that is in accord with the Amontons–Coulomb law, the tangential impulse increases in proportion to the normal impulse *during any period of unidirectional sliding*. Hence for collisions between rough bodies, theorem (3.21) is applicable only during a period of unidirectional sliding. Generally in order to calculate the partial work of a component of impulse using (3.21), the contact period must be separated into a series of discrete periods of unidirectional sliding.

In some cases it can be helpful to recognize that according to (3.21) the work W_i done by the ith component of impulse is

$$W_i = \frac{\tilde{n}_i m_{ij}}{2}[v_j(t_2) - v_j(t_1)]\{\tilde{n}_k [v_k(t_2) + v_k(t_1)]\}$$

so that the total work of three mutually perpendicular components of impulse can be expressed as

$$\bar{W} = \frac{m_{jk}}{2}[v_j(t_2) - v_j(t_1)][v_k(t_2) + v_k(t_1)]$$

$$= \frac{\Delta p_j}{2}[v_j(t_2) + v_j(t_1)]. \qquad (3.23)$$

According to (3.20), however, this total work is independent of whether or not the impulse is unidirectional; i.e., the total work depends solely on the differences between the initial and final states of impulse and relative velocity at the contact point. On the other hand, for eccentric collision configurations ($m_{ij} \neq 0, i \neq j$) Eq. (3.21) for the partial work of any component of impulse is applicable only within periods where the contact force is constant in direction (Stronge, 1992).

As a consequence of (3.23), Thomson and Tait (1879, Art. 309) correctly say "that if any number of impacts be applied to a body, their whole effect will be the same whether they be applied together or successively (provided that the whole time be infinitely short), *although the work done by each particular impact is in general different according to the order in which the several impacts are applied*" (italics added).

3.4.3   Energetic Coefficient of Restitution

A coefficient of restitution relates the normal impulse applied during the restitution phase to that during the compression phase; the sum of these impulses gives the terminal normal impulse p_f when the contact points finally separate. For situations where the direction of slip varies during collision, the only energetically consistent definition of this coefficient is the so-called *energetic coefficient of restitution*. This directly relates the coefficient of restitution to irreversible deformation in the contact region. For analyses employing impulse as an independent variable, this definition of the coefficient of restitution explicitly separates the dissipation due to hysteresis of contact forces from that due to friction between the colliding bodies.

The work W_3 done on the bodies by the normal component of reaction impulse $p = p_3$ is given by

$$W_3(p) = \int_0^{t(p)} F_3 \cdot v_i \, dt = \int_0^p v_3 \, dp.$$

During the compression phase of collision, the work done by the normal component of reaction impulse decreases the sum of the kinetic energies of the colliding bodies and brings the normal component of relative velocity to rest; if tangential compliance is negligible, this work equals the internal energy gained by the bodies during this period. The part of this internal energy that is recoverable is known as elastic strain energy. This strain energy drives the contact points apart during the subsequent restitution phase of collision and thereby restores some of the kinetic energy of relative motion that was absorbed during compression. The difference between the work done to compress the bodies and the work done by the release of strain energy during restitution is the collision energy loss due to internal irreversible deformation. This work is used to define an energetic coefficient of restitution. For an impulse-dependent analysis, this coefficient is independent of friction and the process of slip.

The square of the coefficient of restitution, e_*^2, is the negative of the ratio of the elastic strain energy released during restitution to the internal energy of deformation absorbed during compression,

$$e_*^2 = -\frac{W_3(p_f) - W_3(p_c)}{W_3(p_c)} = \frac{-\int_{p_c}^{p_f} v_3(p) \, dp}{\int_0^{p_c} v_3(p) \, dp} \tag{3.24}$$

---

where a characteristic normal impulse for the compression phase p_c is defined as the impulse that brings the normal component of relative velocity to rest, i.e. $v_3(p_c) = 0$. This characteristic impulse for the compression phase can be calculated from the laws of motion. If the normal component of relative velocity is plotted as a function of impulse as shown in Fig. 3.4, the energetic coefficient of restitution is just the square root of the ratio of areas under the curve before and after p_c. Consequently, if the internal energy loss parameter e_* is known, the terminal impulse at separation p_f can be calculated.

3.4.4 Terminal Impulse p_f for Different Slip Processes

In order to obtain the terminal impulse p_f corresponding to any value of the energetic coefficient of restitution e_*, the work done by the normal reaction impulse is calculated separately for each phase of collision in accord with Eq. (3.21). For planar impact the work done on the colliding bodies by any component of reaction impulse can be calculated most easily by reference to the diagram in Fig. 3.4, which plots changes in relative velocity across the contact point as a function of impulse. Each line segment on this diagram is the solution of the equations of motion (3.4) during a separate period of slip. In the diagram, there is a crosshatched area beneath a relative velocity curve; this area equals the work done by the normal component of reaction impulse. During compression the normal component of impulse does work (on the deformable particle) that increases with increasing impulse until at impulse p_c the work equals $W_3(p_c)$. Subsequently, during restitution, the rate of work is negative, so that finally at separation the part of the kinetic energy of normal relative motion that has been restored by the normal impulse during restitution equals $W_3(p_f) - W_3(p_c)$. The difference in sign for work done during compression and restitution is signified by different crosshatching.

Unidirectional Slip during Contact: $-\frac{v_1(0)}{v_3(0)} \geq (1 + e_*)\frac{\beta_2 + \mu\beta_1}{\beta_3 + \mu\beta_2}$

In the case of dry friction the tangential component of reaction force acts in a direction opposed to sliding in the contact region. For two-dimensional problems, the equations of motion give a constant direction of sliding if either the initial sliding speed is sufficiently large so that slip does not halt during the contact period (impulse) or the initial sliding speed is zero. For this case the characteristic normal impulse for compression, p_c, is obtained from the condition that the normal component of relative velocity vanishes at the termination of compression,

$$v_3(p_c) = 0 = v_3(0) + m^{-1}(\beta_3 + \hat{s}\mu\beta_2)p_c \quad \Rightarrow \quad p_c = \frac{-mv_3(0)}{\beta_3 + \hat{s}\mu\beta_2}. \tag{3.18b}$$

Here, if $\theta < 0$ (i.e. $\beta_2 < 0$) or if the initial direction of sliding is such that $\hat{s} < 0$, then the system jams if the coefficient of friction is sufficiently large, $\mu \geq \beta_3/\beta_2$. Otherwise the contact slides continuously in the initial direction and the normal component of relative velocity at any impulse p is obtained as

$$v_3(p) = v_3(0) + m^{-1}(\beta_3 + \hat{s}\mu\beta_2)p.$$

Hence the partial work done by the normal impulse W_3 during the period of compression p_c can be calculated from (3.21),

$$W_3(p_c) = \int_0^{p_c} v_3(p)\,dp = \frac{-mv_3^2(0)}{\beta_3 + \hat{s}\mu\beta_2}. \tag{3.25}$$

Similarly, for the subsequent period of restitution the partial work done by the normal impulse is

$$W_3(p_f) - W_3(p_c) = \int_{p_c}^{p_f} v_3(p)\,dp = \frac{m^{-1}}{2}(\beta_3 + \hat{s}\mu\beta_2)(p_f - p_c)^2. \tag{3.26}$$

Here we note that the normal reaction does negative partial work during compression, and this reduces the kinetic energy of relative motion. During restitution the partial work of normal impulse is positive, and this restores some of the kinetic energy of relative motion that was transformed to strain energy during compression.

The partial work of the normal impulse during compression and that during restitution are used to evaluate the energetic coefficient of restitution e_*, or rather, to determine the terminal impulse p_f in terms of the coefficient of restitution:

$$e_*^2 \equiv \frac{-[W_3(p_f) - W_3(p_c)]}{W_3(p_c)} = \left(\frac{p_f}{p_c} - 1\right)^2$$

$$\Rightarrow \quad p_f = p_c(1 + e_*) = \frac{-(1 + e_*)mv_3(0)}{\beta_3 + \hat{s}\mu\beta_2}. \tag{3.27}$$

When contact terminates at impulse p_f, the contact points separate with the following ratios for changes in relative speeds during impact:

$$\frac{v_1(p_f)}{v_1(0)} = 1 + (1 + e_*)\left[\frac{\beta_2 + \hat{s}\mu\beta_1}{\beta_3 + \hat{s}\mu\beta_2}\right]\frac{v_3(0)}{v_1(0)}, \quad \frac{v_3(p_f)}{v_3(0)} = -e_*. \tag{3.28}$$

During contact there is also work done by the tangential component of impulse and this always decreases the kinetic energy of relative motion. For monotonous sliding the partial work done by friction is obtained as

$$W_1(p_f) = \int_0^{p_f} v_1(p)\,dp = \frac{-\hat{s}\mu p_f}{2}[2v_1(0) - m^{-1}(\beta_2 + \hat{s}\mu\beta_1)p_f].$$

Slip Reversal during Compression: $0 < \mu < \frac{\beta_2}{\beta_1}, \ -\frac{v_1(0)}{v_3(0)} < \frac{\beta_2 + \mu\beta_1}{\beta_3 + \mu\beta_2}$

For initial sliding in direction $v_1(0) > 0$ that is brought to a halt during compression and then reverses, use first the equation (3.13) for changes in sliding speed in order to determine the normal impulse p_s that brings sliding to a halt:

$$v_1(p_s) \equiv 0 = v_1(0) - m^{-1}(\beta_2 + \mu\beta_1)p_s \quad \Rightarrow \quad p_s = \frac{mv_1(0)}{\beta_2 + \mu\beta_1}. \tag{3.29}$$

At impulse p_s the normal component of relative velocity has changed to

$$v_3(p_s) = v_3(0) + \left(\frac{\beta_3 + \mu\beta_2}{\beta_2 + \mu\beta_1}\right)v_1(0). \tag{3.30}$$

Noting that slip reverses at normal impulse p_s, the characteristic impulse for compression is again found from the condition that $v_3(p_c) \equiv 0$, so that

$$p_c - p_s = \frac{-mv_3(0)}{\beta_3 - \mu\beta_2}\left\{1 + (\beta_2 + \mu\beta_1)\frac{p_s}{mv_3(0)}\right\}$$

or

$$p_c = \frac{-mv_3(0)}{\beta_3 - \mu\beta_2}\left\{1 + \frac{2\mu\beta_2}{\beta_2 + \mu\beta_1}\frac{v_1(0)}{v_3(0)}\right\}. \tag{3.18a}$$

Thus slip reversal occurs at a part of the normal compression impulse that is proportional to

$$\frac{p_s}{p_c} = \frac{-(\beta_3 - \mu\beta_2)v_1(0)/v_3(0)}{\beta_2 + \mu\beta_1 + 2\mu\beta_2 v_1(0)/v_3(0)}.$$

To calculate the partial work done by the normal component of impulse using (3.21), the impulse must be separated into the part occurring during each period of unidirectional slip. Thus at impulse p_s where sliding halts,

$$W_3(p_s) = [2v_3(0) + m^{-1}(\beta_3 + \mu\beta_2)p_s]\frac{p_s}{2}$$

Subsequently, for impulse $p > p_s$ there is a period of reverse sliding if $\mu < \bar{\mu}$. During this period there is additional work done by the normal reaction impulse. This further reduces the kinetic energy of relative motion:

$$W_3(p_c) - W_3(p_s) = \frac{-m^{-1}}{2}(\beta_3 - \mu\beta_2)(p_c - p_s)^2.$$

The sum of these two expressions gives the partial work done by the normal component of impulse during compression, $W_3(p_c)$:

$$W_3(p_c) = \frac{-m^{-1}}{2}(\beta_3 - \mu\beta_2)p_c^2 - \frac{2m^{-1}\mu\beta_2}{2}p_s^2. \tag{3.31}$$

The partial work done by the normal reaction impulse during restitution can be obtained as

$$W_3(p_f) - W_3(p_c) = \frac{v_3(p_f)}{2}(p_f - p_c) = \frac{m^{-1}}{2}(\beta_3 - \mu\beta_2)(p_f - p_c)^2 \tag{3.32}$$

The energetic coefficient of restitution e_* provides a relationship between the partial work $W_3(p_f) - W_3(p_c)$ done by normal impulse during restitution and the partial work $W_3(p_c)$ done during compression. This relationship gives the terminal impulse p_f in terms of the coefficient of restitution. In comparison with the characteristic normal impulse for compression, the terminal impulse is obtained as

$$\frac{p_f}{p_c} = 1 + e_*\left[1 + \frac{2\mu\beta_2}{\beta_3 - \mu\beta_2}\left(\frac{p_s}{p_c}\right)^2\right]^{1/2}. \tag{3.33}$$

This impulse is used to calculate the velocity components at separation,

$$v_1(p_f) = -m^{-1}(\beta_2 - \mu\beta_1)(p_f - p_s)$$

$$v_3(p_f) = m^{-1}(\beta_3 - \mu\beta_2)(p_f - p_c) = \frac{e_*(\beta_3 - \mu\beta_2)p_c}{m}\left[1 + \frac{2\mu\beta_2}{\beta_3 - \mu\beta_2}\left(\frac{p_s}{p_c}\right)^2\right]^{1/2}$$

so that

$$\frac{v_3(p_f)}{v_3(0)} = -e_*\left\{1 + \frac{2\mu\beta_2}{\beta_2 + \mu\beta_1}\frac{v_1(0)}{v_3(0)}\right\}\left[1 + \frac{2\mu\beta_2}{\beta_2 + \mu\beta_1}\left(\frac{p_s}{p_c}\right)^2\right]^{1/2}. \tag{3.34}$$

This expression for the change in the normal component of relative velocity depends on inertia properties, the angle of obliquity at incidence and the coefficient of friction in addition to the energetic coefficient of restitution (Stronge, 1993).

    To calculate dissipation from friction, the partial work of the tangential impulse is separated into that before and that after the impulse p_s where the direction of sliding reverses:

$$W_1(p_s) = \frac{-\mu m^{-1}}{2}(\beta_2 + \mu\beta_1)p_s^2$$

$$W_1(p_f) - W_1(p_s) = \frac{-\mu m^{-1}}{2}(\beta_2 - \mu\beta_1)(p_f - p_s)^2.$$

The initial kinetic energy $T_0 = mv_3^2(0)[1 + v_1^2(0)/v_3^2(0)]$ can be used to nondimensionalize the expressions for work done by the contact force during impact:

$$\frac{W_1(p_f)}{T_0} = \frac{-\mu\left[1 + \dfrac{2\mu\beta_2}{\beta_2 + \mu\beta_1}\dfrac{v_1(0)}{v_3(0)}\right]^2}{(\beta_3 - \mu\beta_2)^2\left[1 + v_1^2(0)/v_3^2(0)\right]}$$

$$\times\left\{(\beta_2 - \mu\beta_1)\left(\frac{p_f}{p_c} - \frac{p_s}{p_c}\right)^2 + (\beta_2 + \mu\beta_1)\left(\frac{p_s}{p_c}\right)^2\right\}$$

$$\frac{W_3(p_f)}{T_0} = \frac{\left[1 + \dfrac{2\mu\beta_2}{\beta_2 + \mu\beta_1}\dfrac{v_1(0)}{v_3(0)}\right]^2}{(\beta_3 - \mu\beta_2)\left[1 + v_1^2(0)/v_3^2(0)\right]}$$

$$\times\left\{\left(\frac{p_f}{p_c}\right)^2 - 2\left(\frac{p_f}{p_c}\right) - \frac{2\mu\beta_2}{(\beta_3 - \mu\beta_2)}\left(\frac{p_s}{p_c}\right)^2\right\}.$$

Slip–Stick Transition during Compression: $\mu > \bar{\mu} = \frac{\beta_2}{\beta_1}, \ -\frac{v_1(0)}{v_3(0)} < \frac{\beta_2 + \mu\beta_1}{\beta_3 + \mu\beta_2}$

The equations of motions indicate that the speed of initial sliding decreases if $\mu > \bar{\mu} \equiv \beta_2/\beta_1$. If the initial speed of sliding is small, so that slip halts during the contact period, then during the remainder of the contact period the contact points stick if the coefficient of friction is sufficiently large ($\mu > \bar{\mu}$); otherwise, the direction of slip reverses. Again taking the case of $\hat{s}(0) > 0$, the normal impulse required to bring slip to a halt and the normal component of velocity $v_3(p_s)$ at this impulse are obtained as

$$p_s = \frac{mv_1(0)}{\beta_2 + \mu\beta_1}, \qquad v_3(p_s) = v_3(0) + \left(\frac{\beta_3 + \mu\beta_2}{\beta_2 + \mu\beta_1}\right)v_1(0). \tag{3.35}$$

For the subsequent period of stick the tangential force is only that which is required to provide $dv_1(p) = 0$, $p > p_s$; i.e., the equations of motion for stick are obtained by replacing the coefficient of friction by $\bar{\mu}$:

$$p_c - p_s = \frac{-mv_3(p_s)}{\beta_3 - \bar{\mu}\beta_2}$$

This gives a normal impulse for compression,

$$p_c = \frac{-mv_3(0)}{\beta_3 - \bar{\mu}\beta_2}\left\{1 + \frac{(\mu + \bar{\mu})\beta_2}{\beta_2 + \mu\beta_1}\frac{v_1(0)}{v_3(0)}\right\}. \tag{3.36}$$

During the parts of the compression period before and after slip halts, the normal impulse does work on the colliding bodies given by

$$W_3(p_s) = \frac{p_s}{2}\left[2v_3(0) + m^{-1}(\beta_3 + \mu\beta_2)p_s\right]$$

$$W_3(p_c) - W_3(p_s) = \frac{-mv_3^2(p_s)}{2(\beta_3 - \bar{\mu}\beta_2)} \tag{3.37}$$

whereas during restitution, if the contact points stick, the normal work will be

$$W_3(p_f) - W_3(p_c) = \frac{m^{-1}(\beta_3 - \bar{\mu}\beta_2)}{2}(p_f - p_c)^2. \tag{3.38}$$

Together with the energetic coefficient of restitution, the normal work during these separate parts of the contact period give a terminal impulse p_f which can be expressed in relation to the normal impulse for compression p_c,

$$\frac{p_f}{p_c} = 1 + e_* \left\{ 1 + \frac{(\mu + \bar{\mu})\beta_2}{\beta_3 - \bar{\mu}\beta_2} \left(\frac{p_s}{p_c}\right)^2 \right\}^{1/2}. \tag{3.39}$$

With this terminal impulse, at the contact point the components of terminal velocity give the following velocity ratios:

$$\frac{v_1(p_f)}{v_3(0)} = 0$$

$$\frac{v_3(p_f)}{v_3(0)} = -e_* \left\{ 1 + \frac{(\mu + \bar{\mu})\beta_2}{\beta_3 - \bar{\mu}\beta_2} \left(\frac{p_s}{p_c}\right)^2 \right\}^{1/2} \left\{ 1 + \frac{(\mu + \bar{\mu})\beta_2}{(\beta_2 + \mu\beta_1)} \frac{v_1(0)}{v_3(0)} \right\}. \tag{3.40}$$

Jam: $\beta_2 < 0$, $\mu > -\beta_3/\beta_2$

Jam (or self-locking) is a process where during an initial period of sliding the normal component of relative velocity increases due to a positive normal acceleration at the contact point – during jam the rate of indentation increases. This acceleration is mostly due to rotational acceleration that is generated by a large friction force. Jam occurs only if there is an eccentric impact configuration with the center of mass at a small negative inclination relative to the common normal $\beta_2 < 0$, a large coefficient of friction $\mu > -\beta_3/\beta_2$, and initial sliding in the positive direction. This process persists until initial sliding is brought to a halt; thereafter, the contact points stick and are driven apart by the normal contact force (Batlle, 1998).

During jam Eq. (3.12) gives

$$dv_1/dp = -m^{-1}(\beta_2 + \mu\beta_1) < 0, \qquad dv_3/dp = m^{-1}(\beta_3 + \mu\beta_2) < 0. \tag{3.41}$$

Since $\beta_1\beta_3 > \beta_2^2$, this gives $\mu > -\beta_3/\beta_2 > -\beta_2/\beta_1$ and $\mu > \bar{\mu}$; consequently, after initial slip is brought to a halt at impulse p_s, the contact points stick. Initial sliding ceases when the normal impulse equals

$$p_s = \frac{mv_1(0)}{\beta_2 + \mu\beta_1}. \tag{3.42}$$

At impulse p_s when initial sliding has been brought to a halt, the normal relative velocity has increased in magnitude to

$$v_3(p_s) = v_3(0) + \left(\frac{\beta_3 + \mu\beta_2}{\beta_2 + \mu\beta_1}\right) v_1(0)$$

Subsequently the contact sticks, since $\mu > \bar{\mu}$, and further impulse accelerates the contact points apart:

$$v_3(p) = v_3(p_s) + m^{-1}(\beta_3 - \bar{\mu}\beta_2)(p - p_s)$$

$$= v_3(0) + m^{-1}(\mu + \bar{\mu})\beta_2 p_s + m^{-1}(\beta_3 - \bar{\mu}\beta_2)p, \qquad p > p_s.$$

The normal impulse for compression, p_c, is obtained from the condition that $v_3(p_c) = 0$, so that

$$p_c = \frac{-mv_3(0)}{\beta_3 - \bar{\mu}\beta_2} - \frac{(\mu + \bar{\mu})\beta_2 p_s}{\beta_3 - \bar{\mu}\beta_2}. \tag{3.43}$$

The terminal normal velocity $v_3(p_f)$ is given by

$$v_3(p_f) = m^{-1}(\beta_3 - \bar{\mu}\beta_2)(p_f - p_c). \tag{3.44}$$

To obtain the terminal impulse, the energetic coefficient of restitution e_* can be used. This requires separate evaluation of the partial work done by the normal impulse during compression, $W_3(p_s) + [W_3(p_c) - W_3(p_s)]$, and that done during restitution, $W_3(p_f) - W_3(p_c)$:

$$W_3(p_c) = \frac{p_c v_3(p_s)}{2} + \frac{p_s v_3(0)}{2}$$

$$= \frac{-1/2}{\beta_3 - \bar{\mu}\beta_2} \left\{ \frac{(\mu + \bar{\mu})\beta_2(\beta_3 + \mu\beta_2)mv_1^2(0)}{(\beta_2 + \mu\beta_1)^2} \right.$$

$$\left. + \frac{2(\mu + \bar{\mu})\beta_2 mv_1(0)v_3(0)}{\beta_2 + \mu\beta_1} + mv_3^2(0) \right\}$$

$$W_3(p_f) - W_3(p_c) = \frac{(p_f - p_c)v_3(p_f)}{2}$$

$$= \frac{m^{-1}}{2}(\beta_3 - \bar{\mu}\beta_2)(p_f - p_c)^2.$$

Hence during restitution the change in normal impulse is obtained from (3.24),

$$p_f - p_c = \frac{e_* mv_3(0)}{\beta_3 - \bar{\mu}\beta_2} \left\{ 1 + \frac{2(\mu + \bar{\mu})\beta_2 v_1(0)}{(\beta_2 + \mu\beta_1)v_3(0)} + \frac{(\mu + \bar{\mu})\beta_2(\beta_3 + \mu\beta_2)v_1^2(0)}{(\beta_2 + \mu\beta_1)^2 v_3^2(0)} \right\}^{1/2}$$

$$\tag{3.45}$$

Figure 3.6 shows the changes in relative velocity and the work done during different

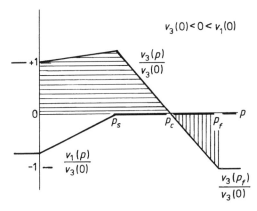

Figure 3.6. Changes in relative velocity for eccentric collision with initial period of jam. The shaded areas indicate the normal work done during compression and that recovered during restitution. Note that if the configuration and coefficient of friction are sufficient to cause jam, the sliding process is jam–stick.

Table 3.2. *Terminal Normal Impulse Ratio p_f/p_c for Planar Impact, $v_1(0) \geq 0$, $v_3(0) < 0$*

Process	Relative velocity at incidence, $v_1(0)/v_3(0)$	Friction coeff. $\mu/|\bar{\mu}|$	Impulse that halts init. slip, $p_s/mv_1(0)$	Normal impulse of compression, $-p_c/mv_3(0)$	Ratio of terminal to comp. normal impulse, p_f/p_c
Continous stick	0	> 1	—	$\bar{\beta}_c^{-1}$	$1+e_*$
Continous slip	$(1+e_*)\dfrac{\beta_a}{\beta_b} < -\dfrac{v_1(0)}{v_3(0)}$	—	—	β_b^{-1}	$1+e_*$
Slip–stick in compression	$-\dfrac{v_1(0)}{v_3(0)} < \dfrac{\beta_a}{\beta_b}$	> 1	β_a^{-1}	$\bar{\beta}_c^{-1}\left\{1+\dfrac{(\mu+\bar{\mu})\beta_2}{\beta_a}\dfrac{v_1(0)}{v_3(0)}\right\}$	$1+e_*\left\{1+\dfrac{(\mu+\bar{\mu})\beta_2}{\bar{\beta}_c}\dfrac{p_s^2}{p_c^2}\right\}^{1/2}$
Slip reversal in compression	$-\dfrac{v_1(0)}{v_3(0)} < \dfrac{\beta_a}{\beta_b}$	< 1	β_a^{-1}	$\beta_c^{-1}\left\{1+\dfrac{2\mu\beta_2}{\beta_a}\dfrac{v_1(0)}{v_3(0)}\right\}$	$1+e_*\left\{1+\dfrac{2\mu\beta_2}{\beta_c}\dfrac{p_s^2}{p_c^2}\right\}^{1/2}$
Slip–stick in restitution	$\dfrac{\beta_a}{\beta_b} < -\dfrac{v_1(0)}{v_3(0)} < (1+e_*)\dfrac{\beta_a}{\beta_b}$	> 1	β_a^{-1}	β_b^{-1}	$\dfrac{p_s}{p_c}+\dfrac{\beta_b}{\bar{\beta}_c}\left(\dfrac{p_s}{p_c}-1\right)\left\{\left[\dfrac{(\mu+\bar{\mu})\beta_2}{\beta_b}+\dfrac{p_c^2e_*^2}{(p_s-p_c)^2}\dfrac{\bar{\beta}_c}{\beta_b}\right]^{1/2}-1\right\}$
Slip reversal in restitution	$\dfrac{\beta_a}{\beta_b} < -\dfrac{v_1(0)}{v_3(0)} < (1+e_*)\dfrac{\beta_a}{\beta_b}$	< 1	β_a^{-1}	β_b^{-1}	$\dfrac{p_s}{p_c}+\dfrac{\beta_b}{\beta_c}\left(\dfrac{p_s}{p_c}-1\right)\left\{\left[\dfrac{2\mu\beta_2}{\beta_b}+\dfrac{p_c^2e_*^2}{(p_s-p_c)^2}\dfrac{\beta_c}{\beta_b}\right]^{1/2}-1\right\}$

$\beta_a \equiv \beta_2 + \mu\beta_1$, $\beta_b \equiv \beta_3 + \mu\beta_2$, $\beta_c \equiv \beta_3 - \mu\beta_2$, $\bar{\beta}_c \equiv \beta_3 - \bar{\mu}\beta_2$, $\bar{\mu} \equiv \beta_2/\beta_1$.

Tangential impulse at separation $p_1(p_f) = -\mu p_s + \bar{\mu}(p_f - p_s)$ if $p_s < p_f$ and $\bar{\mu} \equiv \begin{cases} \mu, & \mu < |\bar{\mu}|, \text{ slip reversal,} \\ \bar{\mu}, & \mu > |\bar{\mu}|, \text{ slip–stick.} \end{cases}$

periods of impulse for an eccentric collision with initial jam. The dynamics of jam have been extended to 3D collisions by Batlle (1998).

Grazing Incidence: $\beta_2 < 0$, $\mu > -\beta_3/\beta_2$, $v_1(0)/v_3(0) \rightarrow \infty$
Jam can result in an impact process in the limiting case of vanishing normal component of velocity $v_3(0) \rightarrow 0$; this is termed *grazing incidence*. In this case the normal impulse that brings initial sliding to a halt p_s is given by (3.42). Subsequently the contact sticks while the normal contact force drives the bodies apart. Hence the normal impulse p_c at the end of compression is obtained from (3.43),

$$p_c = \frac{-(\mu + \bar{\mu})\beta_2 p_s}{\beta_3 - \bar{\mu}\beta_2}$$

while the impulse during restitution $p_f - p_c$,

$$p_f - p_c = \frac{e_* m v_1(0)}{\beta_3 - \bar{\mu}\beta_2} \left\{ \frac{(\mu + \bar{\mu})\beta_2(\beta_3 + \mu\beta_2)}{(\beta_2 + \mu\beta_1)^2} \right\}^{1/2}.$$

Grazing incidence finally results in a positive normal component of relative velocity although the incident relative velocity has no normal component.

Terminal Impulse p_f for Arbitrary Initial Conditions
For various ratios of initial tangential to normal incident speeds, Table 3.2 lists the impulse p_s at which initial slip is halted, the compression impulse p_c and the terminal impulse p_f. The range of angles of incidence wherein each relation applies depends on inertia properties of the bodies and the coefficient of friction as well as the initial ratio of tangential to normal incident speed.

3.5   Friction in Collinear Impact Configurations

Collinear impact configurations always give $\beta_2 = 0$ and $\beta_3 = 1$, so that the equations of relative motion (3.13) simplify to

$$v_1(p) = v_1(0) - \hat{s}m^{-1}\mu\beta_1 p \qquad\qquad (3.46a)$$

$$v_3(p) = v_3(0) + m^{-1}p. \qquad\qquad (3.46b)$$

The sphere shown in Fig. 3.7 has mass M and radius R; it is rotating about a transverse axis when it collides with a stationary half space at a contact point C. At any reaction impulse p during contact the sphere has angular velocity $\boldsymbol{\omega}(p) = (0, \omega, 0)$ and the center

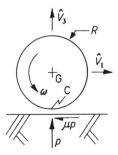

Figure 3.7. Oblique impact of rotating sphere on rough half space.

of mass G has translational velocity $\hat{\mathbf{v}}(p) = (\hat{v}_1, 0, \hat{v}_3)$. At the contact point the friction force satisfies Coulomb's law of friction. For this system and slip at C in the positive direction (i.e. $v_1 > 0$), the differential equations of motion for changes in components of velocity are given in terms of the normal component of impulse p:

$$M d\hat{v}_1 = -\mu\, dp, \quad d\hat{v}_3 = dp, \quad M\hat{k}_r^2\, d\omega = -\mu R\, dp \qquad (v_1 > 0).$$

The components of relative velocity $\mathbf{v}(p) = (v_1, 0, v_3)$ at C are obtained from the following expressions that apply in this case of collinear collision:

$$v_1(p) = \hat{v}_1(p) + R\omega(p), \qquad v_3(p) = \hat{v}_3(p).$$

These translations and the differential equations give differential equations for changes in relative velocity at C that can readily be integrated. On applying the initial conditions $v_1(0) = \hat{v}_1(0) + R\omega(0)$, $v_3(0) = \hat{v}_3(0)$ and noting that $v_3(0) < 0$, these integrations give

$$v_1(p) = v_1(0) - \mu M^{-1}(1 + R^2/\hat{k}_r^2)p, \quad v_3(p) = v_3(0) + M^{-1}p \qquad (v_1 > 0).$$

A transition from compression to restitution occurs at impulse p_c when the normal velocity vanishes [$v_3(p_c) = 0$], so that $p_c = -Mv_3(0)$. Hence the previous expressions can be divided by $v_3(0)$ to give nondimensional expressions,

$$\frac{v_1(p)}{v_3(0)} = \frac{v_1(0)}{v_3(0)} + \mu\left(1 + \frac{R^2}{\hat{k}_r^2}\right)\frac{p}{p_c}, \qquad \frac{v_3(p)}{v_3(0)} = 1 - \frac{p}{p_c}.$$

These changes in relative velocity at C are illustrated in Fig. 3.8. For the planar changes in velocity that result from this collinear configuration, the changes in velocity are linear functions of impulse ratio p/p_c.

The second equation above can be equated to zero to give the impulse ratio p_s/p_c when sliding terminates:

$$\frac{p_s}{p_c} = \frac{-v_1(0)/v_3(0)}{\mu(1 + R^2/\hat{k}_r^2)} = \frac{-[\hat{V}_1(0) + R\omega(0)]/\hat{V}_3(0)}{\mu(1 + R^2/\hat{k}_r^2)}.$$

A collinear collision terminates at impulse $p_f = (1 + e_*)p_c$ when the contact points separate. If sliding halts before separation ($p_s < p_f$), then when sliding halts, the components

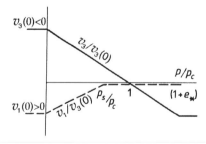

Figure 3.8. Changes in components of velocity and impulse as a function of impulse acting on a rotating sphere. The speed of slip $v_1(p)$ vanishes at impulse p_s and then sticks, whereas the characteristic impulse for compression p_c occurs at the transition from compression to restitution.

of velocity at the center of mass are

$$\frac{\hat{V}_1(p_s)}{\hat{V}_3(0)} = \frac{\hat{V}_1(0)}{\hat{V}_3(0)} + \mu \frac{p_s}{p_c}, \qquad \frac{\hat{V}_3(p_s)}{\hat{V}_3(0)} = 1 + \frac{p_s}{p_c}, \qquad \frac{R\omega(p_s)}{\hat{V}_3(0)} = \frac{R\omega(0)}{\hat{V}_3(0)} + \mu \frac{R^2}{\hat{k}_r^2} \frac{p_s}{p_c}.$$

After sliding halts ($p_s < p < p_f$), there are no additional changes in translational velocity $\hat{V}_1$ or angular speed ω if the collision configuration is collinear. For a collinear configuration there is no friction force or additional tangential impulse required to maintain stick; i.e., $v_1(p) = 0$, $F_1(p) = 0$ and $p_1 = -\mu p_s$ during $p_s < p < p_f$.

Example 3.2   A rigid sphere of radius R is rotating at an initial angular speed $\omega(0)$ about a horizontal axis when it strikes the level surface of a rough elastic half space at an incident speed $\hat{V}_3(0)$. Before impact the horizontal translational velocity of the center of mass is zero: $\hat{V}_1(0) = 0$. Find (i) the angle of rebound Ψ_f and (ii) the coefficient of friction $\tilde{\mu}$ which causes slip to cease at the instant of separation.

Solution   Impulse ratio when slip stops:

$$\frac{p_s}{p_c} = \frac{-R\omega(0)/\hat{V}_3(0)}{\mu\left(1 + R^2/\hat{k}_r^2\right)}.$$

(i) If $p_s < p_f$, any initial slip is brought to a halt during contact, so that the final velocity at G is given by

$$\frac{\hat{V}_1(p_s)}{\hat{V}_3(0)} = \mu \frac{p_s}{p_c}, \qquad \frac{\hat{V}_3(p_s)}{\hat{V}_3(0)} = -e_*, \qquad \frac{R\omega(p_s)}{\hat{V}_3(0)} = \frac{R\omega(0)}{\hat{V}_3(0)} + \mu \frac{R^2}{\hat{k}_r^2} \frac{p_s}{p_c}.$$

Hence the angle of rebound Ψ_f as defined in Fig. 3.9 is given by

$$\Psi_f \equiv \tan^{-1} \frac{\hat{V}_1(p_f)}{\hat{V}_3(p_f)} = \tan^{-1} \frac{-R\omega(0)/\hat{V}_3(0)}{e_*\left(1 + R^2/\hat{k}_r^2\right)}.$$

(ii) If $p_s > p_f$, slip continues throughout contact, giving a final velocity at G

$$\frac{\hat{V}_1(p_s)}{\hat{V}_3(0)} = \mu(1 + e_*), \qquad \frac{\hat{V}_3(p_s)}{\hat{V}_3(0)} = -e_*, \qquad \frac{R\omega(p_s)}{\hat{V}_3(0)} = \frac{R\omega(0)}{\hat{V}_3(0)} + \mu \frac{R^2}{\hat{k}_r^2}(1 + e_*).$$

and an angle of rebound $\Psi_f \equiv \tan^{-1}[-\mu(1 + e_*^{-1})]$. For an initially rotating sphere with

Figure 3.9. Rotating sphere dropped vertically onto a half space, illustrating angle of rebound ψ_f at separation.

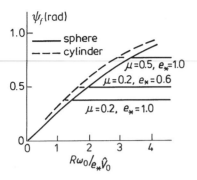

Figure 3.10. Rebound angle for rotating sphere (or cylinder) as function of the angle of incidence at the contact point.

no tangential speed, sliding at the contact point does not stop during impact if

$$\mu < \tilde{\mu} \equiv \left| \frac{R\omega(0)/\hat{V}_3(0)}{(1 + e_*)(1 + R^2/\hat{k}_r^2)} \right|.$$

Figure 3.10 shows the rebound angle Ψ_f of a dropped ball as a function of the rate of rotation. The angle is limited by gross slip vanishing before separation.

Example 3.3 A sphere (or cylinder) of radius R rolls on a level surface without slip before it collides against a ramp inclined at angle $\pi/2 - \Psi_0$ as depicted in Fig. 3.11a. Before impact the center of mass has a horizontal speed $R\omega_0$. Find the maximum coefficient of friction $\tilde{\mu}$ for gross slip and the angle of rebound Ψ_f if $\mu \leq \tilde{\mu}$.

Solution At incidence the normal and tangential components of velocity are as follows:

$$\hat{V}_3(0) = -R\omega_0 \cos \Psi_0, \qquad \hat{V}_1(0) = -R\omega(0) \sin \Psi_0.$$

Initial slip persists until the normal impulse equals p_s, i.e.

$$\frac{p_s}{p_c} = \frac{-[\hat{V}_1(0) + R\omega(0)]/\hat{V}_3(0)}{\mu(1 + R^2/\hat{k}_r^2)} = \frac{(1 - \sin \Psi_0)/ \cos \Psi_0}{\mu(1 + R^2/\hat{k}_r^2)}.$$

Initial slip is stopped during collision if $p_s/p_c < 1 + e_*$; otherwise there is gross slip. The range of values for coefficients of friction and restitution that result in slip–stick is shown in Fig. 3.12. If the coefficient of friction is large enough to cause gross slip, at separation the center of mass of the sphere is traveling at an angle $\tan^{-1}[e_*^{-1} \cot \theta + \mu(1 + e_*^{-1})]$ from normal. This is an upper bound on the change in angle for given values of the coefficients of friction and restitution.

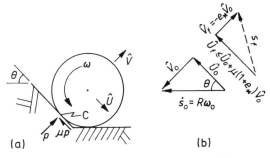

Figure 3.11. (a) Rolling sphere with center of mass traveling at speed s_0 when it collides against inclined bumper, and (b) components of incident and terminal velocities at center of mass G.

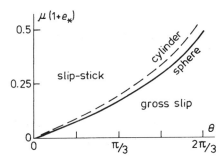

Figure 3.12. Angles of inclination of bumper and coefficient of friction giving either slip–stick or gross slip behavior during impact.

3.6 Friction in Noncollinear Impact Configurations

3.6.1 Planar Impact of Rigid Bar on Rough Half Space

The following is a straightforward example illustrating some effects of friction for oblique impact of eccentric rigid bodies. The example of a slender rigid bar has been chosen for simplicity and clarity of presentation. The reader should recognize, however, that if the eccentricity of the collision configuration is large, rigid body theory is not very accurate for slender bodies, because that theory neglects transverse vibrations, which are important for slender bodies.[5] Here we consider impact of a rigid bar of mass M and length $2L$ with an end that strikes against the surface of a rough half space. At impact the bar is inclined at angle θ from the vertical as shown in Fig. 3.13. The bar has a radius of gyration $\hat{k}_r = L/\sqrt{3}$ about a transverse axis through the center of mass. The center of mass of the bar has an initial velocity $\hat{V}(0) = (\hat{V}_1, \hat{V}_3)$, and the bar is not rotating, so that the initial angular velocity $\omega(0) = 0$. For this impact configuration the effective mass $m = M$ and the elements of the inverse of inertia matrix are

$$\beta_1 = 1 + 3\cos^2\theta = (5 + 3\cos 2\theta)/2$$

$$\beta_2 = 3\sin\theta\cos\theta = (3\sin 2\theta)/2$$

$$\beta_3 = 1 + 3\sin^2\theta = (5 - 3\cos 2\theta)/2$$

---

[5] Experiments by Stoianovici and Hurmuzlu (1996) have shown that the effect of transverse vibrations can be important for impact of a slender bar in an eccentric configuration.

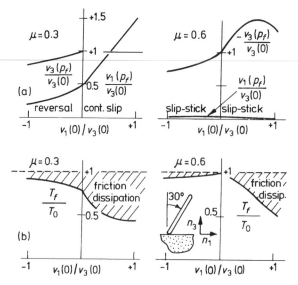

Figure 3.13. Oblique elastic impact of inclined rod ($\theta = \pi/6$) on rough half space ($\bar{\mu} = 0.4$): (a) components for terminal velocity of contact point; (b) terminal kinetic energy $T_f \equiv T(p_f)$ and energy dissipated by friction $T_0 - T_f$.

where $\beta_1 > 0$ and $\beta_3 > 0$. For dry friction with a coefficient of friction μ the equations of motion for frictional impact (3.12) give the following differential equations for changes in relative velocity at the contact point C as a function of normal impulse p:

$$dv_1 = -m^{-1}(\beta_2 + \hat{s}\mu\beta_1)\,dp$$
$$dv_3 = m^{-1}(\beta_3 + \hat{s}\mu\beta_2)\,dp.$$

Integration of these equations requires consideration of the different slip processes which can occur for different parts of the range of angle of incidence $\Psi_0 = \tan^{-1}[v_1(0)/v_3(0)]$.

Terminal Velocity at Contact Point and Energy Dissipation during Impact

Figure 3.13 illustrates calculated values for changes in relative velocity at the contact point and energy loss due to friction of a rigid rod inclined at an angle of $\theta = +30°$ that collides against a stationary half space. The values were calculated from expressions in Sect. 3.3.2. The calculations are for a coefficient of friction equal to unity so that all energy loss is due to friction. Two values of the coefficient of friction have been examined; the graphs on the left are for light friction $\mu < \bar{\mu}$, where small initial slip can reverse during collision, while those on the right are for heavy friction $\mu > \bar{\mu}$, where small initial slip can be halted and then stick.

This example clearly shows that if the coefficient of friction is based on work done by the normal component of contact force (i.e. the energetic coefficient of friction), then the normal component of relative velocity at separation, $v_3(p_f)$, is not directly related to the normal component of relative velocity at incidence, $v_3(0)$. The ratio of these speeds equals the coefficient of restitution only if the direction of slip is constant or if there is no slip during the collision. For light friction the speed of sliding either decreases or increases from its initial value depending on the direction of sliding in relation to the angle of inclination of the center of mass; initial sliding in the direction away from the center of mass decelerates only if friction is heavy ($\mu > \bar{\mu}$). The large coefficient of friction $\mu = 0.6$ brought initial sliding to a halt during collision for $-1 < v_1(0)/v_3(0) < 1$. With heavy friction, small initial sliding results in slip–stick during contact.

On the plots of energy loss in the collision, the dashed line represents the coefficient of restitution. The dependence of frictional dissipation on the direction of slip relative to the angle of inclination of the center of mass is particularly noticeable. There is much more frictional dissipation when the bar strikes with the end that leads the center of mass, much as a javelin does at the end of its flight. If these calculations were done with a coefficient of restitution of less than unity ($e_* < 1$), then the height of the dashed horizontal line would be reduced.

PROBLEMS

3.1 If bodies B and B′ with masses M and M' collide at C, show that the two centers of mass have a relative velocity $\hat{v}_i \equiv \hat{V}_i - \hat{V}'_i$ $i = 1, 3$ which satisfies the differential equation $dv_i = m^{-1} dp_i$. Separate this differential equation for changes in relative velocity into a part for each mass. Do the same for the relative velocity v_i at contact point C. For an eccentric impact between smooth bodies, use the difference between changes in the normal components of relative velocity at C and the center of mass to obtain the change in angular velocity of body B as a function of the coefficient of restitution e_*, the radius of gyration $\hat{k}_r$ and the eccentricity r_1.

3.2 For two colliding bodies, the center of mass of the system has a normal component of translational velocity that is constant. Express this velocity $\hat{V}_3(p_c) = \hat{V}'_3(p_c)$ in terms of the component of initial velocity for bodies with masses M and M' respectively.

3.3 A billiard ball of radius R is initially at rest on a level table. An impulse is imparted to the sphere by striking it with a cue stick. Find that the sphere begins to roll on the table without slip if the line of action of the impulse passes through a point A_C located a distance ξ above the center of mass (G), where $\xi/R = \hat{k}_r^2/R^2$ and $\hat{k}_r$ is the radius of gyration of the sphere about G. Point A_C is termed the center of percussion for the contact point C. An impulse applied through the center of percussion A_C generates no reaction at C in a direction perpendicular to line A_CC.

3.4 For a smooth billiard ball that rolls into an identical ball and collides at an angle of obliquity Ψ_0 as specified in Example 3.1, find the velocity of the center of the initially stationary ball B immediately following a post-impact period of sliding if the limiting coefficient of friction between the ball and the table is μ. Also find an expression for the total sliding distance x_{3B}.

3.5 To obtain the largest angle of deflection for the rolling ball, pool or billiard players frequently aim the centerline of the path of this ball at the lateral surface of the stationary ball which is to be hit. For smooth solid spherical balls, compare the angular deflection of the rolling ball using this tactic with the maximum achievable angular deflection.

3.6 For impact between spherical balls as described in Problem 3.4, calculate the translational momentum parallel and perpendicular to the initial path (a) at the instant of separation from impact and (b) at the instant when sliding terminates and rolling resumes. Explain the difference.

3.7 Prove that after initial slip is brought to a halt at impulse p_s, slip cannot reinitiate in the original direction during $p_s < p < p_f$.

3.8 A golf ball of mass M and radius R is struck by a club of much larger mass. The club has a face inclined at angle $\Psi = \pi/6$ from vertical, and it moves horizontally with speed V_0 when it strikes the ball. Assume the tangential compliance is negligible.

(a) Derive equations of relative motion for the tangential and normal components $v_1(p)$ and $v_3(p)$ of the relative velocity, and show that these can be expressed as

$$v_1 = v_1(0) + \tfrac{7}{2}m^{-1}p_1$$

$$v_3 = v_3(0) + m^{-1}p_3.$$

(b) Obtain the coefficient of Coulomb friction μ that is just sufficient to bring initial sliding to a halt at the transition from compression to restitution.

(c) Write an expression for the work done by the contact force during compression if the coefficient of friction $\mu = 2\sqrt{3}/21$.

3.9 In the case of slip that halts during the restitution phase of contact, obtain an analytical expression for the terminal normal impulse p_f at separation. Show that this is in agreement with the expression given in Table 3.2.

3.10 For a slender rod with one end striking a heavy half space, find the angle of inclination θ where jam requires the smallest coefficient of friction. Evaluate the smallest coefficient of friction that gives jam. Explain why jam necessarily terminates during compression.

3.11 A solid rectangular block has edges of length b parallel to the surface of a level rigid half plane and height a. A bottom edge A of the block initially is in contact with the plane, and the block has an angular velocity ω_0 about A at the instant edge B strikes the plane. Assume that contact with the plane occurs only along the parallel edges A and B.

(a) Assuming edge B does not rebound and there is no slipping during impact, show that after impact the body starts to rotate about B with angular velocity

$$\frac{\omega_0(2a^2 - b^2)}{2(a^2 + b^2)} \qquad \text{for} \qquad a\sqrt{2} > b.$$

(b) Assume instead that the impact at B is perfectly elastic, so that the velocity of the impact point B reverses during the impact. Write down the velocity of B just before and after impact. Calculate the angular velocity of the block just after the impact, using momentum considerations. Confirm that energy is conserved during impact.

(c) From the changes in translational momentum obtain the ratio of tangential to vertical impulse to show that in order to prevent slip during impact, the coefficient of dry friction μ must satisfy $\mu > 0.6$.

3.12 A smooth (frictionless) sphere of mass M and radius R rolls on a level plane at angular speed ω_0 before it collides with a small vertical step of height h. The impact with the step has a coefficient of restitution e_*.

(a) Find an expression for ΔT, the energy dissipated in the collision.

(b) Obtain a lower bound on the initial speed of the center of mass, $R\omega_0$, if the sphere is to pass over the step.

CHAPTER 4

3D Impact of Rough Rigid Bodies

Like a ski resort full of girls hunting for husbands, and husbands
hunting for girls, the situation is not as symmetrical as it might seem.

Alan Lindsay Mackay, Lecture, Birkbeck College,
University of London, 1984

Three-dimensional (3D, or nonplanar) changes in velocity occur in collisions
between rough bodies if the configuration is not collinear and the initial direction of
sliding is not in-plane with two of the three principal axes of inertia for each body. In
collisions between rough bodies, dry friction can be represented by Coulomb's law. If there
is a tangential component of relative velocity at the contact point (sliding contact) this law
relates the normal and tangential components of contact force by a coefficient of limiting
friction. The friction force acts in a direction opposed to sliding. For a collision with planar
changes in velocity, sliding is in either one direction or the other on the common tangent
plane. In general however, friction results in nonplanar changes in velocity. Nonplanar
velocity changes give a direction of sliding that continuously changes, or *swerves*, during
an initial phase of contact in an eccentric impact configuration. This chapter obtains
changes in relative velocity during rigid body collisions as a function of the impulse P
due to the normal component of the reaction force. The changes in velocity depend on two
independent material parameters – the coefficient of friction and an energetic coefficient
of restitution.

During moderate speed collisions between two hard bodies there are continuous
changes in relative velocity which can easily be calculated if we recognize that the bodies
are not entirely rigid – there is a small region of deformation that surrounds the initial
contact point. At the contact point there is a point of contact on each colliding body; these
points are coincident, but they have different velocities. By considering an infinitesimally
small deformable region located between the bodies at the contact point, changes in relative velocity v_i across the contact point can be expressed as a function of the impulse due
to the reaction force. The reaction impulse P_i that develops during a collision depends on
the initial relative velocity $v_i(0)$ and both material and inertia properties of each body at the
impact point. The relative velocity between the colliding bodies and the reaction impulse
can be resolved into components normal and tangential to the common tangent plane
for the contact surfaces at the point of contact. Let the unit vector n be oriented normal
to the tangent plane. The normal component of the impulse, $P \equiv P_i \cdot nn$, monotonically
increases in magnitude during contact because the normal force F is compressive; thus,
changes in relative velocity are obtained as a function of the scalar independent variable P.

For bodies composed of rate-independent materials, the maximum normal force and maximum compression occur at the impulse P_c where the normal component of relative velocity vanishes. During an initial period of contact $P < P_c$, the normal component of contact force $F \equiv dP/dt$ does negative work on the bodies[1]; this work compresses the bodies in a small region around the contact point while decreasing the sum of their kinetic energies. Following maximum compression $P > P_c$, the normal component of contact force decreases as it drives the bodies apart while the compressed region undergoes a release of elastic strain energy. The work done on the bodies by the normal force F is positive during the restitution period $P > P_c$, and this restores that part of the initial kinetic energy which was transformed into elastic strain energy during compression.

In addition to the normal component of relative velocity at the contact point there is also a tangential component of relative velocity called *slip*. If the bodies are rough, a tangential contact force, termed friction, opposes any slip. This friction force complicates the analysis of impact problems, especially if friction and normal force components are interdependent due to coupling in the equations of motion. This coupling occurs if the impact configuration is eccentric or noncollinear, i.e. if the center of mass for each body is not on the common normal line through the contact point. The analytical complications develop if the direction of slip varies during the contact period.

For eccentric collisions with changes in relative velocity that are planar, complications due to friction have been overcome by dividing the total impulse into parts which correspond to successive phases of unidirectional slip. In general, however, collisions between rough bodies result in changes $v_i(P) - v_i(0)$ in relative velocity at the contact point that are nonplanar unless the initial relative velocity $v_i(0)$ lies in the same plane as two principal axes of inertia for each body. We will show that in most cases the direction of slip changes continuously during contact if the collision configuration is eccentric. A general formulation for impact of two unconstrained rough rigid bodies will be developed. Then the effects of friction in 3D (nonplanar) collisions will be illustrated by three examples: the oblique impact of a sphere that has a component of initial rotation about the direction of initial translation, the impact of an inclined rod on a half space and a collision of a spherical pendulum on a half space. These examples represent collinear and two eccentric impact configurations, respectively. The colliding sphere is an example of a collinear impact configuration, while the collision of an inclined rod and that of a spherical pendulum are examples of eccentric impact configurations.

4.1   Collision of Two Free Bodies

Consider two bodies that collide at contact point C; the bodies have no displacement constraints except that they are mutually impenetrable at C. If the surface of at least one of the bodies has continuous curvature at C, there is a common tangent plane that contains point C. First define a common normal direction n that is perpendicular to the common tangent plane. Let $n_i, i = 1,2,3$, be a set of mutually perpendicular unit vectors with n_1 and n_2 in the tangent plane while $n_3 = n$ is normal to this plane, as shown in Fig. 4.1. The bodies have centers of mass located at G and G' respectively. There is a

---

[1] Relative to either body, during compression the normal contact force is opposed to the normal component of relative velocity across the deformable particle at the contact point.

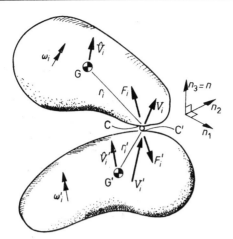

Figure 4.1. Collision between two rough bodies. The rigid bodies have contact points C and C' that are separated by a deformable particle.

position vector r_i from G to the contact point C, while r_i' locates C' from G'. The bodies have masses M and M', inertia tensors $\hat{I}_{ij}$ and $\hat{I}_{ij}'$ for second moments of their masses at G and G' respectively . Let $\hat{V}_i$ and $\hat{V}_i'$ be the velocities of the centers of mass, while ω_i and ω_i' are the corresponding angular velocities for the bodies in reference frame n_i . At the contact points C and C' the bodies are subjected to contact forces F_i and F_i'; these contact forces are reactions that apply an impulse to each body. Denote these impulses as $P_i(t)$ and $P_i'(t)$, where

$$dP_i = F_i\, dt \quad \text{and} \quad dP_i' = F_i'\, dt. \tag{4.1}$$

The first aim is to obtain equations of motion in terms of relative velocities at the contact point, since this is where relative displacement (interpenetration) is resisted by reaction forces.

At the two centers of mass the equations of translational and rotational motion for each body can be expressed as

$$M d\hat{V}_i = dP_i \tag{4.2a}$$

$$\hat{I}_{ij}\, d\omega_j = \varepsilon_{ijk} r_j\, dP_k \tag{4.3a}$$

and

$$M' d\hat{V}_i = dP_i' \tag{4.2b}$$

$$\hat{I}_{ij}'\, d\omega_j' = \varepsilon_{ijk} r_j'\, dP_k' \tag{4.3b}$$

where a repeated index (e.g. j or k) indicates summation and the permutation tensor ε_{ijk} takes the values $\varepsilon_{ijk} = +1$ if the indices are in cyclic order, $\varepsilon_{ijk} = -1$ if the indices are in anticyclic order and $\varepsilon_{ijk} = 0$ for repeated indices. Thus in index notation the vector product $r_j \times dP_k = \varepsilon_{ijk} r_j\, dP_k$. For a body of volume V with density ρ, the elements of the inertia matrix $\hat{I}_{ij}$ for axes through the center of mass are defined as the moments and products of inertia. Typically,

$$\hat{I}_{11} \equiv \int_V \left(r_2^2 + r_3^2\right)\rho\, dV, \qquad \hat{I}_{12} \equiv -\int_V r_1 r_2 \rho\, dV.$$

During collision the reaction forces that act at the contact point are large if the bodies are hard; in particular, these forces are very large in comparison with any body force. Consequently it can be assumed that the only forces acting during a collision are the reactions at C and C' ; the impulse of these reactions depends on changes in relative velocity across a small deforming region at C . The effect of a small deforming region can be represented by assuming that during a collision the contact points C and C' on the colliding bodies are separated by a negligibly small deformable element.[2] The velocity of each contact point, V_i or V_i', can be obtained from the velocity of the respective center of mass and the relationship between velocities of two points on a rigid body,

$$V_i = \hat{V}_i + \varepsilon_{ijk}\omega_j r_k \quad \text{and} \quad V_i' = \hat{V}_i' + \varepsilon_{ijk}\omega_j' r_k'.$$

Let the *relative velocity* v_i between the contact points C and C' be defined as

$$v_i = V_i - V_i'. \tag{4.4}$$

Any incremental changes in reaction impulse acting on the rigid bodies are equal in magnitude but opposite in direction if the infinitesimally small deforming element has negligible mass; i.e.

$$dp_i \equiv dP_i = -dP_i'. \tag{4.5}$$

Changes in relative velocity at C can be related to changes in impulse of the reaction by substituting (4.2) and (4.3) into (4.4) and thence (4.5). This gives an *equation of motion for changes in relative velocity* v_i,

$$dv_i = m_{ij}^{-1} dp_j \tag{4.6}$$

where the elements of the inverse inertia matrix for C are given by

$$m_{ij}^{-1} \equiv \left(\frac{1}{M} + \frac{1}{M'}\right)\delta_{ij} + \varepsilon_{ikm}\varepsilon_{jln}\left(I_{kl}^{-1} r_m r_n + I_{kl}'^{-1} r_m' r_n'\right) \tag{4.7}$$

and δ_{ij} is the Kronecker delta defined as $\delta_{ij} \equiv 1$ if $i = j$ and $\delta_{ij} \equiv 0$ if $i \neq j$. This inverse inertia matrix is symmetric, $m_{ij}^{-1} = m_{ji}^{-1}$. The following are representative elements:

$$m_{11}^{-1} = \left(M^{-1} + r_2^2 I_{33}^{-1} - 2r_2 r_3 I_{23}^{-1} + r_3^2 I_{22}^{-1}\right)$$
$$+ \left(M'^{-1} + r_2'^2 I_{33}'^{-1} - 2r_2' r_3' I_{23}'^{-1} + r_3'^2 I_{22}'^{-1}\right)$$

$$m_{12}^{-1} = \left(r_1 r_3 I_{23}^{-1} - r_3^2 I_{21}^{-1} - r_1 r_2 I_{33}^{-1} + r_2 r_3 I_{31}^{-1}\right)$$
$$+ \left(r_1' r_3' I_{23}'^{-1} - r_3'^2 I_{21}'^{-1} - r_1' r_2' I_{33}'^{-1} + r_2' r_3' I_{31}'^{-1}\right).$$

$$m_{13}^{-1} = \left(r_1 r_2 I_{32}^{-1} - r_2^2 I_{31}^{-1} - r_1 r_3 I_{22}^{-1} + r_2 r_3 I_{21}^{-1}\right)$$
$$+ \left(r_1' r_2' I_{32}'^{-1} - r_2'^2 I_{31}'^{-1} - r_1' r_3' I_{22}'^{-1} + r_2' r_3' I_{21}'^{-1}\right)$$

In these expressions notice that the matrix I_{ij} of moments and products of inertia has an inverse which is denoted by I_{ij}^{-1} ; e.g. $I_{21}^{-1} = (I_{13} I_{23} - I_{12} I_{33})/\det(I_{ij})$.

---

[2] The physical construct of a deformable particle separating contact points on colliding rigid bodies is mathematically equivalent to Keller's (1986) asymptotic method of integrating with respect to time the equations for relative acceleration of deformable bodies and then taking the limit as the compliance (or contact period) becomes vanishingly small.

4.1.1 Law of Friction for Rough Bodies

Dry friction between colliding bodies can be represented by the Amontons–Coulomb law of sliding friction (Johnson, 1985). This law relates the tangential component to the normal component of reaction force at the contact point by introducing a coefficient of limiting friction μ which acts if there is sliding, i.e. $v_1^2 + v_2^2 > 0$. Denoting the magnitude of the normal component of a differential increment of impulse by $dp \equiv dp_3 \equiv dP_n$, this law takes the form

$$\sqrt{(dp_1)^2 + (dp_2)^2} < \mu \, dp \qquad\qquad \text{if } v_1^2 + v_2^2 = 0 \quad (4.8\text{a})$$

$$dp_1 = -\frac{\mu v_1}{\sqrt{v_1^2 + v_2^2}} dp, \quad dp_2 = -\frac{\mu v_2}{\sqrt{v_1^2 + v_2^2}} dp \qquad \text{if } v_1^2 + v_2^2 > 0. \quad (4.8\text{b})$$

Equation (4.8a) expresses an upper bound on the ratio of tangential to normal force for rolling contact; for ratios of tangential to normal contact force that are less than μ the *sliding speed* $s \equiv \sqrt{v_1^2 + v_2^2}$ vanishes. If sliding is present ($s > 0$), the tangential increment of impulse or friction force at any impulse acts in a direction directly opposed to sliding and has a magnitude that is directly proportional to the normal force.[3] The sliding direction can be defined by the angle ϕ measured in the tangent plane from n_1; thus $\phi \equiv \tan^{-1}(v_2/v_1)$ and

$$v_1 = s \cos\phi, \qquad v_2 = s \sin\phi. \tag{4.9}$$

Since the normal contact force must be compressive, the impulse of the normal component of reaction is a monotonously increasing scalar function during the collision period. Hence rates of change for relative velocity at the contact point C can be expressed as a function of the rate of change of impulse for the normal component of reaction; i.e., the collision process can be resolved as a function of the independent variable p.

4.1.2 Equation of Motion in Terms of the Normal Impulse

For sliding in direction $\phi(p)$ the equations of motion can be obtained in terms of the impulse of the normal component of reaction:

$$dv_1/dp = -\mu m_{11}^{-1} \cos\phi - \mu m_{12}^{-1} \sin\phi + m_{13}^{-1} \tag{4.10a}$$

$$dv_2/dp = -\mu m_{21}^{-1} \cos\phi - \mu m_{22}^{-1} \sin\phi + m_{23}^{-1} \tag{4.10b}$$

$$dv_3/dp = -\mu m_{31}^{-1} \cos\phi - \mu m_{32}^{-1} \sin\phi + m_{33}^{-1}. \tag{4.10c}$$

These equations of motion are not separable into independent expressions for each component of velocity unless $\mu = 0$ or $m_{ij}^{-1} = 0$ for $i \neq j$; i.e., either the contact surfaces are perfectly smooth or the impact configuration is collinear and the sliding velocity is in-plane with two principal axes of inertia for the center of mass of each body. Since the inertia terms in (4.10a) and (4.10b) are not proportional, the rates of change are generally different for each tangential component of slip; thus for nonplanar changes in relative velocity the direction of slip $\phi(p)$ continually varies while $s > 0$.

---

[3] To simplify the presentation and reduce the number of parameters, any distinction between static and dynamic coefficients of friction has been neglected.

Alternatively, the equations of motion for slip (4.10a, b) can be expressed in terms of variables (s, ϕ) rather than (v_1, v_2). In this manner, Keller (1986) obtained

$$\frac{ds}{dp} = m_{13}^{-1}\cos\phi + m_{23}^{-1}\sin\phi - \mu m_{11}^{-1}\cos^2\phi - 2\mu m_{12}^{-1}\sin\phi\cos\phi$$
$$- \mu m_{22}^{-1}\sin^2\phi \equiv g(\mu, \phi) \tag{4.11a}$$

$$s\frac{d\phi}{dp} = -m_{13}^{-1}\sin\phi + m_{23}^{-1}\cos\phi + \mu(m_{11}^{-1} - m_{22}^{-1})\sin\phi\cos\phi$$
$$+ \mu m_{12}^{-1}(\sin^2\phi - \cos^2\phi) \equiv h(\mu, \phi). \tag{4.11b}$$

With these definitions, the sliding speed s can be expressed as a function of the current direction of slip ϕ:

$$\frac{s}{s(0)} = \exp\int_{\phi(0)}^{\phi} gh^{-1}\,d\phi'.$$

4.1.3   Sliding that Halts during Collision

If the collision is eccentric and the initial speed of sliding is small enough, slip can halt before separation. After slip halts, the contact patch either sticks or resumes sliding in a new direction. For a sufficiently large coefficient of friction, the contact patch sticks; i.e., after slip halts, $dv_1/dp = dv_2/dp = 0$. This velocity constraint is imposed by a tangential force which corresponds to a differential impulse $(dp_1^2 + dp_2^2)^{1/2} = \bar{\mu}\,dp$ where $\bar{\mu} < \mu$. The constraint force has a direction $\bar{\phi} - \pi$ and a ratio of tangential to normal force $\bar{\mu}$ that can be obtained by equating (4.11a) and (4.11b) to zero:

$$\bar{\phi} = \tan^{-1}\left[\frac{m_{11}^{-1}m_{23}^{-1} - m_{12}^{-1}m_{13}^{-1}}{m_{22}^{-1}m_{13}^{-1} - m_{23}^{-1}m_{12}^{-1}}\right]$$

$$\tag{4.12}$$

$$\bar{\mu} = \frac{\left[\left(m_{11}^{-1}m_{23}^{-1} - m_{12}^{-1}m_{13}^{-1}\right)^2 + \left(m_{22}^{-1}m_{13}^{-1} - m_{12}^{-1}m_{23}^{-1}\right)^2\right]^{1/2}}{m_{11}^{-1}m_{22}^{-1} - m_{12}^{-1}m_{12}^{-1}}.$$

The ratio of tangential to normal contact force, $\bar{\mu}$, is termed the *coefficient for stick*. If the coefficient of friction is larger than the coefficient for stick ($\mu \geq \bar{\mu}$), the contact motion is termed *slip–stick*. In this case, if friction brings initial slip to a halt before separation, subsequently the contact points stick (roll without sliding). On the other hand, if $\mu < \bar{\mu}$, there is a second phase of slip that begins when the initial slip vanishes. In planar collisions this second phase is termed *slip reversal*.

If slip resumes immediately after halting, it does so in a direction $\hat{\phi}(\mu)$ which is a root of $h(\mu, \hat{\phi}) = 0$ as expressed in Eq. (4.11b). In this second phase of slip the direction of slip depends on the coefficient of friction if $\mu < \bar{\mu}$.[4] For a coefficient of friction that is slightly less than the coefficient for stick ($\mu = \bar{\mu} - \varepsilon$), however, the direction of second phase sliding is opposite to the constraint force for stick; i.e. $\lim_{\varepsilon\to 0}\hat{\phi} = \bar{\phi}$. In general, during any second phase of slip, the direction $\hat{\phi}(\mu)$ is constant, since in Eq. (4.11b) h is independent of s. The direction of second phase slip is one of a set of characteristic

---

[4] During any second phase of slip the direction is given also by $\bar{\phi} = \tan^{-1}(dv_2/dv_1)$, since $h = -\sin\phi\,dv_1/dp + \cos\phi\,dv_2/dp$.

directions, termed *isoclinics*, where the direction of slip is constant; in the slip plane v_1, v_2 the isoclinic lines $\hat{\phi}(\mu)$ depend on the impact configuration and the coefficient of friction. Batlle (1996) has shown that if the isoclinic directions are distinct, there is only one along which the relative acceleration is positive $(ds/dp > 0)$. Consequently the direction of second phase slip is unique.

4.1.4   Terminal Normal Impulse from Energetic Coefficient of Restitution

The collision process terminates at a normal impulse p_f that can be obtained from the *energetic coefficient of restitution* that was defined in Eq. (3.24). The terminal impulse p_f is obtained from the energetic coefficient of restitution by first separating the terminal work done by the normal component of force, $W_3(p_f)$, into the work done during compression, $W_c = W_3(p_c)$, and the additional work done during restitution, $W_3(p_f) - W_3(p_c)$. The energetic coefficient of restitution e_* is defined as the ratio

$$e_*^2 = -\frac{W_3(p_f) - W_3(p_c)}{W_3(p_c)}. \tag{4.13}$$

For 3D collisions where the direction of slip is continually varying as a function of impulse, the integration required to calculate these terms is nontrivial because the normal relative velocity $v_3(p)$ is not simply a linear function of normal impulse. In Fig. 4.2 the changes in this normal component of velocity as a function of normal impulse are shown as a curve connecting the initial and final states. According to Eq. (4.10c), at every impulse the slope of this curve depends on the current direction of slip, $\phi(p)$.

Further understanding of the process of slip during collision can be obtained from the following three examples. The first is a central (collinear) collision of a rotating sphere on a rigid half space; because of the initial rotation, the tangential component of translational relative velocity for the mass center is not parallel to the initial direction of slip. The second and third examples are noncollinear collisions where the resultant of forces acting during contact is not in-plane with two of the principal axes of inertia for C, so the changes in velocity are nonplanar.

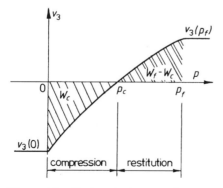

Figure 4.2. Changes in the normal component of relative velocity during collision. The slope of the curve changes when the direction of slip changes. The cross-hatched areas under the curve are equal to the work done by the normal component of force during compression and restitution, respectively.

4.2   Oblique Collision of a Rotating Sphere on a Rough Half Space

Consider a rigid sphere of radius R, mass M and moments of inertia $I_{11} = I_{22} = M\hat{k}_r^2$ about the center of mass G. The center of mass is moving with translational relative velocity $\hat{V}_i(0)$ when it collides with a rough rigid half space at a contact point C . Before the collision the sphere is rotating with angular velocity $\omega_i(0)$. Let the center of mass G have components of velocity $\hat{V}_1$ and $\hat{V}_2$ in the tangent plane and $\hat{V}_3$ in the normal direction n as shown in Fig. 4.3. At the contact point the reaction force gives a normal component of impulse $p_3 \equiv p$ and tangential components p_1 and p_2. The equations of motion for this sphere can be written as

$$d\hat{V}_1 = M^{-1}\,dp_1, \qquad d\hat{V}_2 = M^{-1}\,dp_2, \qquad d\hat{V}_3 = M^{-1}\,dp$$

$$d\omega_1 = M^{-1}R\hat{k}_r^{-2}\,dp_2, \qquad d\omega_2 = -M^{-1}R\hat{k}_r^{-2}\,dp_1, \qquad d\omega_3 = 0$$

(4.14)

where the reaction is assumed to give no couple. Although some experiments on spinning spheres by Horak (1948) have measured a mean reaction couple during collision, this couple is a consequence of development of a finite radius for the contact patch. A time-dependent analysis somewhat like the simulation of Brach (1993) is required to obtain the frictional reaction couple as a function of deformation of the colliding bodies. Hence these experimental results are outside the realm of rigid body impact theory (see Lim and Stronge, 1994).

At the contact point C the relative velocity v_i between the sphere and the surface of the half space has a normal component $v_3 = \hat{v}_3 = \hat{V}_3$ and components of slip

$$v_1 = \hat{V}_1 - R\omega_2, \qquad v_2 = \hat{V}_2 + R\omega_1.$$

The differential equations for change in relative velocity at the contact point are

$$\begin{Bmatrix} dv_1 \\ dv_2 \\ dv_3 \end{Bmatrix} = \frac{1}{M}\begin{bmatrix} 1+R^2/\hat{k}_r^2 & 0 & 0 \\ & 1+R^2/\hat{k}_r^2 & 0 \\ & & 1 \end{bmatrix}\begin{Bmatrix} dp_1 \\ dp_2 \\ dp_3 \end{Bmatrix}.$$

(4.15)

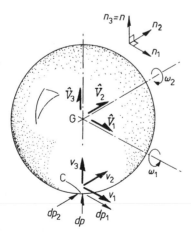

Figure 4.3. Rotating sphere colliding with a rough half space at oblique angle of incidence.

The slip speed $s(p)$ and the angle of slip $\phi(p)$ are defined as follows:

$$s^2 \equiv v_1^2 + v_2^2, \qquad \phi \equiv \tan^{-1}(v_2/v_1)$$

so that

$$v_1 = s \cos \phi, \qquad v_2 = s \sin \phi.$$

If tangential components are related to the normal component of impulse by the Amontons–Coulomb law, then while the contact point is slipping the components of changes in tangential impulse are related to the differential of normal impulse by

$$dp_1 = -\mu \cos \phi \, dp, \qquad dp_2 = -\mu \sin \phi \, dp. \tag{4.16}$$

Instead of expressing slip in terms of velocity components v_1 and v_2 in the tangent plane, the equations of motion for a sphere can be written directly in terms of the slip speed s and angle of slip ϕ. Noting that the moments of inertia about diametral axes of a sphere are $I_{11} = I_{22} \equiv M\hat{k}_r^2$, either Eqs. (4.11) or (4.15) result in differential equations for slip of a sphere,

$$ds/dp = -\mu M^{-1}\left(1 + R^2/\hat{k}_r^2\right), \qquad d\phi/dp = 0. \tag{4.17}$$

Thus, for collinear or central collision of bodies that are axisymmetric about the common normal direction, the direction of slip ϕ does not vary.[5]

From Eq. (4.17) the slip speed can be obtained as a function of impulse:

$$s(p) = s(0) - \mu M^{-1}\left(1 + R^2/\hat{k}_r^2\right)p.$$

The transition impulse p_c when compression terminates and separation begins is given by $p_c = -Mv_3(0)$, whereas in a collinear collision the final impulse at separation is given by $p_f = (1 + e_*)p_c$. This terminal impulse provides an upper limit on the change in slip velocity.

If slip continues throughout the contact period, a solid sphere (with $R^2/\hat{k}_r^2 = 5/2$) has a final speed of slip

$$s(p_f) = s(0) + 3.5\mu(1 + e_*)v_3(0) \tag{4.18}$$

where the coordinate system in Fig. 4.3 gives a normal component of relative velocity that is negative at incidence: $v_3(0) < 0$.

Changes in tangential relative velocity of the contact point C and the center of mass G that develop during collision are illustrated in Fig. 4.4. The slip velocity is directed towards the origin. Slip continues throughout the collision if $s(0)/v_3(0) \geq -3.5\mu(1 + e_*)$, but halts during collision and subsequently sticks if the initial slip speed is smaller than this limiting value. Let the normal impulse where slip halts be denoted by p_s. If the collision configuration is collinear, there is no tangential force after slip halts ($p > p_s$), so there are no further changes in tangential relative velocity. At the center of mass G, the change in tangential velocity is parallel to that for the contact point C, but the change in speed is only 2/7 as large as at C. After slip halts, the sphere rolls in the direction of the tangential component of velocity at the center of mass G; the terminal velocity of the center of

---

[5] For central impact of ellipsoidal bodies with $I_{11} \neq I_{22}$, the angle of slip is constant only if the direction of slip is parallel to a principal axis of inertia; i.e., $\phi(0) = 0, \pi/2, \pi$, where $I_{12} = I_{23} = I_{31} = 0$.

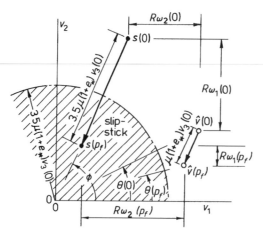

Figure 4.4. Slip trajectory $s_i(p)$ and tangential velocity of center of mass, $\hat{V}_i(p) - \hat{V}_i(p) \cdot nn$, during impact of a solid sphere on a rough half space. The cross-hatched region denotes the domain where initial speed of slip $s(0)$ is halted by friction before separation.

mass has a tangential component $\hat{V}_1(p_s)n_1 + \hat{V}_2(p_s)n_2$. For this collinear configuration, the direction of slip is constant during collision, and all changes in velocity are in this same direction (i.e. planar). Nevertheless, most points in a rotating sphere do not have an initial velocity in the plane that contains both the normal to the common tangent plane and the initial direction of slip; for points where the initial velocity is out of this plane, the velocity continuously changes in direction while the slip speed is changing. The direction of the tangential component of the translational velocity for G changes smoothly between the initial direction $\theta(0)$ and terminal direction $\theta(p_f)$ as shown in Fig. 4.4.

The center of mass has components of final velocity at separation $\hat{V}_{if} \equiv \hat{V}_i(p_f)$ that are given by

$$\frac{\hat{V}_i(p_f)}{\hat{V}_3(0)} = \begin{cases} \dfrac{\hat{V}_i(0)}{\hat{V}_3(0)} - \dfrac{2}{7}\dfrac{v_i(0)}{\hat{V}_3(0)} & \text{if } \left|\dfrac{s(0)}{\hat{V}_3(0)}\right| < 3.5\mu(1+e_*) \\[4mm] \dfrac{\hat{V}_i(0)}{\hat{V}_3(0)} + \mu(1+e_*)\cos\tilde{\phi}_i(0) & \text{if } \left|\dfrac{s(0)}{\hat{V}_3(0)}\right| > 3.5\mu(1+e_*) \end{cases} \quad (i = 1, 2)$$

$$\frac{\hat{V}_3(p_f)}{\hat{V}_3(0)} = -e_* \tag{4.19}$$

where

$$\tilde{\phi}_i \equiv \begin{cases} \phi, & i = 1 \\ \pi/2 - \phi, & i = 2. \end{cases}$$

4.3   Slender Rod That Collides with a Rough Half Space

As a first example of 3D impact, consider a slender rod that is inclined at an angle θ from the normal direction n when it strikes a massive half space as shown in Fig. 4.5. Let the center of mass of the rod be located at the origin of a Cartesian coordinate system

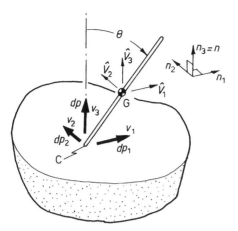

Figure 4.5. Inclined rod colliding with a rough half space. At the contact point C there is initial slip in the tangent plane.

which has unit vectors $n_3 = n$ normal and n_1, n_2 tangential to the surface of the half plane; the rod lies in the n_1, n_3 plane. The rod is assumed to have length $2L$, mass per unit length $M/2L$ and, for the n_2 axis which is transverse to the rod, a radius of gyration $k = L/\sqrt{3}$ about the center of mass. Relative to the center of mass G, the contact point C is located at r_i, where $r_1 = -L \sin\theta$, $r_2 = 0$, $r_3 = -L \cos\theta$. At the contact point there are normal and tangential forces; in an increment of time dt these forces produce a differential of impulse dp_i. Let the center of mass of the rod have translational relative velocity $\hat{V}_i$ and the angular velocity of the rod be denoted by ω_i; then the equations of translational and rotational motion can be expressed as

$$M d\hat{V}_i = M \begin{Bmatrix} d\hat{V}_1 \\ d\hat{V}_2 \\ d\hat{V}_3 \end{Bmatrix} = \begin{Bmatrix} dp_1 \\ dp_2 \\ dp_3 \end{Bmatrix} \qquad (4.20a)$$

$$\hat{I}_{ij} d\omega_j = M k^2 \begin{bmatrix} \cos^2\theta & 0 & -\sin\theta\cos\theta \\ 0 & 1 & 0 \\ -\sin\theta\cos\theta & 0 & \sin^2\theta \end{bmatrix} \begin{Bmatrix} d\omega_1 \\ d\omega_2 \\ d\omega_3 \end{Bmatrix} = \begin{Bmatrix} r_2\,dp_3 - r_3\,dp_2 \\ r_3\,dp_1 - r_1\,dp_3 \\ r_1\,dp_2 - r_2\,dp_1 \end{Bmatrix}.$$

Here the inertia matrix is singular, so it is not possible to obtain the inverse matrix for moments and products of inertia $\hat{I}_{ij}^{-1}$ as required by Eq. 4.7. This is because the rod has been assumed to have a negligible moment of inertia about its longitudinal axis; consequently, in the expression above, the first and last equations are linearly dependent. To eliminate this dependence, we assume that the rod has a constant rate of rotation about its longitudinal axis; i.e.,

$$0 = \sin\theta\,d\omega_1 + \cos\theta\,d\omega_3, \qquad \text{giving} \quad d\omega_3 = -\tan\theta\,d\omega_1 = -\frac{r_1}{r_3}\,d\omega_1. \quad (4.20b)$$

After substitution in the previous expession, the first two equations result in

$$M k^2 \begin{bmatrix} 1 & 0 \\ 0 & 1 \end{bmatrix} \begin{Bmatrix} d\omega_1 \\ d\omega_2 \end{Bmatrix} = \begin{Bmatrix} r_2\,dp_3 - r_3\,dp_2 \\ r_3\,dp_1 - r_1\,dp_3 \end{Bmatrix}. \qquad (4.20c)$$

In a coordinate system with origin at the center of mass, the contact point C is located at r_i. Thus, since only the rod is moving, the relative velocity at C is given by

$$v_i = V_i = \hat{V}_i + \varepsilon_{ijk}\omega_j r_k. \tag{4.20d}$$

Hence substituting (4.20a,b,c) into (4.20d) results in a set of equations identical to (4.6):

$$dv_i = m_{ij}^{-1} dp_j \tag{4.21a}$$

where

$$m_{ij}^{-1} = \frac{1}{M}\begin{bmatrix} 1 + r_3^2/k^2 & 0 & -r_1 r_3/k^2 \\ 0 & 1 + (r_1^2 + r_3^2)/k^2 & 0 \\ -r_1 r_3/k^2 & 0 & 1 + r_1^2/k^2 \end{bmatrix} \tag{4.21b}$$

$$= \frac{1}{2M}\begin{bmatrix} 5 + 3\cos 2\theta & 0 & -3\sin 2\theta \\ 0 & 8 & 0 \\ -3\sin 2\theta & 0 & 5 - 3\cos 2\theta \end{bmatrix}. \tag{4.21c}$$

Because of symmetry about the n_1, n_3 plane, the off diagonal terms $m_{12}^{-1} = m_{23}^{-1} = 0$ and the equations of motion can be expressed as

$$dv_1/dp = -\mu m_{11}^{-1}\cos\phi + m_{13}^{-1} \tag{4.22a}$$

$$dv_2/dp = -\mu m_{22}^{-1}\sin\phi \tag{4.22b}$$

$$dv_3/dp = -\mu m_{13}^{-1}\cos\phi + m_{33}^{-1}. \tag{4.22c}$$

4.3.1 Slip Trajectories or Hodographs

On the relative velocity plane for slip (v_1, v_2) there are lines termed *isoclinics* where the direction of slip is a constant. In general the orientation of the isoclinic lines depends on the inertia properties of bodies and the coefficient of friction; i.e. the orientation of isoclinic lines are characteristic values for the system. These lines are asymptotes for the direction of slip, so that the direction of slip flows towards an isoclinic as the normal impulse increases.

Along an isoclinic, changes in velocity are parallel to the current velocity, so that (4.22) gives

$$\tan\hat{\phi} = \frac{dv_2/dp}{dv_1/dp} = \frac{-\mu m_{22}^{-1}\sin\hat{\phi}}{m_{13}^{-1} - \mu m_{11}^{-1}\cos\hat{\phi}} \tag{4.23}$$

that is either,

$$\sin\hat{\phi} = 0 \quad \text{or} \quad \cos\hat{\phi} = \frac{m_{13}^{-1}}{\mu(m_{11}^{-1} - m_{22}^{-1})}.$$

For the present impact configuration, which is symmetrical with respect to n_2, Eq. (4.23) has roots $\hat{\phi} = 0, \pi$ for all values of the coefficient of friction μ; the symmetry of these roots is due to the inertia being symmetrical, which gives $m_{12}^{-1} = m_{23}^{-1} = 0$. It is worth noting that for the rod, along $\hat{\phi} = \pi$ (i.e. $v_1 < 0, v_2 = 0$) Eq. (4.22a) gives $dv_1/dp > 0$. This is the case where slip speed decreases irrespective of the value of the coefficient of friction μ. On the other hand, slip on the isoclinic $\hat{\phi} = 0$ results in increasing speed ($dv_1/dp > 0$) if $\mu < |m_{13}^{-1}|/m_{11}^{-1}$ and decreasing speed ($dv_1/dp < 0$) if $\mu > |m_{13}^{-1}|/m_{11}^{-1}$. Thus if the

Table 4.1. *Critical Friction Coefficient* μ_* *for Inclined Rod*

Initial Direction of Slip,	Critical Friction Coefficient μ_*		
$\phi(0)$ (deg)	$\theta = 30°$	$\theta = 45°$	$\theta = 60°$
0	1.73	1.0	0.58
30	2.00	1.15	0.67
45	2.45	1.41	0.82
60	3.46	2.0	1.15
90	∞	∞	∞

coefficient of friction is sufficiently large, any initial slip will decrease in magnitude during the contact period; in this case, if the slip vanishes before separation it subsequently sticks rather than beginning a second phase of slip. If however the coefficient of friction is small, the direction of slip always evolves towards π and the speed increases. For the present case of an inclined rod, Eq. (4.12) gives a coefficient for stick of $\bar{\mu} = m_{13}^{-1}/m_{11}^{-1} = (3\sin 2\theta)/(5 + 3\cos 2\theta)$.

If the coefficient of friction satisfies the condition $\mu > \cot\theta$, then Eq. (4.23) has a third root,[6] $\hat{\phi}_* = \cos^{-1}[\mu^{-1} m_{13}^{-1}/(m_{11}^{-1} - m_{22}^{-1})] = \cos^{-1}(\mu^{-1}\cot\theta)$ This again is an isoclinic where the direction of slip remains constant – it will be termed a *separatrix*. A separatrix separates two regions of initial velocity in which the direction of slip asymptotically approaches different isoclinics as the normal impulse increases (or as the sliding speed decreases). For any initial velocity the separatrix passing through this point in phase space is given by the *critical coefficient of friction* $\mu_* \equiv \cot\theta / \cos\phi(0)$. This is the value of the coefficient of friction which causes the sliding speed to continually decrease until slip vanishes simultaneously with termination of the compression period. Typical values for the critical coefficient of friction of an inclined slender rod are listed in Table 4.1.

Figure 4.6 illustrates some flow-lines evolving from various initial velocities of slip for a rod inclined at $45°$ and three different coefficients of friction, $\mu = 0.5$, 1.0 and 1.5. For small friction, $\mu < 0.6$, slip does not stop but tends to an isoclinic direction $\hat{\phi} = \pi$, whereas for larger friction, slip will vanish if the contact period is sufficiently long. In Fig. 4.6 the shaded region indicates initial velocities where slip is brought to a halt during compression. By extrapolation, this gives an indication of the range of initial conditions where slip will vanish before separation. These regions do not appear if $\mu < \bar{\mu}$, because this friction is insufficient to slow initial slip. For a large coefficient of friction, $\mu = 1.5$, there is a separatrix shown as a dashed line; this separates regions of sliding velocity which have different directions of approach as slip vanishes. The flow (or evolution of slip) in more complex examples is described by Mac Sithigh (1996) and Batlle (1996).

4.4 Equilateral Triangle Colliding on a Rough Half Space

A thin plate provides an example of 3D impact which does not suffer from a singular inertia matrix of the type illustrated in Sect. 4.3. Consider a thin uniform plate of equilateral triangular shape which has one corner that collides against the surface of a

---

[6] This lower bound for the coefficient of friction μ is required in order that $-1 \le \cos\hat{\phi} \le 1$.

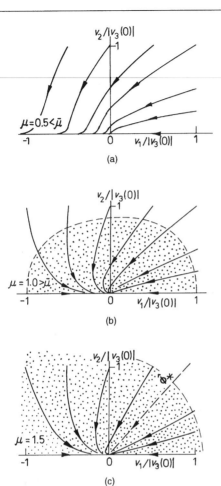

Figure 4.6. Hodograph, phase plane, of slip trajectories for a slender rod inclined at $45°$ and (a) $\mu = 0.5$, (b) $\mu = 1.0$ and (c) $\mu = 1.5$. Isoclines $\hat{\phi} = 0, \pi$ are the same for all μ, while a separatrix $\hat{\phi}_*$ occurs only if $\mu > \cot\theta$. For this symmetrical configuration, the hodograph is symmetrical about the $v_1/|v_3(0)|$ axis, but only one half is shown. For initial slip in the shaded region, slip stops during compression.

massive half space. Let the plate have mass M and sides of length $2L$. At incidence the plate is perpendicular to the surface and the center of mass is inclined at angle θ from the normal to the surface, $n_3 = n$. Axes are chosen such that the plate lies in the n_1, n_3 plane as shown in Fig. 4.7.

In order to obtain the inertia matrix for the plate, consider first the coordinate system ξ, η with an origin at the contact point C and the direction of ξ on an axis of symmetry of the plate. These axes are in the n_1, n_3 plane, and the ξ axis is inclined at angle θ from the normal $n_3 = n$. For these axes the moments and products of inertia are given by

$$I_{\xi\xi} = \tfrac{1}{6}ML^2, \qquad I_{\eta\eta} = \tfrac{1}{6}ML^2, \qquad I_{22} = I_{\xi\xi} + I_{\eta\eta} = \tfrac{1}{3}ML^2, \qquad I_{\xi 2} = I_{2\eta} = I_{\xi\eta} = 0.$$

These principal moments of inertia for C can be transformed to rotated coordinates n_1, n_3 using Mohr's circle or otherwise, to obtain the inertia matrix I_{ij} $(i, j = 1, 2, 3)$ for the

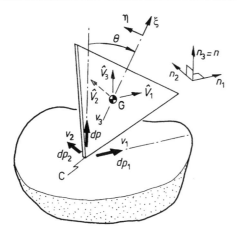

Figure 4.7. Equilateral triangle plate with one corner striking a rough half space.

contact point C of an equilateral triangular plate,

$$I_{ij} = \frac{ML^2}{6} \begin{bmatrix} 1 & 0 & 0 \\ 0 & 2 & 0 \\ 0 & 0 & 1 \end{bmatrix}.$$

The inverse of this matrix for moments of inertia, $\hat{I}_{ij}^{-1}$, as required by (4.7), is

$$\hat{I}_{ij}^{-1} = \frac{3}{ML^2} \begin{bmatrix} 2 & 0 & 0 \\ 0 & 1 & 0 \\ 0 & 0 & 2 \end{bmatrix}.$$

4.4.1 Slip Trajectories and Hodograph for Equilateral Triangle Inclined at $\theta = \pi/4$

In this section slip trajectories are calculated for the example of one corner of an equilateral triangle colliding against a massive body. Calculations are performed for an impact configuration which has a specific angle of eccentricity for the center of mass, $\theta = \pi/4$. For the equilateral triangular plate with side lengths $2L$ the vector r_i from the center of mass to the contact point C has components

$$r_1 = \frac{-2L\sin(\pi/4)}{\sqrt{3}} = -L\sqrt{\frac{2}{3}}, \qquad r_2 = 0, \qquad r_3 = \frac{-2L\cos(\pi/4)}{\sqrt{3}} = -L\sqrt{\frac{2}{3}}.$$

This vector and I_{ij}^{-1} combine to give the inverse of inertia matrix m_{ij}^{-1} for point C according to Eq. (4.7),

$$m_{ij}^{-1} = \frac{1}{M} \begin{bmatrix} 3 & 0 & -2 \\ 0 & 9 & 0 \\ -2 & 0 & 3 \end{bmatrix}.$$

Together with the equations of motion (4.21), this completes the preparation of the differential equations which describe motion during contact between a corner of an equilateral triangular plate and a rough half plane.

As in the case of the inclined rod, slip at the contact point of the triangular plate has isoclinics $\hat{\phi} = 0, \pi$ that are in-plane with the plate. In this case Eq. (4.12) gives a coefficient for stick of $\bar{\mu} = 2/3 \approx 0.66$. An additional isoclinic, the separatrix, is obtained from Eq. (4.11b) only if $\mu \geq m_{13}^{-1}/(m_{11}^{-1} - m_{22}^{-1})$; the separatrix direction in phase space is at an angle $\hat{\phi}_* = \cos^{-1}[\mu^{-1} m_{13}^{-1}/(m_{11}^{-1} - m_{22}^{-1})] = \cos^{-1}(1/3\mu)$. These features of the slip trajectories are illustrated in Fig. 4.8. In both Figs. 4.6 and 4.8 note that for small friction $\mu < \bar{\mu}$, sliding never entirely vanishes unless it approaches the origin along the isoclinic $\hat{\phi} = 0$; even in that case, if sliding vanishes, it reverses immediately and accelerates along an isoclinic pathline. On the other hand, if friction is moderately large ($\mu > \bar{\mu}$), all slip trajectories converge towards zero slip speed. Whether slip vanishes or not in a particular case depends on the initial slip velocity and the distance from the origin along the pathline (i.e. the change in normal impulse during contact). For the present impact configuration the range of initial conditions where slip vanishes during compression is shaded in Fig. 4.8b and c.

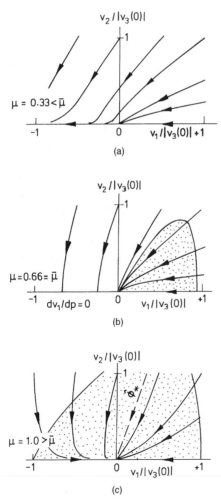

Figure 4.8. Hodograph, or phase plane of slip trajectories for an eccentric triangular plate ($\theta = \pi/4$). For initial slip in the shaded region, friction is sufficient to stop slip during compression.

For slip along isoclinics where the direction of slip remains constant, the *normal impulse at the termination of compression, p_c,* is obtained by setting $v_3(p_c) = 0$ in Eq. (4.22c):

$$\frac{-p_c}{v_3(0)} = \frac{1}{m_{33}^{-1} - \mu m_{13}^{-1}\cos\hat{\phi}} = \frac{M}{3 + 2\mu\cos\hat{\phi}}.$$

Thus along any isoclinic, for a *terminal impulse p_f* there is a change in the component of tangential velocity v_1 that is in-plane with the plate,

$$\frac{-v_1}{v_3(0)} = \frac{-v_1(0)}{v_3(0)} + \left[\frac{m_{13}^{-1} - \mu m_{11}^{-1}\cos\hat{\phi}}{m_{33}^{-1} - \mu m_{13}^{-1}\cos\hat{\phi}}\right]\frac{p_f}{p_c}.$$

Whether or not this is sufficient to bring slip to a halt before separation depends on the ratio $v_1(0)/v_3(0)$ between components of initial velocity and on the impulse ratio p_f/p_c.

4.5   Spherical Pendulum Colliding on a Rough Half Space

As an example of an eccentric impact configuration with a velocity constraint at one point (rather than a free body), consider a spherical simple pendulum that collides on a rough half space. The pendulum is a rigid body of length L with one end pivoted at a stationary point O. At the instant of impact the pendulum is inclined at an angle θ from the normal to a massive half space, so the support is located a perpendicular distance $L\cos\theta$ from the plane surface of the half space as illustrated in Fig. 4.9.

During collision the pendulum is subjected to reactions at both the contact point C and the fixed support O, so the free body impact equations (4.6)–(4.7) are not directly applicable. The impulse of the reaction at O can be eliminated from consideration, however, by obtaining equations of motion in terms of moments about this fixed point. Since the motion of the pendulum is a pure rotation about O, it is convenient to use a cylindrical coordinate system with axis n directed normal to the surface of the half space and a radial unit vector n_1 directed from the projection of O on the surface towards the contact

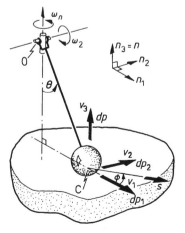

Figure 4.9. Spherical simple pendulum colliding with a rough half space. Initially the pendulum is rotating about both n_2 and n_3.

point C. The differential equations for rotations about these unit vectors can be obtained as follows:

$$\tilde{I}_{ij}\,d\omega_j = \varepsilon_{ijk}r_j\,dp_k$$

where dp_i is the differential of impulse at C, r_i is the position vector of C from the fixed point O, and $\tilde{I}_{ij}$ is the second moment of mass for point O in reference frame n_i. Thence the relative velocity v_i at the contact point can be calculated from

$$v_i = \varepsilon_{ijk}\omega_j r_k$$

and the differential equations for changes in relative velocity can be expressed in terms of the differential impulse of the reaction at C,

$$dv_i = \varepsilon_{ikm}\varepsilon_{jln}\tilde{I}_{kl}^{-1}r_m r_n\,dp_j. \qquad (4.24)$$

In this simple pendulum, however, the bob is assumed to be a particle of mass M at the contact point C. Like the previous example, this idealization has a negligibly small moment of inertia about line OC, so an additional constraint is required to obtain a unique solution (this constrains rotations about OC). Let this constraint on the angular velocity be $\omega_1 = 0$.

This additional constraint reduces the number of equations of motion to the number of degrees of freedom, namely two. Nevertheless, separate equations for the rates of change of components of relative velocity at the contact point as a function of the reaction impulse can be obtained:

$$Mdv_1 = \cos^2\theta\,dp_1 + \sin\theta\,\cos\theta\,dp$$

$$Mdv_2 = dp_2$$

$$Mdv_3 = \sin\theta\,\cos\theta dp_1 + \sin^2\theta\,dp$$

where in reference frame n_i, the spherical pendulum has moments of inertia about the pivot point O given by

$$\tilde{I}_{11} = ML^2\cos^2\theta, \qquad \tilde{I}_{22} = ML^2 \quad \text{and} \quad \tilde{I}_{33} = ML^2\sin^2\theta.$$

If there is slip at the contact point, $v_1^2 + v_2^2 > 0$; in this case the Amontons–Coulomb law of friction gives

$$Mdv_1/dp = \sin\theta\,\cos\theta - \mu\,\cos^2\theta\,\cos\phi \qquad (4.25a)$$
$$Mdv_2/dp = -\mu\sin\phi \qquad (4.25b)$$
$$Mdv_3/dp = \sin^2\theta - \mu\sin\theta\,\cos\theta\,\cos\phi \qquad (4.25c)$$

where $\phi(p) \equiv \tan^{-1}(v_2/v_1)$ is the angle of slip. These equations can be integrated to give the components of velocity as a function of normal impulse p,

$$v_1(p) = v_1(0) + \frac{p}{M}\sin\theta\cos\theta - \frac{\mu\cos^2\theta}{M}\int_0^p \cos\phi(p)\,dp$$

$$v_2(p) = v_2(0) - \frac{\mu}{M}\int_0^p \sin\phi(p)\,dp \qquad (4.26)$$

$$v_3(p) = v_3(0) + \frac{p}{M}\sin^2\theta - \frac{\mu}{M}\sin\theta\cos\theta\int_0^p \cos\phi(p)\,dp$$

where the initial conditions for a spherical pendulum $(0, \omega_2(0), \omega_3(0))$ give the normal and tangential components of initial relative velocity at C,

$$v_1(0) = -L\omega_2(0)\cos\theta, \qquad v_2(0) = L\omega_3(0)\sin\theta, \qquad v_3(0) = -L\omega_2(0)\sin\theta.$$

Additional information about behavior during impact can be gained by considering the slip trajectory dv_2/dv_1 at any impulse p,

$$\frac{dv_2}{dv_1} = -\frac{\mu\sec^2\theta\sin\phi}{\tan\theta - \mu\cos\phi}. \tag{4.27}$$

Suppose that the angle of slip $\phi(p)$ is a constant throughout the compression phase of collision and that slip vanishes simultaneously with the normal component of relative velocity at impulse p_c, i.e. that the incident angle of slip is coincident with a separatrix $\phi(0) = \hat{\phi}_*$. For $0 < \phi < \pi$ this condition can be expressed as

$$\tan\phi(p) = \frac{dv_2(p)/dp}{dv_1(p)/dp} = \tan\phi(0), \qquad 0 \le p < p_c.$$

After substitution from (4.25), this constant direction of slip is associated with a critical value μ_* for the coefficient of friction,

$$\mu_* = -\cot\theta/\cos\phi(0). \tag{4.28}$$

Note that the spherical pendulum has initial conditions $v_1(0) < 0$ and $v_2(0) > 0$, which give $-1 < \cos\phi(0) < 0$ and consequently a critical coefficient of friction which is positive $(\mu_* > 0)$.

At this point we can anticipate a few results. If the coefficient of friction is large $(\mu > \mu_*)$, the circumferential component of slip vanishes at impulse p_c where compression terminates, while if $\mu < \mu_*$ and $\mu < \tan\theta$, the circumferential component of slip asymptotically approaches zero but never completely vanishes during the collision. In either case the direction of slip is continuously changing unless the coefficient of friction equals the critical value $\mu = \mu_*$.

For the spherical pendulum the kinematic constraints give

$$v_3(0)/v_1(0) = \tan\theta, \qquad v_3(0)/v_2(0) = \tan\theta\cot\phi(0).$$

With these constraints, the components of velocity (4.26) can be expressed as a function of the normal component of reaction impulse p:

$$\frac{v_1(p)}{v_3(0)} = \left\{1 + \frac{p\sin^2\theta}{Mv_3(0)} + \frac{\mu}{\mu_*}\frac{\cos^2\theta}{Mv_3(0)}\int_0^p \frac{\cos\phi(p)}{\cos\phi(0)}\,dp\right\}\cot\theta$$

$$\frac{v_2(p)}{v_3(0)} = \left\{1 + \frac{\mu}{\mu_* Mv_3(0)}\int_0^p \frac{\sin\phi(p)}{\sin\phi(0)}\,dp\right\}\cot\theta\tan\phi(0) \tag{4.29}$$

$$\frac{v_3(p)}{v_3(0)} = 1 + \frac{p\sin^2\theta}{Mv_3(0)} + \frac{\mu}{\mu_*}\frac{\cos^2\theta}{Mv_3(0)}\int_0^p \frac{\cos\phi(p)}{\cos\phi(0)}\,dp.$$

Here it is worth noting that an alternative derivation of the equations of motion in terms of generalized coordinates can result in an expression directly in terms of angular speeds rather than the normal and slip velocity components at the contact point. For the spherical simple pendulum this gives

$$\tilde{I}_{22}\begin{bmatrix} 1 & 0 \\ 0 & \sin^2\theta \end{bmatrix}\begin{Bmatrix} d\omega_2 \\ d\omega_3 \end{Bmatrix} = \begin{bmatrix} -L\cos\theta & 0 & -L\sin\theta \\ 0 & L\sin\theta & 0 \end{bmatrix}\begin{Bmatrix} dp_1 \\ dp_2 \\ dp \end{Bmatrix}. \quad (4.30)$$

These equations relating changes in generalized momentum with differential increments of impulse can also be obtained from Smith's (1991) general formulation of equations of impulsive motion for colliding bodies with velocity constraints. If the rates of change of tangential components of impulse are related to the normal component by the Amontons–Coulomb law, then $dp_1 = -\mu\,\cos\phi\,dp$ and $dp_2 = -\mu\,\sin\phi\,dp$. Hence,

$$\frac{d\omega_2}{\omega_2(0)} = -\left[1 + \frac{\mu}{\mu_*}\frac{\cos\phi(p)}{\cos\phi(0)}\cot^2\theta\right]\frac{dp}{\bar{p}_c}$$

$$\frac{d\omega_3}{\omega_3(0)} = -\frac{\mu}{\mu_*}\frac{\sin\phi(p)}{\sin\phi(0)}\csc^2\theta\,\frac{dp}{\bar{p}_c} \quad (4.31)$$

where a characteristic normal impulse for compression $\bar{p}_c$ that brings the normal component of relative velocity to a halt in the absence of friction is obtained as

$$\bar{p}_c = \frac{M\hat{k}_2^2\omega_2(0)}{L\sin\theta} = -\frac{Mv_3(0)}{\sin^2\theta}. \quad (4.32)$$

To integrate these differential equations, the angle of slip $\phi(p)$ must be expressed in terms of the rotation rates,

$$\sin\phi = \frac{\omega_3\sin\theta}{\sqrt{\omega_2^2\cos^2\theta + \omega_3^2\sin^2\theta}}, \qquad \cos\phi = \frac{-\omega_2\cos\theta}{\sqrt{\omega_2^2\cos^2\theta + \omega_3^2\sin^2\theta}}.$$

The components of slip are directly related to the angular velocities, so a comparison of (4.31) with the combination of (4.25), (4.29) and (4.32) gives

$$\frac{dv_3}{v_3(0)} = \frac{d\omega_2}{\omega_2(0)}, \qquad \frac{dv_2}{v_3(0)} = \frac{-d\omega_3}{\omega_2(0)}.$$

The terminal impulse p_f that corresponds to any specific value of the energetic coefficient of restitution e_* can be calculated either by integrating the last equation in (4.29) or by double integration of the first equation in (4.31). This terminal impulse is the upper limit of integration for each component of relative velocity; it depends on any variations in the angle of slip.

4.5.1   Numerical Results for $\theta = \pi/3$ and $\pi/4$

The values of the critical friction coefficient μ_* that give a steady direction of slip during compression are listed in Table 4.2 for a variety of initial conditions and impact configurations. For most of these conditions the critical value is rather large in comparison with typical friction measurements.

During collision the normal and tangential components of velocity are continuously changing as a function of impulse. The changes in components of slip are illustrated in Fig. 4.10; unless $\mu/\mu_* = 1.0$ or the slip direction is an isoclinic, the direction of slip is continuously varying. In Fig. 4.10 the radial component of slip v_1 is negative during the

Table 4.2. *Critical Friction Coefficient μ_* for Spherical Pendulum*

Inclination at Impact, θ (deg)	Initial Direction of Slip, $\phi(0)$ (deg)	Ratio of Initial Components of Slip, $v_2(0)/v_1(0)$	Ratio of Initial Angular Speeds $\omega_3(0)/\omega_2(0)$	Critical Coeff. of Friction, μ_*
45	165	−0.27	0.27	1.03
	154	−0.50	0.50	1.12
	150	−0.58	0.58	1.16
	135	−1.00	1.00	1.42
	120	−1.75	1.75	2.01
	117	−2.00	2.00	2.24
60	165	−0.28	0.16	0.59
	150	−0.59	0.34	0.67
	139	−0.86	0.50	0.76
	135	−1.00	0.58	0.82
	120	−1.74	1.01	1.16
	106	−3.47	2.0	2.10

compression and positive during the restitution phase of collision. The curvature of the slip trajectories indicates that the pendulum bob in this system has a radial acceleration that is larger than the circumferential acceleration during the compression phase, but the circumferential acceleration is the largest component during the restitution phase. If $\mu < \mu_*$ and $\mu < \tan\theta$, the circumferential component of velocity monotonically decreases throughout the collision, whereas if $\tan\theta < \mu < \mu_*$, the circumferential slip vanishes during restitution. After circumferential slip vanishes, the slip velocity is radial and decreasing. Furthermore, if $\mu \geq \mu_*$, the circumferential and radial components of slip vanish simultaneously at the end of the compressive phase of collision. In the latter case, Eqs. (4.25a) and (4.27) show that friction is sufficient to prevent the radial component of slip from reversing. Consequently, if $\mu \geq \mu_*$, there is no final separation, as the pendulum is stuck in the compressed state.

The present analysis gives ratios of angular speeds at separation to those at incidence (final/initial angular speed) that can be obtained from Fig. 4.10 for several initial conditions. The separation velocity is diminished if either the coefficient of friction increases or the coefficient of restitution decreases. If friction dissipates some of the initial kinetic energy of relative motion, then at separation both the radial and the circumferential slip speeds are smaller than the initial values even if the coefficient of restitution $e_* = 1$.

4.6   General 3D Impact

Some of the results presented in this chapter have been known for a long time. It was Coriolis in his *Jeu de Billiard* (1835) who first proved that during a collision between two rough spheres the contact points on the two bodies have the same direction of slip throughout any collision. Later Ed. Phillips generalized this result in *Liouville's Journal*, Vol. 14 (1849). He showed that in a collinear collision between two rough bodies, the direction of slip is constant if each body has a principal axis of inertia in the direction of

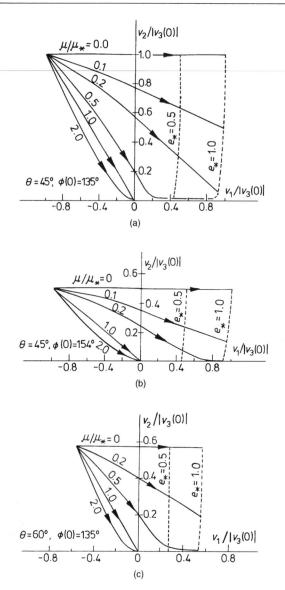

Figure 4.10. Slip trajectories for spherical pendulum: (a) $\theta = 45°$, $\phi(0) = 135°$; (b) $\theta = 45°$, $\phi(0) = 154°$; (c) $\theta = 60°$, $\phi(0) = 135°$.

initial slip. In all other cases, however, the contact points have a rate of change of relative tangential velocity with impulse, $d(v_i - v_i \cdot nn)/dp$, that is not parallel to the direction of slip and therefore the direction of slip changes continuously. While the direction of slip swerves only if the impact configuration is eccentric, the tangential velocity of the center of mass swerves whenever there is friction and the body has a component of initial angular velocity parallel to the initial tangential velocity of the center of mass. These changes in direction occur smoothly as a function of impulse. Slip can halt before separation only if the coefficient of friction is larger than a characteristic value. Slip that halts can immediately resume in a different direction only if the impact configuration is eccentric.

In this chapter, changes in relative velocity that occur during an instant of impact have been obtained from ordinary differential equations of motion using the normal impulse of the reaction at the contact point as an independent variable. These equations represent collisions between hard bodies where deformation is limited to an infinitesimally small region around the contact point. By considering that contact points on two colliding bodies are separated by a deformable particle, we have been able to follow the process of slip. This gives a method of dividing the energy loss into separate parts due to friction and irreversible deformations, e.g., by calculating separately the work done by normal and tangential components of the reaction impulse. In fact, the advance of the present analytical method, compared with a semigraphical method presented by Routh (1905) and an analytical method given by Keller (1986), is in the use of an *energetic coefficient of restitution* e_* to calculate the terminal impulse at final separation. This coefficient of restitution represents hysteresis of contact force due to internal irreversible deformation. It is calculable from the kinetic equations for rigid body motions. This method of obtaining changes in velocity as a function of impulse represents collisions between slightly deformable solids where the contact patch remains small.

PROBLEMS

4.1 A rigid rod of mass M and length $2L$ is inclined at angle θ from the vertical when it strikes a massive half space. For a small coefficient of friction μ, find the *isoclinics* (lines along which slip has a constant direction). For these isoclinic lines find the limiting coefficient of friction which, if slip vanishes during collision, prevents a second phase of slip. On the isoclinic for initial slip that approaches the origin, find the largest ratio of slip to normal velocity at incidence which results in slip vanishing simultaneously with the termination of compression.

4.2 For the spherical pendulum in Sect. 4.5, find the isoclinic lines. Also find the limiting coefficient of friction which prevents second phase slip. For $\mu = 0.5$ sketch flow lines for $\phi(0) = 45, 90$, and $135 \deg$.

4.3 A homogeneous sphere of radius R rolls on a level table at an initial angular speed ω_0 before striking a rough vertical wall. At incidence $\hat{\psi}_0$ is the angle between the velocity of the center of mass and the normal to the wall. Let μ be the coefficient of friction and e_* the coefficient of restitution between the sphere and the wall.
   (a) Find that at the impact point, $R\omega_0$ is the initial slip speed and the angle of incidence is inclined at $\hat{\psi}_0$ from vertical.
   (b) Show that in order for slip to stop during compression, the coefficient of friction must satisfy

$$\mu > 2/7 \qquad \text{if} \quad \hat{\psi}_0 < \pi/4$$

$$\mu > 2\tan(\hat{\psi}_0)/7 \qquad \text{if} \quad \hat{\psi}_0 < \pi/4.$$

   (c) Show that the center of mass has a rebound angle $\hat{\psi}_f$ that satisfies

$$\tan(\hat{\psi}_f) = \frac{5\tan(\hat{\psi}_0)}{7e_*} \qquad \text{if} \quad 0 < p_s/p_c < 1 + e_*$$

$$\tan(\hat{\psi}_f) = \frac{\tan(\hat{\psi}_0)}{e_*} - \frac{\mu(1+e_*)}{e_*} \qquad \text{if} \quad p_s/p_c > 1 + e_*.$$

   (d) Sketch regions of slip–stick and continuous slip for $e_* = 0, 1$ on a plot of $\hat{\psi}_0$ vs μ.

CHAPTER 5

Rigid Body Impact with Discrete Modeling of Compliance for the Contact Region

In ancient days two aviators procured to themselves wings. Daedalus flew safely through the middle air and was duly honoured on his landing. Icarus soared upwards to the sun till the wax melted which bound his wings and his flight ended in fiasco ... The classical authorities tell us, of course, that he was only 'doing a stunt'; but I prefer to think of him as the man who brought to light a serious constructional defect in the flying-machines of his day.

So, too, in science. Cautious Daedalus will apply his theories where he feels confident they will safely go; but by his excess of caution their hidden weaknesses remain undiscovered. Icarus will strain his theories to the breaking-point till the weak joints gape. For the mere adventure? Perhaps partly, that is human nature. But if he is destined not yet to reach the sun ... we may at least hope to learn from his journey some hints to build a better machine.

Sir Arthur Eddington, *Stars and Atoms*, 1927.

In this chapter lumped parameter models for compliance of the deforming region are used to examine the influence of factors which previously in this book were assumed to be negligibly small – namely the effects of (a) a viscoelastic or rate-dependent normal compliance relation and (b) tangential compliance. Because these factors depend on the interaction force and not simply the impulse, the analysis of their effects necessarily uses time rather than normal impulse as an independent variable. Thus these analyses are closely akin to problems of vibration of one and two degree of freedom systems where the dependent variables (i.e. displacements and velocities) depend on the initial conditions as well as the system parameters.

5.1   Direct Impact of Viscoelastic Bodies

Although most of this book describes the dynamics of collision between elastic or elastic–plastic bodies, nevertheless there are many examples of impact between nonmetallic bodies where the contact force is rate-dependent or viscoelastic. This is certainly true of golf balls, and it may be the best representation for other hard balls that are hit by a bat, e.g. baseballs and cricket balls. Impact mechanics of bodies represented by nonlinear viscoelastic relations has been investigated by Hunter (1956) and by Hunt and Crossley (1975). Here a simpler development is based on a linear viscoelastic compliance relation for the deformable element in order to illustrate some consequences of material rate dependence.

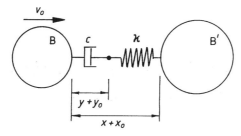

Figure 5.1. Collinear collision of bodies separated by a Maxwell linear viscoelastic element.

5.1.1 Linear Viscoelasticity – the Maxwell Model

The *Maxwell model* is the simplest viscoelastic element which represents the contact force arising from mutual compression of colliding bodies[1]; this element has a linear spring and dashpot in series as shown in Fig. 5.1. For this model the compliance of the deforming region gives a normal force which increases smoothly with normal compression and some kinetic energy of normal relative motion that is restored during restitution; i.e., the model gives coefficients of restitution in a range from 0 to 1. Let bodies B and B′ be separated by a Maxwell element; the spring has a spring constant κ and uncompressed length x_0, while the dashpot has a damping force constant c and uncompressed length y_0. The relative displacement of the bodies x and a part of this displacement that is due to the compression of the dashpot y give the normal force F acting on body B′; the same force acts in both the spring and dashpot,

$$F = -\kappa(x - y) = -c\dot{y}. \tag{5.1}$$

If the colliding bodies B and B′ have masses M and $M′$ respectively, then the effective mass m can be obtained from the definition $m^{-1} \equiv M^{-1} + M'^{-1}$. Thus there is an equation of relative motion,

$$m\ddot{x} = -\kappa(x - y).$$

By differentiating Eq. (5.1) and adding this to the equation of motion, another equation of motion is obtained in terms of the spring extension $z \equiv x - y$,

$$\ddot{z} + 2\zeta\omega_0\dot{z} + \omega_0^2 z = 0, \qquad \omega_0^2 \equiv \kappa/m, \quad \zeta \equiv m\omega_0/2c. \tag{5.2}$$

To represent a collision between bodies with an *initial normal relative speed* v_0, the initial conditions are

$$x(0) = y(0) = z(0) = 0$$
$$\dot{x}(0) = \dot{z}(0) = -v_0, \qquad \dot{y} = 0.$$

Equation (5.2) gives simple harmonic motion for spring extension during the contact period, i.e. the period in which the spring is compressed:

$$z = -\omega_d^{-1} v_0 e^{-\zeta\omega_0 t} \sin(\omega_d t), \qquad \omega_d t \le \pi \tag{5.3}$$

---

[1] The Kelvin–Voight solid is an alternative elementary viscoelastic model that has a linear spring and dashpot in parallel. With an initial discontinuity in relative velocity at the contact point, however, this model gives a normal force which jumps to a finite value at the instant of incidence.

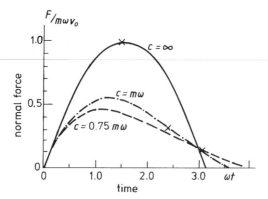

Figure 5.2. For Maxwell model the maximum force occurs during compression, before the normal relative velocity vanishes at time t_c. (On each curve the nondimensional time ωt_c is indicated by a cross.)

where the damped natural frequency $\omega_d \equiv \omega_0\sqrt{1 - \zeta^2}$. Consistent with the usual assumption of no tensile normal force, separation is assumed to occur at time $t_f = \pi/\omega_d$. At separation the normal relative velocity $\dot{x}_f \equiv \dot{x}(t_f)$ is obtained as

$$\dot{x}_f = \dot{z}(t_f) + \frac{\kappa}{c}z(t_f) = v_0 e^{-\zeta\pi/\sqrt{1-\zeta^2}}. \tag{5.4}$$

The normal force between the two bodies during collision is given by either the force in the spring or that in the dashpot – these forces are the same:

$$F = -\kappa z = (1 - \zeta^2)^{-1/2}m\omega_0 v_0 e^{-\zeta\omega_0 t}\sin(\omega_d t), \qquad \omega_d t \leq \pi. \tag{5.5}$$

Figure 5.2 illustrates this force for both an elastic collision $c = \infty$ (or $\zeta = 0$) and a collision with a large damping ratio $c = m\omega$ (or $\zeta = 0.5$). The Maxwell model gives an asymmetrical force with a contact period t_f that increases with the damping ratio ζ. The maximum force is reduced as a result of the compliance of the dashpot. This force has a normal impulse

$$p(t) = mv_0\{1 - e^{-\zeta\omega_0 t}[\cos(\omega_d t) + \zeta(1 - \zeta^2)^{-1/2}\sin(\omega_d t)]\}.$$

The transition from the compression to restitution phases of impact occurs at time t_c when the normal impulse has eliminated the initial normal relative momentum; i.e., $p_c \equiv p(t_c) = mv_0$ at time $t_c = \omega_d^{-1}[\pi - \tan^{-1}(\zeta^{-1}\sqrt{1 - \zeta^2})]$. Thus at time t_f when the bodies separate, the final impulse p_f imparted to body B' is given by the ratio

$$p_f/p_c = 1 + e^{-\zeta\pi/\sqrt{1-\zeta^2}}. \tag{5.6}$$

For a direct impact this gives a coefficient of restitution e_* that agrees with that obtained from the ratio of normal components of relative velocity at incidence and separation,

$$e_* = e^{-\zeta\pi/\sqrt{1-\zeta^2}}. \tag{5.7}$$

The Maxwell viscoelastic model results in a coefficient of restitution e_* that is *independent of the incident relative velocity.*

In a collision between viscoelastic bodies, the compliance relation is rate-dependent; consequently the transition from the compression to the restitution phase of contact does not occur at the instant of maximum force when the spring compression is maximum. With the Maxwell model the dashpot continues to compress throughout the entire contact period. The analysis above indicates that the transition from compression to restitution occurs when the normal impulse is equal in magnitude to the initial momentum of relative motion mv_0 and that this impulse causes the normal relative velocity to vanish; i.e. $\dot{x}_c = \dot{x}(t_c) = 0$.

5.1.2 Simplest Nonlinear Viscoelastic Deformable Element

The linear viscoelastic model gives a coefficient of restitution that is independent of the normal component of relative velocity at incidence; i.e. it is independent of the normal impact speed. For many collision pairs a coefficient of restitution that decreases with increasing relative velocity at incidence is required in order to represent experimental measurements. A nonlinear viscoelastic model is a means of achieving this velocity dependence.

One possibility is a linear spring in parallel with a nonlinear dashpot; in this material model the dashpot provides a force that depends on both the relative velocity and the displacement. The proposed material model gives a normal force F that depends on the normal relative displacement x and the relative velocity $\dot{x}$ across the deformable element,

$$F = -c|x|\dot{x} - \kappa x. \tag{5.8}$$

This is a simple example of a more general nonlinear model that was proposed by Walton (1992). Similar linear-spring–nonlinear-dashpot models have been employed by Stoianovici and Hurmuzlu (1996) and Chatterjee (1997). For a collinear collision between bodies B and B' with points of contact C and C' that are separated by this viscoelastic element, the equation of relative motion is

$$m\ddot{x} - cx\dot{x} + \kappa x = 0, \qquad \dot{x}(0) \equiv -v_0 < 0 \tag{5.9}$$

where m is the effective mass, $m^{-1} \equiv M^{-1} + M'^{-1}$. In (5.9) the absolute value sign has been omitted after taking into account the initial conditions $\dot{x}(0) \equiv -v_0 < 0$ and $x(0) = 0$. To analyze the dynamic response of this system it is useful to change the independent variable from time t to displacement x and rewrite the equation of motion as

$$\frac{m}{\kappa}\dot{x}\frac{d\dot{x}}{dx} + x\left(1 - \frac{c\dot{x}}{\kappa}\right) = 0.$$

A nondimensional displacement X and velocity Z are defined as follows:

$$X \equiv \frac{cx}{\sqrt{m\kappa}} = \frac{c\omega x}{\kappa}, \quad Z \equiv \frac{c\dot{x}}{\kappa}, \quad \omega^2 = \frac{\kappa}{m}. \tag{5.10}$$

In terms of these nondimensional variables, the equation of relative motion becomes

$$Z\frac{dZ}{dX} + X(1 - Z) = 0.$$

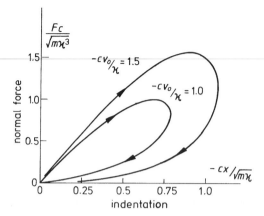

Figure 5.3. Nonlinear viscoelastic element gives a hysteresis loop for contact force as a function of deflection. The size of the loop increases with incident relative velocity.

This equation is separable and can be integrated to give

$$Z - Z_0 + \ln\left(\frac{1 - Z}{1 - Z_0}\right) = \frac{X^2}{2} \tag{5.11}$$

where initially, the displacement $X(0) = 0$ and the relative velocity $Z(0) \equiv Z_0 = -c\kappa^{-1}v_0 < 0$.

For this model the nondimensional force across the deformable element is given by

$$c(m\kappa^3)^{-1/2} F = -X(1 - Z). \tag{5.12}$$

Figure 5.3 plots this force for two different initial velocities. Because of the velocity-dependent element the maximum force occurs substantially before the maximum relative deflection. This maximum force increases roughly in proportion to the incident relative velocity. For loading followed by unloading the force–deflection curve forms a loop. With increasing impact speed there is an increase in the area within this hysteresis loop relative to the area under the curve during loading, i.e. in the part of the initial kinetic energy of relative motion that is dissipated increases with impact speed. Consequently this nonlinear viscoelastic element gives a coefficient of restitution that decreases as the normal incident relative velocity increases.

For a collinear collision the coefficient of restitution can be calculated from the ratio of normal components of relative velocity at separation and at incidence, i.e. the kinematic coefficient of restitution. For any initial velocity Z_0 separation occurs when the force vanishes, i.e. $X = 0$. Hence the separation velocity Z_f is a root of (5.11) with the right side of the equation set equal to zero; i.e., Z_f is obtained from

$$Z_f - Z_0 = -\ln\left(\frac{1 - Z_f}{1 - Z_0}\right), \qquad 0 \le Z_f \le -Z_0 \tag{5.13}$$

and the coefficient of restitution e_* is given by

$$e_* = -Z_f/Z_0.$$

The coefficient of restitution for this nonlinear viscoelastic element is shown in Fig. 5.4.

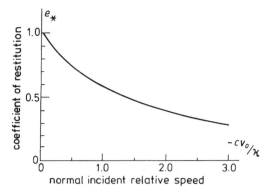

Figure 5.4. Nonlinear viscoelastic element gives a coefficient of restitution that decreases with increasing normal component of incident relative velocity.

Chatterjee has identified a close approximation for this curve,

$$e_* \approx (-Z_0 + e^{0.4Z_0})^{-1}.$$

Thus for large impact speeds $Z_0 \ll -1$ the coefficient of restitution is inversely proportional to the normal relative speed at incidence; i.e. $e_* \sim -Z_0^{-1}$.

5.1.3   Hybrid Nonlinear Viscoelastic Element for Spherical Contact

For elastic spheres the Hertz contact relation gives a normal force proportional to the 3/2 power of indentation or relative displacement. In cases where during the cycle of loading and unloading the elastic restoring force is dominant but there also is a small velocity-dependent dissipation of energy, Simon (1967) suggested a hybrid relation between contact force F and relative displacement x,

$$F = -\kappa |x|^{1/2}(x + c|x|\dot{x}), \qquad \dot{x}(0) \equiv -v_0 < 0 \tag{5.14}$$

where the elastic stiffness κ can be obtained from the Hertz relation $\kappa = 4E_* R_*^{1/2}/3$ [see Eq. (6.8)] and c is a damping coefficient.

For collinear collision between two bodies with spherical contact surfaces and $x < 0$ the Simon relation gives an equation of relative motion,

$$\ddot{x} = \dot{x}\, d\dot{x}/dx = -m^{-1}\kappa |x|^{1/2}(x - cx\dot{x}). \tag{5.15}$$

The analytical results can be expressed most compactly in terms of nondimensional variables, viz. a nondimensional relative displacement X, velocity Z, and time τ:

$$X \equiv x/R_*, \qquad Z \equiv dX/d\tau = c\dot{x}, \qquad \tau \equiv (cR_*)^{-1}t.$$

With these variables the response of the system depends on only a single nondimensional composite parameter ζ which represents the energy loss.[2] The ordinary differential equation for relative displacement can then be expressed in nondimensional form as

$$Z\frac{dZ}{dX} = \zeta |X|^{3/2}(1 - Z), \qquad \zeta \equiv \frac{\kappa c^2 R_*^{5/2}}{m} \tag{5.16}$$

---

[2] The principal advantage of nondimensionalization is that the parameter ζ is itself nondimensional and thus independent of the set of units used for measurement.

where indentation results in a force of positive sign. After integrating and applying initial conditions $X(0) = 0$ and $Z_0 \equiv Z(0) < 0$, the nondimensional velocity is obtained as

$$Z - Z_0 + \ln \frac{1 - Z}{1 - Z_0} = \frac{2}{5} \zeta X^{5/2}.$$

This can be rearranged to obtain the displacement at any nondimensional velocity,

$$X = -\left\{ \frac{5}{2\zeta} \left[Z - Z_0 + \ln \frac{1 - Z}{1 - Z_0} \right] \right\}^{2/5} \qquad (5.17)$$

which has a negative sign for displacement to indicate indentation. The corresponding nondimensional expression for the contact force is

$$F/\kappa R_*^{3/2} = |X|^{3/2}(1 - Z). \qquad (5.18)$$

The latter can be compared with Eq. (5.12) from the previous nonlinear viscoelastic model.

The contact force as a function of displacement X, shown in Fig. 5.5, displays a hysteresis loop similar to that in Fig. 5.3. The size of this loop in comparison with the area under the loading curve is equal to the part of the normal partial work that is dissipated by the damper, i.e. to the coefficient of restitution.

The terminal velocity at separation Z_f or the coefficient of restitution $e_* = -Z_f/Z_0$ can be obtained from Eq. (5.17) by recognizing that when $X = 0$ the contact force vanishes. This condition, however, gives an expression identical to (5.11). Thus the relation between coefficient of restitution and nondimensional incident speed for this viscoelastic model is identical to that shown in Fig. 5.4; only the nondimensionalization of velocity is different. Hence for simulation of experiments, the damping coefficient c is obtained from the curve in Fig. 5.4 with measurements of the impact speed v_0 and the coefficient of restitution e_*.

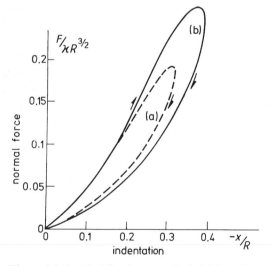

Figure 5.5. Load–deflection curve for hybrid nonlinear viscoelastic model and two different initial velocities: (a) $Z_0 = -0.38$, $\zeta = 2.5$ (solid curve) and (b) $Z_0 = -0.50$, $\zeta = 2.5$ (dashed curve).

Table 5.1. *Viscoelastic Parameters for Two-Piece Golf Ball from Dynamic Measurements*

v_0 (m s^{-1})	F_{max} (N)	t_f (μs)	e_*	c (s m^{-1})	κ (10^6 N m$^{-3/2}$)	Z_0	ζ	$F_{max}/\kappa R_*^{3/2}$
36.6	12,260	450	0.81	0.0105	20.8	0.38	2.5	0.189
42.7	14,980	—	0.81	0.0089	20.9	0.38	1.8	0.230

Note: Size of golf ball: radius $R = 0.0213$ m, mass $M = 61$ g.
Source: Experimental data from Johnson and Lieberman (1996).

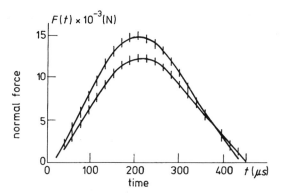

Figure 5.6. Contact force $\mathbf{F}(t)$ for a two-piece golf ball struck by a heavy body with normal incident speed $\dot{x}(0) = v_0 = -36.6$ m s^{-1} (lower curve) or -42.7 m s^{-1} (upper curve).

5.1.4   Parameters of the Hybrid Nonlinear Element for Impact on a Golf Ball

There are two types of golf balls in common use: (a) two-piece balls with a solid core of cross-linked rubber and an ionomer cover, and (b) wound balata balls with either a durable ionomer or a balata cover. The results in Table 5.1 are for two-piece balls.[3] The dynamic behavior of these balls is accurately represented by the hybrid nonlinear viscoelastic relation for spherical contacts as shown in Fig. 5.6. This figure compares calculated contact forces for two collisions with experimental measurements.

While this analysis has focused on impact between a golf ball and a club head, the impact behavior of the ball bouncing off the turf also has been related to viscoelastic parameters. Haake (1991) measured the incident and separation velocities of rotating golf balls striking a turf green at different speeds and used these measurements to derive constants for a linear three parameter viscoelastic model. Data for similar lumped parameter models have been obtained by Ujihashi (1994) and Johnson and Liebermann (1996).

5.2   Tangential Compliance in Planar Impact of Rough Bodies

For oblique impact between rough bodies there are both normal and tangential components of contact force; the tangential force (friction) opposes the tangential relative velocity, which is termed sliding or slip. Effects of dry friction on changes in velocity during impact can be obtained from Coulomb's law only by summation of changes during

---

[3] Wound balata balls have a more nonlinear compliance that is not so well represented by the Simon two parameter hybrid viscoelastic relation.

successive stages of unidirectional slip (Goldsmith, 1960; Stronge, 1990; Brach, 1993). In previous chapters of this book the analyses have assumed that tangential compliance of the bodies is negligible in the contact region. While this assumption is an asymptotic limit that represents large initial slip, it may not be accurate for small initial slip where the direction of slip can change during contact. Maw, Barber and Fawcett (1976, 1981) performed a dynamic analysis of oblique impact between rough elastic spheres using Hertz contact theory to obtain the normal tractions in the contact area. For deformable bodies they found that at small sliding speeds the contact area had an outer annulus which was sliding. The annulus surrounded a central area where there was no tangential component of relative velocity, i.e., the central area was sticking. This combined state of slip and stick was termed *microslip*. Both the analysis and experiments of Maw, Barber and Fawcett showed that if initial slip was small, the direction of slip could be reversed during collision by *tangential compliance*. With negligible tangential compliance such reversals are not possible if the impact configuration is collinear (i.e. if the centers of mass of the colliding bodies are on the line of the common normal passing through the contact point).

There have been several attempts to develop approximations which produce the effect of slip reversal for small angles of incidence in collinear as well as noncollinear collisions. Bilbao, Campos and Bastero (1989) defined a kinematic tangential coefficient of restitution which varies exponentially with the coefficient of friction and the ratio of normal to tangential components of incident velocity. Smith (1991) defined a kinetic coefficient of restitution that relates the tangential impulse to the coefficient of friction, the normal impulse and an average velocity of sliding during collision. Brach (1989) used two linear relations for changes in tangential velocity; these employed a kinematic coefficient of restitution (negative) at very small angles of incidence and a kinetic coefficient of restitution at larger angles. All of these approaches were designed to produce at large angles of incidence a ratio of tangential to normal impulse equal to the coefficient of friction; they do not represent Coulomb's law of friction in general.

In contrast to the elastic continuum approach of Maw, Barber and Fawcett (1976) and the approximations above, the present section uses a lumped parameter representation for compliance of the contact region. Equations of motion are developed from a few physical laws, and system characteristics are expressed in terms of coefficients that are independent of the angle of incidence. This model yields either slip or stick at the contact point, depending on whether the ratio of tangential to normal contact force is as large as the coefficient of friction. During stick the tangential force depends on relative displacement, so a time-dependent analysis is required to resolve the changes in velocity that occur in a collision; nevertheless, we assume that the total period of contact is so brief that there is no change in configuration during collision. Comparison will reveal that the present lumped parameter modeling gives velocity changes that are almost the same as those in experiments by Maw, Barber and Fawcett (1981) and contact forces that are similar to measurements by Lewis and Rogers (1988, 1990). In addition the present analysis yields results for a model of inelastic collisions with tangential compliance.

5.2.1   Dynamics of Planar Collision for Hard Bodies

To focus on the effects of tangential compliance during collision, consider two bodies with masses M and M' that collide at contact point C as shown in Fig. 5.7.

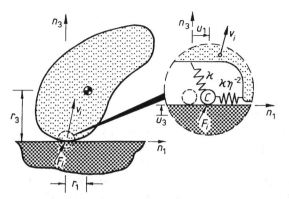

Figure 5.7. Discrete parameter model for impact of a rough, compliant body on a half space.

At C the contact areas of the colliding bodies have a common tangent plane. Let unit vectors n_1 and n_3 be oriented in directions tangent and normal to this plane respectively. At incidence (time $t = 0$) a point on each body comes into contact; at incidence these points have a relative velocity $v_i(0)$ with tangential and normal components $v_1(0)$ and $v_3(0)$ respectively. The orientation of the coordinate system is defined such that at incidence both normal and tangential components of relative velocity are negative: $v_1(0) < 0$ and $v_3(0) < 0$. The collision period is separated into an initial period of normal compression and a subsequent period of separation. The compression period terminates at time t_c when the normal component of relative velocity vanishes, $v_3(t_c) = 0$. The contact points separate at time t_f when the final relative velocity has a normal component $v_3(t_f) \geq 0$. Hence during compression, kinetic energy is absorbed by deformation of the bodies, while during restitution, elastic strain energy generates the force that drives the bodies apart and restores some of the kinetic energy that was absorbed during compression.

To simplify the dynamic analysis we assume that both bodies are rigid except for an infinitesimally small deformable region that separates the bodies at the contact point. Let the bodies have masses M and M' and radii of gyration $\hat{k}_r$ and $\hat{k}_r'$ about their respective centers of mass. From the center of mass of each colliding body, the contact point is located by a position vector r_i or r_i' with components in directions n_i, $i = 1, 3$. Thus the colliding bodies have an inverse of inertia matrix for C

$$m_{ij}^{-1} \equiv m^{-1} \begin{bmatrix} \beta_1 & -\beta_2 \\ -\beta_2 & \beta_3 \end{bmatrix}$$

where

$$\beta_1 = 1 + mr_3^2/M\hat{k}_r^2 + mr_3'^2/M'\hat{k}_r'^2$$
$$\beta_2 = mr_1 r_3/M\hat{k}_r^2 + mr_1'r_3'/M'\hat{k}_r'^2$$
$$\beta_3 = 1 + mr_1^2/M\hat{k}_r^2 + mr_1'^2/M'\hat{k}_r'^2.$$

Notice that if the collision configuration is collinear ($\beta_2 = 0$), then in the inverse of the inertia matrix the terms off the principal diagonal all vanish. Hence for a collinear collision configuration, the effects of normal and tangential impulse on changes in the components of relative velocity at C are decoupled.

The infinitesimal deforming region around C is modeled by assuming that one of the colliding bodies is connected to a massless particle located at C as illustrated in Fig. 5.7. The connection is via two independent compliant elements – one tangential and one normal to the tangent plane. This particle has a tangential component of displacement $u_1(t)$ relative to the body that depends on the tangential force at C and the tangential compliance. The contact force F_i acts at the contact point; it has components in directions n_i. This force applies a differential impulse $dp_i = F_i\, dt$ in an increment of time dt. Thus the equations of planar motion for the rigid body can be expressed as

$$dv_i = m_{ij}^{-1}\, dp_j .$$

To progress further it is necessary to be specific about the compliance in the contact region so that contact force can be calculated. The components of this force are needed in order to distinguish between periods of slip and stick at C. In order to calculate the changes in relative velocity that occur while the contact sticks, it is necessary first to obtain the components of contact force.

Linear Compliance Model

Here we assume that both normal and tangential compliant elements are piecewise linear. During compression let the normal element have an arbitrary stiffness κ while the tangential element has stiffness κ/η^2 as shown in Fig. 5.8. During compression the parameter η^2 is a ratio of normal to tangential stiffness at the contact point; this ratio depends on the structure of the colliding body in the deforming region. These spring constants yield components of force at the contact point which, together with the inverse m_{ij}^{-1} of the inertia matrix, give an equation of motion in terms of relative displacements u_1 and u_3 at the contact point. This equation of motion is expressed as follows:

$$\begin{Bmatrix} \ddot{u}_1 \\ \ddot{u}_3 \end{Bmatrix} = -m^{-1}\kappa \begin{bmatrix} \beta_1 \eta^{-2} & -\beta_2 \\ -\beta_2 \eta^{-2} & \beta_3 \end{bmatrix} \begin{Bmatrix} u_1 \\ u_3 \end{Bmatrix}. \tag{5.19}$$

This system has two natural frequencies, ω and Ω, which are the eigenvalues of the

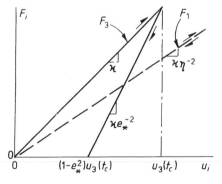

Figure 5.8. Stiffness of normal and tangential compliant elements during loading and unloading. The energy dissipated by irreversible internal deformations is proportional to the ratio of the area under the curve during unloading to that during loading.

frequency equations,

$$\begin{Bmatrix} \omega^2/\omega_0^2 \\ \Omega^2/\omega_0^2 \end{Bmatrix} = \frac{\beta_1 + \beta_3\eta^2}{2\eta^2} \left\{ 1 \pm \sqrt{1 - \frac{4\left(\beta_1\beta_3 - \beta_2^2\right)\eta^2}{\left(\beta_1 + \beta_3\eta^2\right)^2}} \right\}, \qquad \omega_0^2 = \frac{\kappa}{m}.$$

For a collinear impact configuration and linear stiffness, the colliding bodies undergo independent simple harmonic motions (SHMs) in the normal and tangential directions while the contact point sticks. These motions have frequencies that depend on the stiffness and the components β_i of the inverse of inertia matrix that were given in (5.14). For a collinear impact configuration the natural frequency of normal motion (Ω) and that of tangential motion (ω) are given by

$$\Omega \equiv \sqrt{\frac{\beta_3\kappa}{m}} = \frac{\pi}{2t_c}, \qquad \omega \equiv \sqrt{\frac{\beta_1\kappa}{\eta^2 m}} = \frac{\pi}{2\eta t_c}\sqrt{\frac{\beta_1}{\beta_3}} \tag{5.20}$$

where time t_c is the instant when the compression period terminates. At this instant the normal compliant element has a maximum compression $\dot{u}_3(t_c) = 0$.

For normal indentation by a rigid circular punch on an elastic half space, Johnson (1985) gives expressions for the normal stiffness $\kappa = Ea/(1 - v^2)$ and the tangential stiffness $\kappa/\eta^2 = 2Ea/(2 - v)(1 + v)$, where a is the radius of the punch. These relations give $\eta^2 = (2 - v)/2(1 - v)$.

Coefficient of Restitution

The coefficient of restitution e_* can be defined as the square root of the ratio of the elastic strain energy released at the contact point during restitution to the kinetic energy absorbed by internal deformation during compression. For negligible tangential compliance, the loss of kinetic energy due to irreversible internal deformations in the contact region can be obtained from the work done on the bodies by the normal component of contact force. Equation (3.24) defined the energetic coefficient of restitution e_* according to this work. For negligible tangential compliance, this is the only part of the work done on the bodies that goes into deformation. Consequently, it is the only part that can be stored as elastic strain energy if tangential compliance is negligible – the work done by the tangential component of force is all dissipated by friction. If tangential compliance is not negligible, however, there is also energy absorbed by tangential deformations u_1. Here it is assumed that this deformation is entirely elastic. This assumption is based on considering the coefficient of restitution as representing nonfrictional energy losses that are due primarily to *contained plastic deformation*. In an initial range of elastic–plastic deformation the region of plasticity is contained beneath the surface of a deforming body; this contained, or subsurface, plasticity has very little effect on tangential compliance. Hence for elastic–plastic bodies which collide at low speeds, the energetic coefficient of restitution still applies; i.e.

$$e_*^2 = \frac{-\int_{t_c}^{t_f} F_3 v_3\, dt}{\int_0^{t_c} F_3 v_3\, dt}$$

With linear compliance the normal component of force does work $W_3(t_c)$ on the bodies during compression; this is simply the area under the normal compression line in Fig. 5.8,

$W_3(t_c) = F_3(t_c)u_3(t_c)/2$. In the present model the effect of the coefficient of restitution e_* is obtained by changing the stiffness of the normal compliant element at the transition time t_c when compression terminates. Hence at the instant of maximum compression, t_c, the stiffness of the normal element increases from κ to κ/e_*^2. For changes in the normal component of relative velocity, this change in stiffness makes the frequency of SHM larger during restitution than it was during compression. The collision terminates and separation occurs at a final time t_f; for collinear collisions $t_f = (1 + e_*)t_c$. At separation the normal component of relative displacement has a terminal value $u_3(t_f) = (1 - e_*^2)u_3(t_c)$.

Normal Components of Velocity and Force in Collinear Collision

Henceforth in this section the analysis is limited to collinear collisions (i.e. $\beta_2 = 0$), so that the normal and tangential equations of motion (5.19) are decoupled. For a collinear collision the normal force (and hence changes in the normal component of relative velocity) is independent of the process of slip or stick at the contact point. For a linear compliant element, the colliding bodies undergo separate stages of SHM during compression and restitution periods of the collision. Thus at any time during the collision, the normal component of relative velocity is as follows:

$$v_3(t) = v_3(0)\cos\Omega t, \qquad\qquad\qquad 0 \le t \le t_c \qquad (5.21\text{a})$$

$$v_3(t) = e_*v_3(0)\cos\left(\frac{\Omega t}{e_*} + \frac{\pi}{2}(1 - e_*^{-1})\right), \qquad t_c < t \le t_f. \qquad (5.21\text{b})$$

This normal component of velocity is continuous at the time of maximum compression t_c when the frequency of SHM increases from Ω to Ω/e_*. The normal component of impulse that causes these changes in velocity can be obtained directly, and this impulse can be differentiated to obtain the normal component of contact force. The expressions for these variables are listed in Table 5.2.

Notice that to maintain contact without interpenetration (or separation), we must have $v_3(t) + \dot{u}_3(t) = 0$, where normal compression gives $u_3 > 0$, and where $\dot{u}_3(t) \equiv du_3/dt$.

Tangential Velocity and Force During Stick

During the period in which the contact point sticks, SHM applies also to tangential changes in velocity if the tangential compliant element is linear. The tangential oscillations during stick occur with frequency ω. It is convenient to express the velocity and force during stick in terms of an initial displacement $u_1(t_2)$ and an initial velocity $v_1(t_2)$ at time $t = t_2$

Table 5.2. *Normal Displacement, Velocity, Force and Impulse during Collision*

	Formula	
Quantity	Compression, $0 \le t \le t_c$	Restitution, $t_c \le t \le t_f$
Displacement	$u_3(t) = -\Omega^{-1}v_3(0)\sin\Omega t$	$u_3(t) = -e_*^2\Omega^{-1}v_3(0)\sin\left(\dfrac{\Omega t}{e_*} + \dfrac{\pi}{2}(1 - e_*^{-1})\right)$
Velocity	$v_3(t) = v_3(0)\cos\Omega t$	$v_3(t) = e_*v_3(0)\cos\left(\dfrac{\Omega t}{e_*} + \dfrac{\pi}{2}(1 - e_*^{-1})\right)$
Force	$F_3(t) = -\dfrac{m\Omega v_3(0)}{\beta_3}\sin\Omega t \ge 0$	$F_3(t) = -\dfrac{m\Omega v_3(0)}{\beta_3}\sin\left(\dfrac{\Omega t}{e_*} + \dfrac{\pi}{2}(1 - e_*^{-1})\right) \ge 0$
Impulse	$p(t) = -\dfrac{mv_3(0)}{\beta_3}(1 - \cos\Omega t)$	$p(t) = -\dfrac{mv_3(0)}{\beta_3}\left[1 - e_*\cos\left(\dfrac{\Omega t}{e_*} + \dfrac{\pi}{2}(1 - e_*^{-1})\right)\right]$

when stick begins. While the contact point sticks, there is no sliding [$v_1(t) + \dot{u}_1(t) = 0$, where $\dot{u}_1 \equiv du_1/dt$]; thus the tangential displacement, the velocity and the contact force on C can be written as

$$u_1(t) = u_1(t_2)\cos\omega(t - t_2) - \omega^{-1}v_1(t_2)\sin\omega(t - t_2) \tag{5.22a}$$

$$v_1(t) = \omega u_1(t_2)\sin\omega(t - t_2) + v_1(t_2)\cos\omega(t - t_2) \tag{5.22b}$$

$$F_1(t) = m\beta_1^{-1}\omega^2 u_1(t_2)\cos\omega(t - t_2) - m\beta_1^{-1}\omega v_1(t_2)\sin\omega(t - t_2), \qquad t \geq t_2. \tag{5.22c}$$

The state of stick persists while the ratio between tangential and normal components of contact force satisfies $|F_1|/F_3 < \mu$.

Velocity Changes during Sliding with Friction

For dry friction that can be represented by Coulomb's law, sliding occurs if the ratio between the tangential and normal components of force is equal to the coefficient of friction μ. Thus sliding depends on the tangential compliance and relative displacement $u_1(t)$ between the particle at C and the body; if there is sliding contact, the particle is sliding at velocity $v_1(t) + \dot{u}_1(t)$. On the other hand, if $|F_1|/F_3 < \mu$ then the contact is sticking, so $v_1(t) = -\dot{u}_1(t)$. For simplicity, coefficients of static and dynamic friction are assumed to be equal. While the contact slides in direction $\hat{s} \equiv \mathrm{sgn}(v_1 + \dot{u}_1)$, the contact force applies a differential impulse

$$dp_j = \{-\mu\hat{s} \quad 1\}^T dp, \qquad j = 1, 3$$

where $dp \equiv dp_3$. Consequently, in a collinear collision the differential equation for sliding can be expressed in terms of a different independent variable – the normal component of impulse p rather than time:

$$\left\{ \begin{array}{c} dv_1/dp \\ dv_3/dp \end{array} \right\} = m^{-1} \begin{bmatrix} \beta_1\eta^{-2} & 0 \\ 0 & \beta_3 \end{bmatrix} \left\{ \begin{array}{c} -\mu\hat{s} \\ 1 \end{array} \right\}. \tag{5.23}$$

Initial Stick or Slip?

With the present model, the contact point C either sticks or slips, depending on the ratio between the components of the contact force, viz. whether or not the contact force is inside the cone of friction.[4] First suppose that stick begins at the initial instant of contact and then test whether this satisfies the limiting force ratio. An initial period of stick terminates at a time t_1 when the ratio of tangential to normal force first becomes as large as μ; i.e., time t_1 is obtained from

$$\frac{|F_1|}{\mu F_3} = \frac{1}{\eta^2}\frac{v_1(0)}{\mu v_3(0)}\frac{\Omega}{\omega}\frac{\sin\omega t_1}{\sin\Omega t_1} = 1, \qquad 0 \leq t_1 < t_c \tag{5.24a}$$

$$\frac{|F_1|}{\mu F_3} = \frac{1}{\eta^2}\frac{v_1(0)}{\mu v_3(0)}\frac{\Omega}{\omega}\frac{\sin\omega t_1}{\sin\left(\dfrac{\Omega t_1}{e_*} + \dfrac{\pi}{2}(1 - e_*^{-1})\right)} = 1, \qquad t_c < t_1 \leq t_f \tag{5.24b}$$

---

[4] This model gives a period of initial stick if the incident tangential velocity is small – initial stick occurs because the particle at C has negligible mass. Mindlin and Deresiewicz (1953) have shown that during an early stage of collision for elastic bodies, any finite tangential velocity results in central stick plus a peripheral annulus of slip around the small contact patch.

The process of initial stick takes place if $t_1 > 0$, i.e. if in the limit as $t_1 \to 0$ the force ratio is inside the cone of friction. This requires an angle of incidence such that

$$\frac{v_1(0)}{v_3(0)} < \mu\eta^2.$$

Thus initial stick occurs if the angle of incidence is small, i.e. if at incidence the ratio of the tangential to normal component of relative velocity at C (which equals the tangent of the angle of incidence ψ_0) is within a range $0 < v_1(0)/v_3(0) < \mu\eta^2$ that is bounded by the product of the coefficient of friction and the ratio of normal to tangential stiffness. For collisions with an angle of incidence at C which is larger in magnitude than $\tan^{-1}(\mu\eta^2)$, the contact point begins sliding when contact initiates.

5.2.2   Slip Processes

Small Angle of Incidence, $v_1(0)/v_3(0) < \mu\eta^2$
For small angles of incidence, stick initiates at initial contact and continues until time t_1, when slip begins. Thereafter slip continues until separation at time t_f. Figure 5.9 sketches

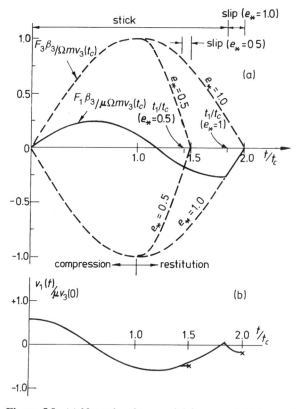

Figure 5.9. (a) Normal and tangential forces and (b) tangential velocity during collision for small angle of incidence $v_1(0)/v_3(0) < \mu\eta^2$ and frequency ratio $\omega/\Omega = 1.7$. Forces are illustrated for both $e_* = 0.5$ and 1.0. For each coefficient of restitution, the final velocity at separation is indicated by a cross $\times$.

the components of force for this case of initial stick and terminal slip. At time t_1 when slip begins, the relative velocity at C has a tangential component

$$v_1(t_1) = v_1(0) \cos \omega t_1.$$

Thus from (5.24) the terminal tangential relative velocity can be obtained:

$$v_1(t_f) = v_1(t_1) - \frac{\mu \hat{s} \beta_1}{m} \left[p(t_f) - p(t_1) \right] \qquad (5.25)$$

where the change in normal impulse during the final period of slip is given by

$$p(t_f) - p(t_1) = -\gamma_1 m v_3(0)/\beta_3.$$

The impulse ratio during slip, γ_1, is the ratio of the normal impulse applied during period $t_f - t_1$ to the normal impulse during compression,

$$\gamma_1 \equiv e_* \left\{ 1 + \cos\left(\frac{\Omega t_1}{e_*} + \frac{\pi}{2}(1 - e_*^{-1}) \right) \right\}.$$

This ratio depends on initial conditions. In the subsequent analysis we consider the ratio of compliances $\omega / \Omega > 1$, since this applies to most elastic bodies.[5] For this case slip begins during restitution at a time $t_1 > t_c$ when $u_1(t_1) > 0$, so that $\hat{s} = +1$ and the terminal velocity can be expressed as

$$\frac{v_1(t_f)}{\mu v_3(0)} = \frac{v_1(t_1)}{\mu v_3(0)} + \frac{\beta_1}{\beta_2} \gamma_1. \qquad (5.26)$$

Intermediate Angle of Incidence $\mu \eta^2 < v_1(0)/v_3(0) < \mu(1 + e_*)\beta_1/\beta_3$
If the angle of incidence is larger than $\tan^{-1}(\mu \eta^2)$, initially there is sliding at the contact point as illustrated in Fig. 5.10. During the initial period of sliding the body has a tangential component of relative velocity at C that is given by

$$v_1(t) = v_1(0) - \frac{\mu \hat{s} \beta_1}{m} p(t). \qquad (5.27)$$

This initial sliding terminates and stick begins at time t_2 when subsequent sliding and stick give the same rate of change for the tangential force; i.e., the transition to stick occurs when

$$\lim_{\varepsilon \to 0} \left[\left| \frac{d F_1(t_2 + \varepsilon)}{d\varepsilon} \right| = \mu \frac{d F_3(t_2 + \varepsilon)}{d\varepsilon} \right].$$

Differentiation of Eq. (5.22c) gives for the period $t > t_2$,

$$\frac{d F_1(t)}{dt} = \frac{m \omega^3 u_1(t_2)}{\beta_1} \sin \omega(t - t_2) - \frac{m \omega^2 v_1(t_2)}{\beta_1} \cos \omega(t - t_2)$$

where the dynamics for sliding during $t < t_2$ give transition values at the beginning of

---

[5] For elastic spheres with Poisson's ratio $\nu = 0.3$, the ratio of stiffnesses $\eta^2 = 1.21$ and elements of the inverse of inertia matrix give $\beta_1/\beta_3 = 3.5$, so that $\omega/\Omega = 1.7$ (e.g. see Johnson, 1985), whereas for incompressible elastic spheres, $\eta^2 = 1.5$ and $\omega/\Omega = 1.53$. Furthermore, in the asymptotic limit of negligible tangential compliance, $\eta^2 \to 0$, so that $\omega \to \infty$.

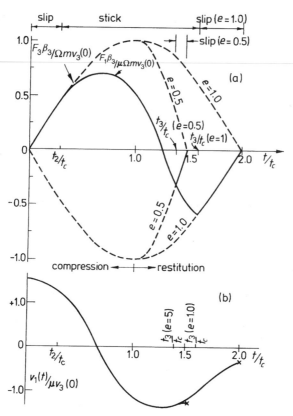

Figure 5.10. (a) Normal and tangential forces and (b) tangential velocity during collision for intermediate angle of incidence $\mu\eta^2 < v_1(0)/v_3(0) < \mu(1 + e_*)\beta_1/\beta_3$ and frequency ratio $\omega/\Omega = 1.7$. Forces are illustrated for both $e_* = 0.5$ and 1.0. For each coefficient of restitution, the final velocity at separation is indicated by a cross $\times$.

stick,

$$u_1(t_2) = \frac{\eta^2 F_1(t_2)}{\kappa} = \mu \frac{\beta_1 \Omega v_3(0)}{\beta_3 \omega^2} \sin \Omega t_2$$

$$v_1(t_2) = v_1(0) - \mu \frac{\beta_1}{\beta_3} v_3(0)[1 - \cos \Omega t_2], \qquad t_2 \le t_c$$

$$u_1(t_2) = \frac{\eta^2 F_1(t_2)}{\kappa} = \mu \frac{\beta_1 \Omega v_3(0)}{\beta_3 \omega^2} \sin\left(\frac{\Omega t_2}{e_*} + \frac{\pi}{2}(1 - e_*^{-1})\right)$$

$$v_1(t_2) = v_1(0) - \mu \frac{\beta_1}{\beta_3} v_3(0)\left[1 - e_* \cos\left(\frac{\Omega t_2}{e_*} + \frac{\pi}{2}(1 - e_*^{-1})\right)\right], \qquad t_2 > t_c.$$

The normal force during restitution is given in Table 5.2; hence by differentiation we obtain

$$\frac{d F_3(t_2)}{dt} = -\beta_3^{-1}\Omega^2 m v_3(0) \cos \Omega t_2, \qquad t_2 \le t_c$$

$$\frac{d F_3(t_2)}{dt} = -\frac{\Omega^2 m v_3(0)}{\beta_3 e_*} \cos\left(\frac{\Omega t_2}{e_*} + \frac{\pi}{2}(1 - e_*^{-1})\right), \qquad t_2 > t_c.$$

Thus after equating the rates of change for components of force, we obtain the expression to be solved for Ωt_2 by taking the limit as $t \to t_2$,

$$\Omega t_2 = \cos^{-1}\left(\frac{v_1(0)/\mu v_3(0) - \beta_1/\beta_3}{\eta^2 - \beta_1/\beta_3}\right), \qquad \frac{v_1(0)}{v_3(0)} \le \mu\frac{\beta_1}{\beta_3}$$

$$\frac{\Omega t_2}{e_*} = -\frac{\pi}{2}(1 - e_*^{-1}) + \cos^{-1}\left(\frac{v_1(0)/\mu v_3(0) - \beta_1/\beta_3}{\eta^2 e_*^{-1} - e_*\beta_1/\beta_3}\right), \qquad \frac{v_1(0)}{v_3(0)} > \mu\frac{\beta_1}{\beta_3}.$$

The limits of applicability for these equations have been expressed in terms of the ratio of velocity components by noting that the transition time $t_2 = t_c$ occurs if $v_1(0)/v_3(0) = \mu\beta_1/\beta_3$.

For intermediate angles of incidence the contact point sticks at time t_2 and then the tangential compliant element begins a period of SHM. During the period of sticking the tangential components of velocity and force are given by Eq. (5.22):

$$v_1(t) = \omega u_1(t_2)\sin\omega(t - t_2) + v_1(t_2)\cos\omega(t - t_2)$$

$$F_1(t) = m\beta_1^{-1}\omega^2 u_1(t_2)\cos\omega(t - t_2) - m\beta_1^{-1}\omega v_1(t_2)\sin\omega(t - t_2).$$

This period of stick terminates and slip begins again at time t_3 when the ratio of components of force next becomes as large as the coefficient of friction, $|F_1|/F_3 = \mu$. This gives a final phase of slip which begins at time t_3 – a time that can be obtained from

$$\left|\frac{\Omega u_1(t_2)}{\mu v_3(0)}\cos\omega(t_3 - t_2) - \frac{\Omega v_1(t_2)}{\omega\mu v_3(0)}\sin\omega(t_3 - t_2)\right| = \eta^2\sin\left[\frac{\Omega t_3}{e_*} + \frac{\pi}{2}(1 - e_*^{-1})\right].$$

Sticking ceases at time t_3 when a second phase of slip begins. For this second phase of slip the initial tangential components of velocity and force, $v_1(t_3)$ and $F_1(t_3)$, are given above.

The second phase of slip terminates at separation. During this final phase of slip the particle at the contact point is sliding as the elastic strain energy in the compliant elements is decreasing to zero. Thus changes in velocity during this phase are given by (5.23). The final tangential velocity of the contact point at time $t_f = (1 + e_*)t_c$ is given by

$$v_1(t_f) = v_1(t_3) - \mu\beta_1 m^{-1}[p(t_f) - p(t_3)]$$

where $\hat{s}(t_f) = +1$, since the final direction of slip is opposite to the initial direction as illustrated in Fig. 5.10. The nondimensional tangential velocity at separation is

$$\frac{v_1(t_f)}{\mu v_3(0)} = \frac{v_1(t_3)}{\mu v_3(0)} + \frac{\beta_1}{\beta_3}\left[1 + e_* - \frac{p(t_3)}{p(t_c)}\right]. \tag{5.28}$$

Large Angle of Incidence, $v_1(0)/v_3(0) > \mu(1 + e_*)\beta_1/\beta_3$

If initial slip does not cease before separation, the transition time $t_2 > (1 + e_*)t_c$; i.e., slip continues in the initial direction throughout the entire contact period – a contact process that is sometimes termed *gross slip*. For this case, at separation Eq. (5.23) gives the tangential velocity $v_1(t_f)$ as follows:

$$v_1(t_f) = v_1(0) + \mu\beta_1 m^{-1}(1 + e_*)p(t_c)$$

where $\hat{s}(t_f) = \hat{s}(0) = -1$. This can be rewritten as

$$\frac{v_1(t_f)}{\mu v_3(0)} = \frac{v_1(0)}{\mu v_3(0)} - \frac{\beta_1}{\beta_3}(1 + e_*).\tag{5.29}$$

For gross slip, the direction of sliding is constant – friction merely reduces the speed.

5.2.3   Oblique Impact of an Elastic Sphere on a Rough Half Space

Combined effects of friction and tangential compliance have been evaluated for oblique impact of a sphere on a half space. For a solid sphere composed of material with Poisson's ratio $v = 0.3$, the ratio of stiffnesses is $\eta^2 = 1.21$ while the ratio of elements of inertia is $\beta_1/\beta_3 = 3.5$, so that the ratio of frequencies is $\omega/\Omega = 1.7$. These values are used in the following examples of oblique impact of an elastic sphere against a massive half space.

Tangential Velocity at Separation
For a sphere striking a half space at angle of incidence $\psi_0 = \tan^{-1}[v_1(0)/v_3(0)]$, Eqs. (5.26), (5.28) and (5.29) were used to calculate the tangential relative velocity for contact point C at separation. In Fig. 5.11, the results from the present discrete parameter model are compared with the elasticity solution given by Maw, Barber and Fawcett (1976) and experimental measurements by K.L. Johnson (1983) for a rubber sphere (Poisson's ratio $v = 0.5$) striking a heavy steel plate at a small speed. The elastic solution and the discrete parameter model each have similar processes that develop at the contact point in three parts of the range of angle of incidence. The predictions of these two models are most different for small and intermediate angles of incidence where the discrete parameter model has a final period of slip that is prolonged by elastic strain energy stored in the tangential compliant element. Throughout most of the range of small to intermediate angles of incidence, both the elastic continuum and the discrete parameter models of sphere impact have a tangential relative velocity at separation that is in the opposite direction to the incident tangential velocity. For a collinear collision this velocity reversal

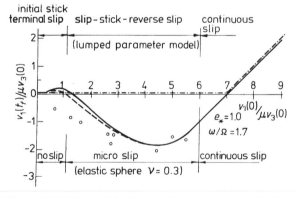

Figure 5.11. Tangential velocity of contact point on elastic sphere at instant of separation as a function of the angle of incidence. Solid curve, lumped parameter model; dashed curve, elastic continuum analysis; dot–dash curve, analysis for negligible tangential compliance; circles, experiments with a rubber ball.

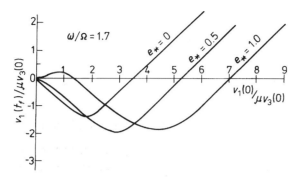

Figure 5.12. Tangential velocity of contact point on sphere at separation as function of the angle of incidence for coefficients of restitution $e_* = 0, 0.5$ and 1.0.

at C is entirely due to tangential compliance. In almost all respects the results of these models are practically identical. Figure 5.11 also shows the final velocities calculated for negligible tangential compliance; in this case the tangential component of relative velocity at C is zero unless the angle of incidence is large enough to cause gross slip, i.e. $v_1(0)/v_3(0) > \mu(1 + e_*)\beta_1/\beta_3$.

The effect of coefficient of restitution e_* on the change in the tangential component of relative velocity at C is shown in Fig. 5.12. The angle of incidence for gross slip decreases with increasing internal dissipation. The coefficient of restitution e_* affects only the impulse imparted during restitution; the restitution impulse (and changes in velocity during restitution) decreases with decreasing e_*. This causes the shift in the curve for separation velocity that is apparent in Fig. 5.12.

Angle of Incidence for Maximum Friction
Experiments using repeated impacts on steel tubes at oblique angles of incidence were performed by Ko (1985). He showed that for relatively small normal impact speeds, $v_3 < 1 \text{ m s}^{-1}$, the wear rate of steel tubes is closely correlated with the maximum tangential force $F_{1\text{max}}$ and that for any colliding missile this force varies with angle of obliquity. For the present model, the tangential force $F_1(t)$ can be calculated as a function of the angle of incidence, $\tan^{-1}[v_1(0)/v_3(0)]$. Irrespective of the angle of incidence, the largest value of friction, $F_{1\text{max}}$, occurs during the compression period if $\omega/\Omega \geq 1$; consequently, the maximum tangential force is independent of the coefficient of restitution. For any impact speed $v(0) \equiv [v_1^2(0) + v_3^2(0)]^{1/2}$ the tangential component of force can be compared with the largest normal force $F_{3\text{max}} = F_3(t_c)$, where

$$\frac{F_{3\text{max}}}{\Omega m v(0)} = \frac{\beta_3^{-1}}{[1 + v_1^2(0)/v_3^2(0)]^{1/2}}. \tag{5.30}$$

Expressions for the maximum tangential force are given below for different ranges of the angle of incidence.

Small Angle of Incidence, $v_1(0)/v_3(0) \leq \mu\eta^2$
For small angles of obliquity the contact point initially sticks and only begins to slide during the restitution period. The maximum tangential force occurs during compression when the tangential velocity reverses in direction at time $\tau \equiv \Omega t_c/\omega$. At this time the

contact is sticking, so the maximum tangential force can be obtained from Eq. (5.22c):

$$F_{1\max} = -\omega m v_1(0)/\beta_1.$$

This maximum tangential component of force can be expressed as a nondimensional ratio $|F_{1\max}|/\Omega m v(0)$. Thus for small angles of obliquity

$$\frac{F_{1\max}}{\Omega m v(0)} = \frac{\omega \beta_1^{-1}|v_1(0)|}{\Omega v(0)} = \frac{\beta_3^{-1}}{\eta} \sqrt{\frac{\beta_3}{\beta_1}} \frac{v_1(0)/v_3(0)}{\left[1 + v_1^2(0)/v_3^2(0)\right]^{1/2}}.$$

Intermediate Angle of Incidence, $\mu\eta^2 < v_1(0)/v_3(0) \le \mu(1 + e_*)\beta_1/\beta_3$

At intermediate angles of obliquity there is initial sliding, but then stick begins at time t_2 during the compression period. When stick begins, the contact point is still moving in the initial direction; i.e., $v_1(t_2) > 0$. Maximum friction develops shortly after the period of stick begins and before the instant of maximum compression. The friction force during sticking can be expressed as

$$\frac{F_1(t)}{\Omega m v(0)} = \frac{\mu \beta_3^{-1}}{\eta^2 v(0)} \left\{ \frac{\Omega u_1(t_2)}{\mu v_3(0)} \cos \omega(t - t_2) + \eta \sqrt{\frac{\beta_3}{\beta_1}} \frac{v_1(t_2)}{\mu v_3(0)} \sin \omega(t - t_2) \right\}.$$

At the transition from sliding to stick the displacement $u_1(t_2)$ and velocity $v_1(t_2)$ depend on the coefficient of friction. The transition velocity $v_1(t_2)$ is obtained from (5.23), while the displacement $u_1(t_2)$ is calculated from the friction law $F_1 = -\mu \hat{s} F_2$ and the force F_2 given in Table 5.2. Thus

$$\frac{\Omega u_1(t_2)}{\mu v_3(0)} = \eta^2 \sin \Omega t_2, \qquad \frac{v_1(t_2)}{\mu v_3(0)} = \frac{v_1(0)}{\mu v_3(0)} - \frac{\beta_1}{\beta_3}(1 - \cos \Omega t_2).$$

Large Angle of Incidence, $v_1(0)/v_3(0) > \mu(1 + e_*)\beta_1/\beta_3$

If the direction of slip is constant throughout the collision period, then the maximum friction force is directly proportional to the normal force and the coefficient of friction:

$$\frac{F_{1\max}}{\Omega m v(0)} = \frac{\mu \beta_3^{-1}}{\left[1 + v_1^2(0)/v_3^2(0)\right]^{1/2}}.$$

For gross sliding this maximum tangential force occurs simultaneously with the largest normal force; i.e., the tangential force is a maximum at time t_c when the compression period terminates.

These expressions for the maximum values of components of contact force have been used to calculate the largest normal and tangential forces which occur during oblique collisions. The largest values vary with the angle of incidence and the coefficient of friction as shown in Fig. 5.13. The largest values for the peak force occur for collisions at intermediate angles of incidence. The angle of incidence where the peak force is largest increases from about 20° for a coefficient of friction $\mu = 0.1$ to almost 60° for $\mu = 1.0$.

Maximum Friction for Negligible Tangential Compliance

If tangential compliance is negligible, oblique collision always results in an initial period of sliding; this sliding is halted before separation, and there is a subsequent final period of stick unless the angle of incidence is large enough to cause gross slip. Since we are considering central or collinear collisions, slip reversal does not occur. Consequently, if

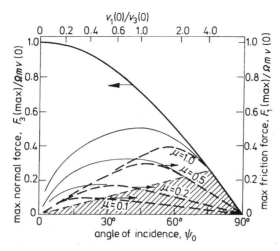

Figure 5.13. Maximum normal component of force (bold solid curve) and tangential component of force (dashed curves) during oblique impact of a sphere. The light solid curves show the maximum force if tangential compliance is negligible.

initial slip comes to a halt during compression, the peak tangential force occurs at the instant t_2 when sliding terminates:

$$\frac{F_{1\max}}{\Omega m v(0)} = \frac{\mu \beta_3^{-1} \sin \Omega t_2}{\left[1 + v_1^2(0)/v_3^2(0)\right]^{1/2}}, \qquad t_2 < t_c$$

where

$$\Omega t_2 = \cos^{-1}\left[1 - \frac{\beta_3}{\beta_1}\left(\frac{v_1(0)}{\mu v_3(0)}\right)\right].$$

On the other hand, if the bodies are still sliding when compression terminates at time t_c, then the largest friction force occurs at this instant simultaneously with the largest normal force:

$$\frac{F_{1\max}}{\Omega m v(0)} = \frac{\mu \beta_3^{-1}}{\left[1 + v_1^2(0)/v_3^2(0)\right]^{1/2}}, \qquad t_2 > t_c.$$

In Fig. 5.13 the dashed lines show the maximum friction force for a compliant solid sphere, whereas the light extensions to the left of these curves are results for similar collisions between spheres with negligible tangential compliance. For solid spheres, the largest tangential force for a compliant body is substantially less than the largest force calculated with the assumption of negligible tangential compliance. In either case *the largest tangential force occurs in the range of small to intermediate angles of incidence*. At large angles of incidence there is gross slip; in this case the maximum tangential force is independent of tangential compliance.

Comparison with Measurement of Peak Force during Oblique Impact
Lewis and Rogers (1988) performed impact experiments in which a 25.4 mm diameter steel sphere collided against a heavy steel plate at angles of incidence that varied between $0°$ and $85°$ from normal. The impact speeds were small, being in the range 0.01–0.05 m s^{-1}. The sphere was attached at the free end of a 1.8 m long pendulum by a steel "ball holder".

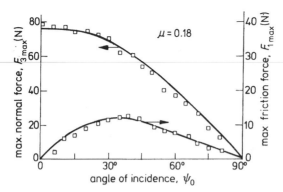

Figure 5.14. Comparison of the maximum normal and tangential components of contact force with experimental measurements by Lewis and Rogers (1988). At each angle of incidence, the normal and friction forces on a 101 g sphere colliding against a half space were calculated for an incident speed 0.048 m s^{-1} and coefficient of friction $\mu = 0.18$.

Piezoelectric force transducers were used to make separate measurements of normal and tangential components of contact force during impact. For gross slip or continuous sliding of this ball on the plate, Lewis and Rogers reported a coefficient of dynamic friction $\mu = 0.179$.

In Fig. 5.14 experimental data taken from collisions at an impact speed of 0.048 m s^{-1} are compared with normal and tangential components of force calculated by the present theory (using a coefficient of friction $\mu = 0.18$). The calculations depend on an estimate of the mass of the "ball holder". The agreement between experiment and theory shown in Fig. 5.14 was achieved by increasing the mass of the ball by 50% in order to allow for inertia of the support system. In addition the calculations used a relative compliance ratio $\eta^2 = 1.21$ and ratio of elements of inertia $\beta_1/\beta_3 = 3.5$ that are representative of solid spheres. This resulted in both qualitative and quantitative agreement between the calculations and the experiments for the full range of possible angles of incidence. Four series of tests using different impact speeds each gave a largest measurement of tangential force at an angle of incidence of about $40°$. For $\mu = 0.18$ the present lumped parameter model gives a largest value of peak tangential force at about $35°$ irrespective of impact speed. [For any angle of incidence, the ratio between peak tangential and normal components of force can only be as large as the coefficient of friction if these peak values occur simultaneously, i.e. if $v_1(0)/v_3(0) \geq \mu\beta_1/\beta_3$ or the angle of incidence $\psi_0 \geq 32°$ for a rough solid sphere with $\mu = 0.18$.]

Comparison with Measurements of Impulse Ratio at Separation
Using thin pucks on an air table, Chatterjee (1997) and colleagues have measured angles of incidence and reflection for uniform circular disks colliding against a heavy steel bar. The disks were made from a polymer, Delrin. They struck the fixed bar at angles of incidence from normal in the range $0 < \psi_0 < 85°$. At impact the disks were translating at speeds in the range $0.25 < v(0) < 0.75 \text{ m s}^{-1}$, and they had negligible angular velocity. For incident angles in the range $0 < \psi_0 < 50°$ two similar pucks gave measurements of the coefficient of restitution $e_* = 0.95 \pm 0.3$. For larger angles of incidence $\psi_0 > 50°$, the coefficient of restitution was slightly larger.

Measurements of the angle of incidence and the angle of rebound for the center of mass were used to obtain the ratio of the tangential to the normal component of impulse

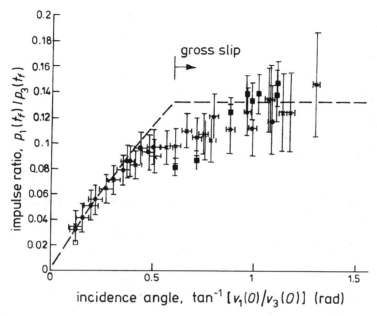

Figure 5.15. Ratio of tangential to normal impulse for a circular Delrin puck colliding against a steel half space at angle of incidence $\tan^{-1}[v_1(0)/v_3(0)]$. Vertical bars describe the range of experimental values. The cross and circle symbols designate two pucks which are identical. The dashed line indicates the calculation using the bilinear spring model.

at separation. If the coefficient of friction is independent of sliding speed and the angle of incidence is large enough to induce gross sliding, this ratio should be a constant that is equal to the coefficient of sliding friction.

Figure 5.15 compares the measurements of Chatterjee (1997) with calculations based on the bilinear spring model in this chapter. The calculations presented here assume a coefficient of sliding friction[6] $\mu = 0.13$ and a coefficient of restitution $e_* = 0.95$, which are consistent with Chatterjee's measurements. For the thin uniform disk, a ratio of normal to tangential stiffness $\eta^2 = 1$ was used on the basis of an estimate from the 2D finite element calculations of Lim (1996); together with the radius of gyration for a circular disk, this gives a frequency ratio $\omega/\Omega \approx \sqrt{3}$.

The bilinear spring model accurately represents the experimental results other than near the transition to gross sliding; this transition occurs at an angle of incidence $\psi_0 = \tan^{-1}[v_1(0)/v_3(0)] = 0.66$ rad. Near this transition the tangential impulse is smaller than the calculated value, indicating that there is stick for more of the contact period than is calculated on the basis of the bilinear spring model. Chatterjee's experiments showed no dependence of the coefficient of friction on either the normal force or the sliding speed.

5.2.4 Dissipation of Energy

Internal Dissipation from Hysteresis of Normal force
During partly elastic collisions ($e_* < 1$) there is always irreversible internal deformation that dissipates a part of the initial kinetic energy T_0. In the present model this internal

---

[6] Crude measurements indicated that the coefficient of sliding friction $\mu < 0.2$.

dissipation D_3 is entirely due to hysteresis of the normal component of force. It can be obtained as the negative of work done by this component of force,

$$D_3(t_f) = -\int_0^{t_f} F_3 v_3 \, dt = (1 - e_*^2)\frac{mv_3^2(0)}{2\beta_3}$$

since in collinear collisions ($\beta_3 = 1$) the part of the initial kinetic energy that is dissipated internally by irreversible deformations can be expressed as

$$\frac{2D_3(t_f)}{mv_3^2(0)} = 1 - e_*^2. \tag{5.31}$$

Frictional Dissipation

Friction dissipates energy only during periods of slip. During these periods the tangential force does some work that changes the tangential strain energy and also some that is dissipated by friction D_1. Whereas the total work done on the body by the tangential force depends on the tangential relative velocity v_1 the frictional energy loss depends only on the sliding speed $v_1 + \dot{u}_1$. The remainder of the work done by the tangential force is stored as elastic strain energy in the tangential compliant element and later recovered as the normal force decreases before separation. Thus for small angles of incidence where the contact point slides only after an initial period of sticking, the tangential dissipation during sliding is calculated from[7]

$$D_1(t_f) = -\int_{t_1}^{t_f} F_1 v_1 \, dt = -\int_{p(t_1)}^{p(t_f)} \left\{ v_1(t_1) + \frac{\mu\beta_1}{m}[p(t) - p(t_1)] \right\} \mu \, dp.$$

Using Eq. (5.20), this gives

$$\frac{2D_1(t_f)}{mv_3^2(0)} = -\frac{2\mu^2\gamma_1}{\beta_3} \left\{ \frac{v_1(0)}{\mu v_3(0)} \cos \omega t_1 + \frac{\gamma_1}{2}\frac{\beta_1}{\beta_3} \right\}. \tag{5.32a}$$

On the other hand, if the angle of incidence is intermediate, the contact point slides prior to time t_2 and again slides after time t_3. Thus for an intermediate angle if $t_2 < t_c$,

$$\frac{2D_1(t_f)}{mv_3^2(0)} = \frac{\mu^2}{\beta_3}\frac{\beta_3}{\beta_1} \left\{ \frac{v_1^2(0)}{\mu^2 v_3^2(0)} - \frac{v_1^2(t_2)}{\mu^2 v_3^2(0)} + \frac{v_1^2(t_3)}{\mu^2 v_3^2(0)} - \frac{v_1^2(t_4)}{\mu^2 v_3^2(0)} \right\}. \tag{5.32b}$$

Finally, if the angle of incidence is large so that there is gross sliding, the part of the energy dissipation due to friction can be expressed as

$$\frac{2D_1(t_f)}{mv_3^2(0)} = (1 + e_*)\frac{2\mu^2}{\beta_3} \left\{ \frac{v_1(0)}{\mu v_3(0)} - \frac{\beta_1}{2\beta_3}(1 + e_*) \right\}. \tag{5.32c}$$

Equations (5.32a–c) were used to evaluate the part of the initial kinetic energy of relative motion that is dissipated by friction $2D_1(t_f)/mv^2(0)$. In Fig. 5.16 this frictional dissipation is plotted as a function of angle of incidence for coefficients of friction $\mu = 0.1$ and 0.5. If the contact region has nonnegligible tangential compliance there is almost no frictional dissipation if the angle of incidence is small; i.e., not much of the kinetic energy T_0 is

---

[7] During any period $t_b - t_a$ *with a constant direction of sliding* the part of the total energy dissipation $D(t_b) - D(t_a)$ that is due to a component in direction $\bar{n}_i$ of the contact force F_j can be calculated from theorem (3.20), $D_i(t_b) - D_i(t_a) = \bar{n}_i[p_j(t_b) - p_j(t_a)][v_j(t_b) + v_j(t_a)]/2$, where at any time t the force has provided an impulse $p_j(t)$. For a *collinear impact configuration* this gives $D_i(t_b) - D_i(t_a) = m_{ij}^{-1}[v_j^2(t_b) - v_j^2(t_a)]/2$ if the direction of sliding is constant.

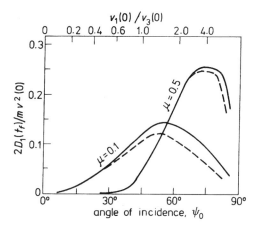

Figure 5.16. The part of the initial kinetic energy that is dissipated by friction during impact of a sphere for various angles of incidence and coefficients of restitution $e_* = 1$ (solid curves) and $e_* = 0.5$ (dashed curves). For gross slip the frictional dissipation is independent of tangential compliance.

required to bring small initial slip to a halt before separation. On the other hand, if there is gross slip due to a large angle of incidence, tangential compliance has no effect on frictional dissipation. The fraction of the initial kinetic energy which is dissipated by friction is maximum at an angle of incidence slightly larger than the smallest angle giving gross slip.

These results are based on the supposition that the coefficient of friction is a parameter that is constant during impact. For impact speeds $v_3(0) > 50\,\mathrm{m\,s^{-1}}$ where indentation of metals results from uncontained plastic deformation, Sundararajan (1990) has pointed out that the tangential force can be increased by finite indentation and decreased by frictional heating.

5.2.5 Effects of Tangential Compliance

For almost all angles of incidence, the response of this simple lumped parameter model is identical with that of the quasistatic (Hertz) elastic analysis. The microslip present in the continuum analysis has no significant effect on changes in velocity of the colliding bodies. Both the elastic continuum analysis and the lumped parameter model show that if slip is brought to a halt during collision, tangential compliance can subsequently reverse the direction of slip. Slip can be brought to a halt during collision only if the tangential component of incident velocity is not too large, i.e. if $v_1(0)/v_3(0) < \mu(1 + e_*)\beta_1/\beta_3$.

For oblique impacts, the largest tangential force generated by friction during collision occurs during compression. This maximum frictional force is independent of the coefficient of restitution. In a compliant body the largest force is somewhat smaller than that which occurs if the contact region has negligible tangential compliance. Although the present calculation of the largest force is based on a model with normal compliance equal to a constant rather than the nonlinear compliance suggested by Hertz type analysis, the details of normal compliance have only a very small effect on the largest tangential force in a collision. In this respect a more significant factor is the ratio of the tangential to the normal compliance, η^2.

Although the present analysis has considered collinear impact configurations (and consequently planar changes in velocity), the same framework can be used to analyze noncollinear collisions. In a noncollinear collision, the normal and tangential motions are coupled so components of relative displacement u_i and relative velocity v_i do not undergo SHM; nevertheless, the equations of motion can be integrated numerically to obtain changes in contact force and velocity during separate phases of stick or slip.

In this model only the normal compliant element is irreversible; consequently, energy losses due to internal hysteresis and those due to friction remain decoupled. While this is representative of dissipation due to contained elastic–plastic deformation where indentation is barely perceptible, it is unlikely to be accurate at higher impact speeds. In elastic–plastic bodies, if the impact energy is large enough to develop significant permanent indentation (uncontained plastic deformation), the inelastic internal deformation depends on both normal and tangential components of contact force. Consequently, for impact energies that produce significant indentation, sources of dissipation are no longer assignable to separate components of force, nor representable by coefficients which are independent of angle of incidence.

5.2.6   Bounce of a Superball

A Superball is a solid rubber ball that is highly elastic. When such a ball is gently launched in a horizontal direction and simultaneously given backspin about a transverse axis, when the falling ball strikes a level floor the horizontal component of velocity at the center of mass can reverse direction simultaneously with a reversal in the direction of rotation. This behavior is illustrated in Fig. 5.17. For a spherical ball these reversals are solely due to tangential compliance of the contact region; in Fig. 5.11 they occur in the region where the tangential velocity is negative. K.L. Johnson (1983) analyzed the bounce of a Superball as an example of an observable effect entirely attributable to tangential compliance and friction; this effect cannot be obtained from rigid body impact theory.

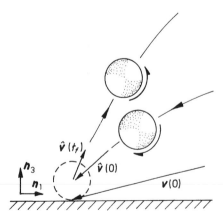

Figure 5.17. Spherical rubber ball with backspin. During collision, friction and tangential compliance combine to reverse directions of both translation and spin.

For each bounce the changing velocity of the ball can be plotted on a graph of the incident horizontal (tangential) velocity of the center of mass, $\hat{v}_1(0)$, and the incident angular speed about the transverse axis, $\omega(0)$; in Fig. 5.18 each of these variables has been nondimensionalized by the product of the coefficient of Coulomb friction μ and the normal component $\hat{v}_3(0)$ of the incident velocity of the center of mass. The moment of momentum h_c about the contact point C on the lower surface of the ball during impact can be expressed as

$$h_C(t) = mR\hat{v}_1(t) + m\hat{k}_r^2\omega(t) = mR\hat{v}_1(t) + \frac{2}{5}mR^2\omega(t).$$

If the contact area remains small in comparison with the radius R, the reaction force at C has no resultant moment; hence the initial and terminal moments of momentum about the contact point are equal:

$$\frac{h_C}{\mu mR\hat{v}_3(0)} = \frac{\hat{v}_1(0)}{\mu\hat{v}_3(0)} + \frac{2}{5}\frac{R\omega(0)}{\mu\hat{v}_3(0)} = \frac{\hat{v}_1(t_f)}{\mu\hat{v}_3(0)} + \frac{2}{5}\frac{R\omega(t_f)}{\mu\hat{v}_3(0)}. \tag{5.33}$$

During impact the velocity changes from an initial state $(\hat{v}_1(0), \omega(0))$ to a final state $(\hat{v}_1(t_f), \omega(t_f))$; these changes occur along lines of constant moment of momentum h_C. Constant moment of momentum h_C is represented by lines of slope $-5/2$ that are labeled across the bottom of Fig. 5.18.

Any initial state $(\hat{v}_1(0), \omega(0))$ gives a specific tangential velocity $v_1(0)$ at contact point C:

$$v_1(0) = \hat{v}_1(0) - R\omega(0)$$

or

$$\frac{v_1(0)}{\mu\hat{v}_3} = \frac{\hat{v}_1(0)}{\mu\hat{v}_3} - \frac{R\omega(0)}{\mu\hat{v}_3} \tag{5.34}$$

Lines on the graph with slope $+1$ have constant values of $v_1(0)/\mu\hat{v}_3(0)$; these have been labeled along the left side of the figure. The initial and terminal tangential velocities at C are directly related by Fig. 5.12 for three different values of the coefficient of restitution. In Fig. 5.18, the terminal tangential velocities at C for a coefficient of restitution $e_* = 1$ are labeled along the right hand side of the figure. Thus for any incident tangential velocity at C the terminal tangential velocity at C is given at the right hand end of this diagonal line; this terminal velocity equals the incident tangential velocity for the succeeding impact.

In Fig. 18 a numbered series of impacts (1,2,3,...) is shown; these begin with initial conditions $[\hat{v}_1(0)/\mu\hat{v}_3(0) = -1, R\omega(0)/\mu\hat{v}_3(0) = 4]$ which result in the tangential velocity at C changing from $v_1(0)/\mu\hat{v}_3(0) = -5$ to $v_1(t_f)/\mu\hat{v}_3(0) = 1.7$. In this case, when the ball undergoes the first impact, the velocity jumps from the fourth to the second quadrant; thus both tangential and angular velocities change sign. In the second impact the direction of rotation changes once again, but the direction of the tangential velocity does not change. This second bounce gives, for the third impact, an incident tangential velocity at C of $v_1(0)/\mu\hat{v}_3(0) = -0.3$. The shaded field in this figure represents the range of incident speeds which give simultaneous reversal of direction for both the tangential and rotational velocities. For most initial conditions there are only a few simultaneous reversals before the ball bounces off in one direction.

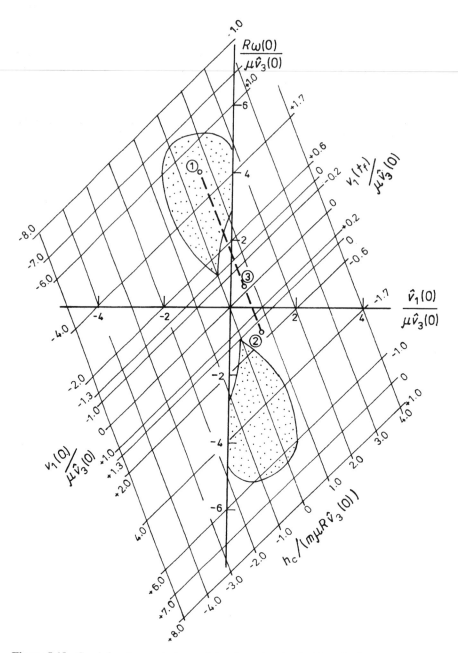

Figure 5.18. Graph for changes in tangential and rotational velocities of sphere during impact where $e_* = 1$. The incident tangential velocity of the center of mass is on the horizontal axis while that of the contact point C is along the left border. The dashed line represents a series of three collisions which occur on the line $h_C = 0.6$. Incident velocities in the shaded teardrop-shaped regions give reversal for both translational and rotational velocities.

PROBLEMS

5.1 For the Maxwell solid, find the relative velocity $\dot{x}_c \equiv \dot{x}(t_c)$ at the instant when the normal impulse is equal to the initial difference in translational momenta, mv_0. (This proves that the transition from compression to restitution occurs simultaneously with maximum relative indentation.)

5.2 For the Maxwell solid, find the following expression for the nondimensional time ωt_c when the normal component of relative velocity vanishes [$\dot{x}(t_c) = 0$]:

$$\omega t_c = \frac{1}{\sqrt{1 - \zeta^2}} \tan^{-1}\left(-\frac{\sqrt{1 - \zeta^2}}{\zeta}\right)$$

Hence show that for $c = m\omega$ ($\zeta = 0.5$), the relative velocity vanishes at a time $\omega t_c = 2.42$.

CHAPTER 6

Continuum Modeling of Local Deformation Near the Contact Area

Only those bodies which are absolutely hard are exactly reflected according to these rules. Now the bodies here amongst us (being an aggregate of smaller bodies) have a relenting softnesse and springynesse, which makes their contact be for some time and in more points than one. And the touching surfaces during the time of contact doe slide one upon another more or lesse or not at all according to their roughnesse. And few or none of these bodyes have a springynesse soe strong as to force them one from another with the same vigor that they came together.

Isaac Newton, Laws of Motion Paper, MS. Add 3958,
Cambridge University.

In practice the bodies that are colliding are composed of elastic, elastic–plastic or viscoplastic materials, so that the large contact forces acting during a collision induce both local deformations near the contact point and global deformations (vibrations) of the entire body. This chapter focuses on the local deformations in a contact region that can be represented as an elastic–perfectly plastic solid; the additional effect of global deformations will be introduced in Chapter 7.

For collisions between hard bodies, the analysis of changes in velocity during collision is simplified by assuming that the initial point of contact is surrounded by an infinitesimally small deforming region. For other purposes, however, it is necessary to consider deformations in the small region surrounding a finite area of contact. One such purpose is to relate the coefficient of restitution e_* to energy dissipated by plastic deformation in the contact region. Here we analyze details of deformation in the contact region and relate these to interface *pressure* between the bodies. The aim is to express hysteresis of contact forces as a function of impact parameters and properties of the colliding bodies, i.e. to obtain a theory for estimating nonfrictional energy losses in collisions between elastoplastic solids. In the first instance this theory is based on the assumption that nonfrictional energy loss is entirely due to plastic deformation. Such a theory is useful for identifying the range of impact speeds where a particular form of material behavior is representative of the physics of deformation. Subsequently, effects of friction and additional energy losses associated with elastic waves are considered.

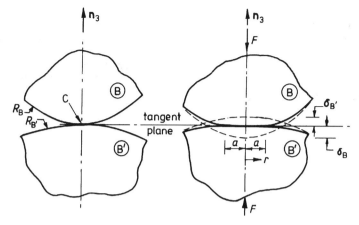

Figure 6.1. Compression and indentation of spherical contact surfaces.

6.1   Quasistatic Compression of Elastic–Perfectly Plastic Solids

6.1.1   Elastic Stresses – Hertzian Contact

A continuum analysis of contact forces and the deformations that arise from quasistatic compression of elastic, elastic–plastic or perfectly plastic bodies can be used to develop a theory of impact for hard bodies composed of rate-independent materials. In this theory deformations are negligible outside a small contact region, and the deforming region acts as a nonlinear inelastic spring between two rigid bodies; the mass of the deforming region is assumed to be negligible. Hertz[1] (1882) first developed this quasistatic theory for elastic deformation localized near the contact patch and applied it to the collision of solid bodies with spherical contact surfaces. Hertz's theory provides a very good approximation for collisions between hard compact bodies where the contact region remains small in comparison with the size of either body.

Let nonconforming elastic bodies B and B' come into contact at a point C; in a neighborhood of C the surfaces of the bodies have radii of curvature R_B and $R_{B'}$, as described in Fig. 6.1. If these bodies are compressed by force $F \equiv F_3$ in the normal direction, Hertz showed that the contact region spreads to radius a and within the contact area there is an elliptical distribution of contact pressure

$$p(r) = p_0\left(1 - r^2/a^2\right)^{1/2}, \qquad r \leq a \tag{6.1}$$

where r is a radial coordinate originating at the center and $p_0 \equiv p(0)$ is the pressure at the center of the contact area. This contact pressure generates local elastic deformations and surface displacements that cause initially nonconforming surfaces to touch or conform within a contact area. This pressure distribution results in a compressive reaction force F on each body,

$$F = \int_0^a p(r)\, 2\pi r\, dr = \frac{2\pi}{3} p_0 a^2. \tag{6.2}$$

---

[1] Hertz developed this theory during Christmas vacation 1880; he was 23 at the time and studying with Kirchoff. Although this theory was initially dismissed by Kirchoff, subsequently it has proven to be extremely useful.

The mean pressure $\bar{p}$ is two-thirds of the pressure at the center of the contact circle, $\bar{p} = 2p_0/3$. For the pressure distribution given in Eq. (6.1), Hertz obtained the normal displacement $w_i(r)$ at the surface of body i ($i = $ B, B$'$) from the Boussinesq solution for a force applied normal to the surface of an elastic half space (Timoshenko and Goodier, 1970):

$$w_i(r) = 0.25\left(1 - v_i^2\right)\pi a p_0 E_i^{-1}(2 - r^2/a^2), \qquad r \leq a \tag{6.3}$$

where compressive displacements are positive. In this expression the elastic moduli of body i are given as Young's modulus E_i and Poisson's ratio v_i.

The compression of each body δ_i is equivalent to the relative displacement between the initial contact point C and the center of mass, $\delta_i = w_i(0)$. Thus for axisymmetric bodies with convex contact surfaces of curvature R_i^{-1}, if the contact area is small in comparison with the cross-section, the radial distribution of the normal displacement can be expressed as

$$w_i(r) \approx \delta_i - r^2/2R_i.$$

The total indentation from compression $\delta = \delta_B + \delta_{B'}$ can be related to the pressure magnitude p_0 at the center of the contact area by summing the individual effects expressed by Eq. (6.3):

$$\delta = \pi a p_0/2E_* \tag{6.4}$$

where an effective radius R_* and modulus E_* have been defined as

$$R_* = \left(R_B^{-1} + R_{B'}^{-1}\right)^{-1}$$

$$E_* = \left[\left(1 - v_B^2\right)E_B^{-1} + \left(1 - v_{B'}^2\right)E_{B'}^{-1}\right]^{-1}.$$

The size of the contact area can be determined from Eqs. (6.3) and (6.4); the contact radius a is then related to contact force F using Eq. (6.2):

$$\frac{\delta}{R_*} = \frac{a^2}{R_*^2} = \frac{3F}{4aE_*R_*}.$$

Rearranging, we obtain

$$\frac{a}{R_*} = \left(\frac{3F}{4E_*R_*^2}\right)^{1/3} \tag{6.5}$$

$$\frac{\delta}{R_*} = \frac{a^2}{R_*^2} = \left(\frac{3F}{4E_*R_*^2}\right)^{2/3} \tag{6.6}$$

$$\frac{p_0}{E_*} = \frac{3F}{2\pi a^2 E_*} = \left(\frac{6F}{\pi^3 E_*R_*^2}\right)^{1/3} \tag{6.7}$$

The mean pressure in the contact region $\bar{p}$ and a compliance relation for interaction force $F(\delta)$ are obtained from Eqs. (6.6) and (6.7):

$$\frac{\bar{p}}{E_*} = \frac{4}{3\pi}\sqrt{\frac{\delta}{R_*}}, \qquad \frac{F}{E_*R_*^2} = \frac{4}{3}\left(\frac{\delta}{R_*}\right)^{3/2} \tag{6.8}$$

where R_* is the effective radius of curvature in the contact area before compression. This force can be integrated to obtain the work W done by the normal contact force in compressing the small deforming region to any indentation δ,

$$\frac{W}{E_* R_*^3} = \int_0^{\delta/R_*} \frac{F(\delta'/R_*)}{E_* R_*^2} d(\delta'/R_*) = \frac{8}{15}\left(\frac{\delta}{R_*}\right)^{5/2} \tag{6.9}$$

6.1.2 Indentation at Yield of Elastic–Plastic Bodies

Elastic indentation continues until some point in the contact region has a state of stress satisfying the yield criterion of a constituent material. If plasticity (i.e. irreversible deformation) initiates at a uniaxial yield stress Y, the elliptical (Hertzian) contact pressure distribution for a spherical contact surface gives solely elastic deformation if the mean pressure $\bar{p} < 1.1Y$ (Johnson, 1985). The transition pressure $\bar{p}_Y \equiv 1.1Y \equiv \vartheta_Y Y$ results in yield at a point beneath the contact surface for either von Mises or Tresca yield criteria.[2] This transition pressure occurs at a limiting indentation for elastic deformation δ_Y that can be obtained from

$$\frac{\bar{p}_Y}{Y} \equiv \vartheta_Y = \frac{4}{3\pi} \frac{E_*}{Y} \sqrt{\frac{\delta_Y}{R_*}}.$$

Thus the nondimensional *indentation* δ_Y/R, *normal force* F_Y/YR^2 and *work* W_Y/YR^3 required to initiate yield are *material properties*,

$$\frac{\delta_Y}{R_*} = \left(\frac{3\pi}{4}\right)^2 \left(\frac{\vartheta_Y Y}{E_*}\right)^2, \tag{6.10a}$$

$$\frac{F_Y}{\vartheta_Y YR_*^2} = \pi\left(\frac{3\pi}{4}\right)^2 \left(\frac{\vartheta_Y Y}{E_*}\right)^2, \tag{6.10b}$$

$$\frac{W_Y}{\vartheta_Y YR_*^3} = \frac{2\pi}{5}\left(\frac{3\pi}{4}\right)^4 \left(\frac{\vartheta_Y Y}{E_*}\right)^4. \tag{6.10c}$$

With this definition of the indentation at yield δ_Y, the contact radius, normal force and work done by normal force during elastic deformation $\delta < \delta_Y$ can be expressed as

$$\frac{a}{a_Y} = \left(\frac{\delta}{\delta_Y}\right)^{1/2}, \qquad \frac{F}{F_Y} = \left(\frac{\delta}{\delta_Y}\right)^{3/2}, \qquad \frac{W}{W_Y} = \left(\frac{\delta}{\delta_Y}\right)^{5/2}$$

Contours of maximum shear stress (Tresca yield criterion) are illustrated in Fig. 6.2 for an elastic solid compressed by a spherical indenter. The location of the maximum shear stress where yield initiates is substantially beneath the surface of the body.

---

[2] The Hertz pressure distribution causes yield to initiate at a point below the contact surface at a nondimensional depth $x_3/a = 0.45$. The plastically deforming region expands from this point in a lenticular shape as the mean pressure increases above the yield pressure $\bar{p}_Y$. Nevertheless, the plastically deforming region remains contained below the surface until the mean pressure is as large as $\bar{p} = 2.8\bar{p}_Y$; consequently, after loading into this elastoplastic range and subsequent unloading, there is very small final deflection at the surface.

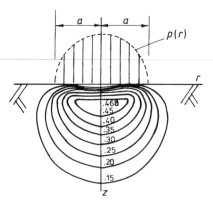

Figure 6.2. Contours of maximum shear stress beneath spherical indenter.

6.1.3 Quasistatic Elastic–Plastic Indentation

For colliding bodies with spherical contact surfaces which are composed of material that can be represented as an elastic–perfectly plastic solid with uniaxial yield stress Y, plastic deformation initiates beneath the contact surface when the mean contact pressure equals $\bar{p}_Y = 1.1Y$; at this pressure plastic flow begins at a nondimensional depth $x_3/a = 0.45$ beneath the contact surface. This depth is less than the contact radius a. Although the plastically deforming region enlarges as contact pressure increases, it remains confined below the surface for pressures throughout most of the range $1.1 < \bar{p}/Y < 2.8$; this state is termed *contained plastic deformation*. In this elastoplastic range the observable permanent indentation of the surface is small because plastic deformation is incompressible and the plastically deforming region is encased within an otherwise elastic body. For contact pressures in the elastoplastic range, Fig. 6.3 shows the development of the plastic region and the evolution of the distribution of contact pressure. The analytical solutions shown in this figure are contours for the second stress invariant J_2 (equivalent to the von Mises yield criterion) at the perimeter of the plastically deforming region for each load $F > F_Y$.

While the shape of the evolving plastic zone can be calculated using the finite element method, it is useful to have an analytical approximation that can estimate this behavior. Following observations by Mulhearn (1959) that any blunt indenter (pyramid, cone or sphere) produced roughly spherical displacements below the surface, K.L. Johnson (1985) suggested a simplified spherical expansion model for elastoplastic indentation. This model consists of an incompressible hemispherical core of radius a beneath the indenter; within this core the state of stress is hydrostatic pressure $\bar{p}$. Surrounding the core is a plastically deforming thick hemispherical shell wherein the radial stress decreases. The outer surface of the shell is at radius $\bar{c}$, where stresses satisfy the yield condition.

The model gives the radial and tangential components of stress in the plastically deforming zone as

$$\frac{\sigma_r}{Y} = -2\ln\left(\frac{\bar{c}}{r}\right) - \frac{2}{3}, \quad \frac{\sigma_\theta}{Y} = -2\ln\left(\frac{\bar{c}}{r}\right) + \frac{1}{3}, \qquad a \leq r \leq \bar{c} \qquad (6.11a)$$

and an effective stress $\bar{\sigma} = (\sigma_r + 2\sigma_\theta)/3$ which can be expressed as

$$\frac{\bar{\sigma}}{Y} = -2\ln\left(\frac{\bar{c}}{r}\right), \qquad a \leq r \leq \bar{c}. \qquad (6.11b)$$

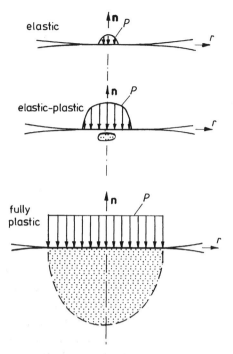

Figure 6.3. Contact pressure distribution and region of plastic deformation for indentation, giving elastic, contained plastic or uncontained fully plastic deformations.

Outside the plastic region $r > \bar{c}$ there is an elastic zone where

$$\frac{\sigma_r}{Y} = -\frac{2}{3}\left(\frac{\bar{c}}{r}\right)^3, \quad \frac{\sigma_\theta}{Y} = \frac{1}{3}\left(\frac{\bar{c}}{r}\right)^3, \quad r > \bar{c}. \tag{6.11c}$$

Beneath the indenter the core pressure $\bar{p}$ is assumed to be uniform, so it is given by the radial component of stress at the hemispherical surface of the core,

$$\frac{\bar{p}}{Y} = -\frac{\sigma_r(a)}{Y} = 2\ln\left(\frac{\bar{c}}{a}\right) + \frac{2}{3}. \tag{6.12}$$

The rigid core and the plastically deforming region both increase in size as the indentation increases. From elastic compressibility of the plastic region $a \le r \le \bar{c}$, Hill (1950, p. 101) obtained a ratio between incremental changes in the radius of contact a and the radius of the elastic–plastic interface $\bar{c}$,

$$\frac{da}{d\bar{c}} = \frac{Y}{E_*}\left\{\frac{3(1-v)\bar{c}^2}{a^2} - \frac{2(1-2v)a}{\bar{c}}\right\}.$$

Within the plastic region a similar ratio relates differential increments of radial displacement $u(r)$ to incremental changes in the elastic–plastic interface,

$$\frac{du}{d\bar{c}} = \frac{Y}{E_*}\left[3(1-v)\frac{\bar{c}^2}{r^2} - 2(1-2v)\frac{r}{\bar{c}}\right].$$

If the material is rigid–plastic, the core and plastically deforming region are incompressible; for a rigid–plastic material the previous equation gives

$$\frac{\bar{c}}{a} \approx \left[\frac{3}{4}\left(\frac{E_* a}{3YR_*}\right)\right]^{1/3} = 0.865\left(\frac{a}{a_Y}\right)^{1/3} \tag{6.13}$$

where a equals the contact radius. Hence the core pressure is obtained as

$$\frac{\hat{p}}{Y} = \frac{2}{3}\left[0.569 + \ln\left(\frac{a}{a_Y}\right)\right]. \tag{6.14}$$

In fact the stresses in the core are also at yield and therefore not hydrostatic. From Eq. (6.11a) a better estimate for the mean contact pressure is given by the tangential component of stress in the elastoplastic region, $\bar{p} \approx -\sigma_r(a) = \hat{p} + 2Y/3$; this is discussed by Johnson (1985, p. 175). Hence for an elastic–plastic boundary located at $r = \bar{c}$, the equilibrium of forces on the core requires a mean pressure $\bar{p}$ in the contact area given by

$$\frac{\bar{p}}{\bar{p}_Y} = 1 + \frac{2}{3\vartheta_Y}\ln\left(\frac{a}{a_Y}\right). \tag{6.15}$$

Yield begins at a mean pressure $\bar{p}/Y \equiv \vartheta_Y$ where an initially spheroidal contact surface has $\vartheta_Y = 1.1$ at a contact radius $a(\delta_Y) \equiv a_Y$. Notice that the change in indentation model at yield results in a slight discontinuity in indentation force F at the transition from elastic to elastic–plastic behavior.

During elastic–plastic indentation the contact force F increases with indentation:

$$F/F_Y = (a/a_Y)^2\left[1 + 0.67\vartheta_Y^{-1}\ln(a/a_Y)\right]. \tag{6.16}$$

The normal relative displacement of the colliding bodies (indentation) for this phase is obtained by assuming that there is negligible elastic deformation in the material surrounding the contact patch; i.e., at the edge of the contact area the surface neither sinks in nor piles up. Then the total indentation is given by

$$\delta/\delta_Y = 0.5\left(a^2/a_Y^2 + 1\right). \tag{6.17}$$

(Recall that in the elastic range $\delta/\delta_Y = a^2/a_Y^2$.) Although approximation (6.17) results in the contact area being a discontinuous function of indentation at yield, both indentation δ and contact force F are continuous functions of the nondimensional contact radius a/a_Y. Thus for the range of indentation where deformations are elastic–plastic, Eq. (6.17) gives a ratio of the contact radius a to the contact radius at yield a_Y as a function of the indentation δ,

$$a/a_Y = (2\delta/\delta_Y - 1)^{1/2}$$

Hence (6.15) can be expressed also as

$$\bar{p}/\bar{p}_Y = \left[1 + (3\vartheta_Y)^{-1}\ln(2\delta/\delta_Y - 1)\right]. \tag{6.18}$$

This gives the following expression for the normal contact force F:

$$F/F_Y = (2\delta/\delta_Y - 1)\left\{1 + (3\vartheta_Y)^{-1}\ln(2\delta/\delta_Y - 1)\right\}.$$

The total work done by the normal contact force during indentation into the elastoplastic range is obtained by integrating the product of this force and the differential increment of normal relative displacement:

$$\frac{W}{W_Y} = 1 + \int_1^{\delta/\delta_Y}\left(\frac{2\delta'}{\delta_Y} - 1\right)\left[1 + \frac{1}{3\vartheta_Y}\ln\left(\frac{2\delta'}{\delta_Y} - 1\right)\right]d\left(\frac{\delta'}{\delta_Y}\right).$$

After integration, substitution from (6.16) and recognizing that $W_Y = \frac{2}{5}F_Y\delta_Y$, the work during indentation into the elastic–plastic range is obtained as

$$\frac{W}{W_Y} = 1 + \frac{5}{8}\left\{0.85\left[\left(\frac{2\delta}{\delta_Y} - 1\right)^2 - 1\right] + 0.30\left(\frac{2\delta}{\delta_Y} - 1\right)^2 \ln\left(\frac{2\delta}{\delta_Y} - 1\right)\right\}. \quad (6.19)$$

The elastic–plastic phase continues until the mean contact pressure satisfies $\bar{p}/Y = 2.8$. This contact pressure results in a nondimensional contact radius $a = a_p$ where the contact pressure is fully plastic. Indentation experiments show that thereafter $(a > a_p)$ any additional indentation occurs without further increase in mean pressure $\bar{p}$.

At $a/a_Y \approx 12.9 \equiv a_p/a_Y$ or $\delta_p/\delta_Y \approx 84$ Eq. (6.15) or (6.18) gives a mean pressure equal to that for fully plastic indentation, $\bar{p}/Y = 2.8$. Further indentation $a > a_p$ is in the regime of uncontained plastic deformation, where there is a different load–deflection relation.

6.1.4   Fully Plastic Indentation

In the previously discussed range of contained plastic deformation the mean contact pressure $\bar{p}$ increases with increasing indentation. This behavior has an upper limit where the plastic deformation is no longer contained beneath the contact surface; for spheroidal contact surfaces this occurs at a contact pressure of about $\bar{p} = 2.8Y$. Throughout the range of fully plastic indentation the mean contact pressure $\bar{p}$ is constant; this pressure is the same as that measured in a Brinell hardness test.

Uncontained plastic deformation or *fully plastic indentation* begins at a contact radius $a/a_Y \approx 12.9$ where the contact pressure $\bar{p} = 2.8Y$. The force F_p at the transition from elastoplastic to fully plastic indentation is given by

$$F_p/F_Y \approx 424 \quad (6.20)$$

so the elastoplastic range of force spans more than two orders of magnitude, $1 < F/F_Y < 424$.[3]

The transition from contained to uncontained plastic deformation that occurs at $\delta_p/\delta_Y \approx 84$ requires a large amount of plastic work in comparison with the work to initiate yield:

$$W_p/W_Y = 41.5 \times 10^3. \quad (6.21)$$

In the fully plastic range the contact pressure is uniform and remains constant $\bar{p} = 2.8$ while the contact force increases as the contact area continues to increase:

$$\frac{F}{F_Y} = \frac{2.8}{\vartheta_Y}\left(\frac{2\delta}{\delta_Y} - 1\right), \quad F > F_p. \quad (6.22)$$

In the fully plastic range the indentation $\delta/\delta_Y = 0.5(a^2/a_Y^2 + 1)$ is given by the same condition as that used in the range of elastoplastic indentation, Eq. (6.17). This force results in the work done during compression given by

$$\frac{W}{W_Y} = \frac{W_p}{W_Y} + \frac{7.0}{\vartheta_Y}\left[\left(\frac{\delta^2}{\delta_Y^2} - \frac{\delta}{\delta_Y}\right) - \left(\frac{\delta_p^2}{\delta_Y^2} - \frac{\delta_p}{\delta_Y}\right)\right], \quad \delta > \delta_p. \quad (6.23)$$

---

[3] Finite element analysis of elastic–perfectly plastic bodies with initially spherical contact surfaces gives fully plastic indentation initiating at $F_p/F_Y \approx 650$ and $\delta_p/\delta_Y \approx 140$.

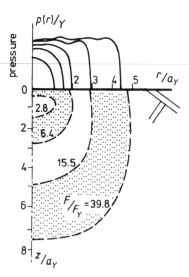

Figure 6.4. Finite element calculation of contact pressures and region of plastic deformation for a sphere indenting an elastic–perfectly plastic solid (Hardy, Baronet and Tordion, 1971).

While this model gives reasonable estimates of final indentation and energy dissipated by indentation, the transition from elastic–plastic to fully plastic indentation occurs at a much larger force and indentation than are indicated by some numerical simulations (Hardy, Baronet and Tordion, 1971; Follansbee and Sinclair, 1984). These numerical analyses show that $F_p/F_Y \approx 20$ rather than 420.

Figure 6.3 illustrates the distribution of pressure on the surface and the extent of the plastically deforming region for the elastic, elastic–plastic and fully plastic ranges of indentation by a spherical indenter. The ranges of applicability for contained elastic–plastic and uncontained fully plastic deformations have been obtained from the analytical approximations. These conceptual images for the stress distribution can be compared with results from a finite element analysis of indentation in an elastic–perfectly plastic solid by a spherical indenter that are shown in Fig. 6.4 (Hardy, Baronet and Tordion, 1971).

6.1.5   Elastic Unloading from Maximum Indentation

The work done on the deforming region by contact force during compression goes into deformation; part of this work is absorbed by elastic strain energy, and part is dissipated by plastic deformation. Immediately following the period of compression, the elastic strain energy sustains the normal contact force that drives the bodies apart during the period of restitution.

During unloading from maximum compression the compliance relation for the contact region is elastic. Complete unloading from a maximum compressive force F_c or indentation δ_c that is in the plastically deforming range results in a change in indentation δ_r, so that when the compressed bodies separate there is a final indentation $\delta_f = \delta_c - \delta_r$.

To obtain an expression for the change in indentation we recognize that as a consequence of plastic deformation during loading, the contact area has an effective curvature that has changed from the initial value R_*^{-1} to a new unloaded curvature $\bar{R}_*^{-1}$. The transition is assumed to occur at maximum indentation. This curvature of the deformed surface depends on whether the deformed bodies are both convex or whether one has become concave:

$$\bar{R}_* = \begin{cases} \bar{R}\,\bar{R}'/(\bar{R} + \bar{R}'), & \bar{R} \geq 0, \quad \bar{R}' \geq 0 \\ \bar{R}\,\bar{R}'/(\bar{R} - \bar{R}'), & -\bar{R} > \bar{R}' \geq 0. \end{cases} \tag{6.24}$$

During elastic unloading the changes in the contact region are geometrically similar to the changes that occur during loading; thus

$$\delta_Y/R_* = \delta_r/\bar{R}_*. \tag{6.25}$$

Unloading results in a change in indentation δ_r from the maximum indentation δ_c. For contact forces in the plastically deforming range $F > F_Y$ these indentations are related to the respective contact radii by

$$\delta_r/\bar{R}_* = a_r^2/\bar{R}_*^2, \qquad 2\delta_c/R_* = a_c^2/R_*^2 + \delta_Y/R_*.$$

The assumption used to derive the ratio between indentation δ_r recovered during unloading and maximum indentation δ_c is that during unloading the change in contact radius a_r equals the contact radius at maximum indentation a_c; hence

$$\bar{R}_*/R_* = \delta_r/\delta_Y = (2\delta_c/\delta_Y - 1)^{1/2}. \tag{6.26}$$

This expression applies to unloading from either elastoplastic or fully plastic indentation.

During unloading (i.e. the period of restitution), strain energy of elastic deformation provides the power that is transformed into relative kinetic energy of the colliding bodies. This energy transformation is achieved by means of work done on the bodies by the contact force. The work done on the colliding bodies by the contact force during unloading can be obtained by integrating the unloading normal force $F = \frac{4}{3}E_*\bar{R}_*^{1/2}(\delta - \delta_f)^{3/2}$ over the range $\delta_f \leq \delta \leq \delta_c$. This work is negative, since the deforming region is expanding in the normal direction during restitution and this expansion is opposed by the normal component of contact force. Since $\delta_r = \delta_c - \delta_f$, this integration results in

$$\frac{W_r}{YR_*^3} = -\int_0^{\delta_r/\bar{R}_*} \left(\frac{4}{3}\frac{E_*}{Y}\right)\left(\frac{\bar{R}_*}{R_*}\right)^3 \left(\frac{\delta}{\bar{R}_*}\right)^{3/2} d\left(\frac{\delta}{\bar{R}_*}\right) = -\frac{8}{15}\frac{E_*}{Y}\left(\frac{\delta_r}{\delta_Y}\right)^3 \left(\frac{\delta_Y}{R_*}\right)^{5/2}$$

By recalling the geometric relation for unloading (6.25) and expressions for indentation and loading work at yield, we obtain

$$\frac{\delta_Y}{R_*} = \frac{\delta_r}{\bar{R}_*} = \left(\frac{3\pi}{4}\right)^2 \left(\frac{\vartheta_Y Y}{E_*}\right)^2 \quad \text{and} \quad \frac{W_Y}{YR_*^3} = \frac{8}{15}\frac{E_*}{Y}\left(\frac{\delta_Y}{R_*}\right)^{5/2}$$

These expressions can be substituted into the relation for work done by the contact force during unloading W_r to obtain the recovered energy as[4]

$$\frac{W_r}{W_Y} = -\left(\frac{\delta_r}{\delta_Y}\right)^3 = -\left(\frac{2\delta_c}{\delta_Y} - 1\right)^{3/2} \tag{6.27}$$

---

[4] An alternative expression that at maximum indentation satisfies continuity of force (but not contact area) is the following:

$$\frac{\delta_r}{\delta_Y} = \left(\tfrac{3}{2}\right)^{1/2}\left(\tfrac{2\delta_c}{\delta_Y} - 1\right)^{1/2}, \qquad \text{which gives} \qquad \frac{W_r}{W_Y} = -\left(\tfrac{3}{2}\right)^{1/2}\left(\tfrac{2\delta_c}{\delta_Y} - 1\right)^{3/2}$$

The outcome of the difference from (6.22) is only a small decrease in the normal impact speed to initiate yield v_Y, so the former relation is retained in order to provide continuity at the elastic limit.

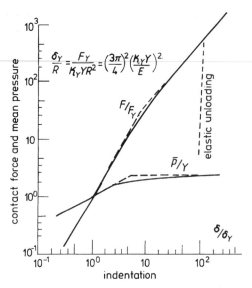

Figure 6.5. Indentation force as a function of indentation for contact of elastoplastic spherical bodies. (Dashed line is for an elastic–perfectly plastic approximation that neglects the intermediate range of contained plastic deformation.)

The relationship between the indentation and the normal force can be represented as being within one of three successive ranges on indentation – elastic, elastic–plastic or fully plastic, where the latter two represent contained and uncontained plastic deformation respectively. Figure 6.5 illustrates the normal contact force as a function of indentation. In this figure elastic unloading is illustrated also. Notice that because the contact surface curvature for unloading, $\bar{R}_*$, does not equal the initial contact radius curvature R_*, the unloading line is not parallel to the elastic loading line.

6.2   Resolved Dynamics of Planar Impact

6.2.1   Direct Impact of Elastic Bodies

At each instant during collision the rate of change of the normal component of relative velocity depends on the interaction force F and hence on the current relative displacement δ. Accelerations during a collision depend on the relative displacement or interference between the colliding objects. Here only the normal component of translational relative velocity is considered. Assuming that the deforming region surrounding the initial contact point C is sufficiently small so that it has negligible mass in comparison with the remainder of the body, the mass of body i moves uniformly in the normal direction $\mathbf{n}_3$. As the contact region is compressed during a collision, the approach of one center of mass relative to the other, $\delta = \delta_B + \delta_{B'}$, results in a reaction force F at C that opposes the approach of the two centers of mass,

$$F = -m_B\ddot{\delta}_B = -m_{B'}\ddot{\delta}_{B'} = \kappa_s\delta^{3/2} \tag{6.28}$$

where κ_s is a stiffness parameter for nonconforming spherical contact,

$$\kappa_s = \tfrac{4}{3}E_*R_*^{1/2}$$

that depends on material properties in the deforming (contact) region. Here another assumption has been tacitly introduced – namely, that the compliance relation is the same as that obtained from quasistatic compression. This assumption is valid if the small deforming region is composed of material with a rate-independent constitutive relation.[5]

For a central or collinear collision, an effective mass m is defined in terms of the individual masses m_i by

$$m^{-1} \equiv m_{\rm B}^{-1} + m_{\rm B'}^{-1} \quad \text{or} \quad m \equiv \frac{m_{\rm B} m_{\rm B'}}{m_{\rm B} + m_{\rm B'}}$$

This gives an equation of relative motion

$$m\ddot{\delta} = m\dot{\delta}\,d\dot{\delta}/d\delta = -\kappa_s \delta^{3/2}.$$

Integration of this equation and subsequent application of the initial conditions $\dot{\delta}(0) = -v(0) \equiv -v_0$ and $\delta(0) = 0$ result in a relative velocity given by

$$\dot{\delta}^2 = v_0^2 - \frac{4\kappa_s}{5m}\delta^{5/2}. \tag{6.29}$$

The contact period is separated into a period of compression and a period of restitution. The compression phase of collision terminates at time t_c when $\dot{\delta}(t_c) = 0$. At this time the compressive relative displacement between the centers of mass has its largest value, $\delta_c = \delta(t_c)$; likewise, for these rate-independent materials, the interaction force $F_c = F(t_c)$ is a maximum at the time when the normal relative velocity vanishes:

$$\delta_c/R = R^{-1}\left(5mv_0^2/4\kappa_s\right)^{2/5} = \left(15mv_0^2/16E_*R_*^3\right)^{2/5} \tag{6.30a}$$

$$\frac{F_c}{E_*R_*^2} = \frac{4}{3}\left(\frac{\delta_c}{R_*}\right)^{3/2} = 1.94\left(\frac{T_0}{E_*R_*^3}\right)^{3/5} \tag{6.30b}$$

where at incidence the kinetic energy of normal relative motion is $T_0 = mv_0^2/2$. By numerical integration of the relative velocity (6.29) Deresiewicz (1968) obtained the period of elastic compression as

$$t_c = \int_0^{\delta_c} \left(v_0^2 - \frac{4\kappa_s}{5m}\delta^{5/2}\right)^{-1/2} d\delta = 1.47\frac{\delta_c}{v_0} = 1.43\left(\frac{m^2}{E_*^2 R_* v_0}\right)^{1/5} \tag{6.31}$$

Alternatively this integral can be expressed in terms of gamma functions.

The variation of the relative displacement $\delta(t)$ with respect to time is obtained from the nonlinear relation (6.28); this is not very different from that given by the compression of a linear spring between two rigid bodies that collide with an initial difference in momentum $-mv_0$ (see Fig. 5.1). The linear approximation involves a spring force $F = F_c \sin(0.5\pi t/t_c)$. Using the compression period obtained in Eq. (6.31), this linear approximation gives a largest reaction force $F_c \approx 1.48\kappa_s^{2/5} T_0^{3/5}$; i.e. roughly 14% less than the force F_c calculated with Eq. (6.30b). Since the Hertz theory and the approximation obtained with linear compliance have the same normal impulse during compression, this comparison of maximum force F_c implies that the linear approximation gives a contact period that is slightly longer than the contact period for collision between elastic bodies.

---

[5] Wagstaff (1924) and Andrews (1931) conducted experiments that demonstrated the validity of the Hertz theory for contact force during impact. Additional validation is shown in Table 6.2 below.

An expression for the impact speed which is just sufficient to initiate yield v_Y can be obtained from (6.10a) with $\delta_c = \delta_Y$ and (6.30a). This gives

$$v_Y^2 = \frac{4\pi}{5} \left(\frac{3\pi}{4} \right)^4 \left(\frac{\vartheta_Y Y}{E_*} \right)^4 \frac{\vartheta_Y Y R_*^3}{m}; \tag{6.32}$$

i.e., for direct collisions the normal impact speed v_Y where yielding initiates is a material property that is directly related to the indentation δ_Y that initiates yield.

Example 6.1 Suppose a steel sphere with Young's modulus $E_B = 210 \times 10^9$ N m^{-2} and Poisson's ratio $\nu_B = 0.3$ is dropped onto a flat steel anvil from a height $h = 50$ mm. If the sphere has radius $R_B = 10$ mm, it will have mass $m_B = 32.4$ g, an impact speed $v_0 = 1.0$ m s^{-1} and kinetic energy at impact $T_0 = m_B v_0^2/2 = 0.016$ J. Assuming elastic deformations, what are the largest contact force F_c, the maximum size of the contact region a_c and the elastic contact period $2t_c$?

Solution:

$$m = \left(m_B^{-1} + m_{B'}^{-1} \right)^{-1} = 0.0324 \text{ kg} \qquad E_* = 0.5 E_B \left(1 - \nu_B^2 \right)^{-1}$$

$$= 115 \times 10^9 \text{ N m}^{-2}.$$

$$R_* = \left(R_B^{-1} + R_{B'}^{-1} \right)^{-1} = 0.01 \text{ m} \qquad \kappa_s = \tfrac{4}{3} E_* R_*^{1/2} = 1.51 \times 10^{10} \text{ N m}^{-3/2}$$

Hence the largest force, contact radius, indentation and pressure are given by

$$F_c = 1.73 \kappa_s^{2/5} T_0^{3/5} = 1.71 \text{ kN} \qquad a_c/R_* = (F_c/\kappa_s)^{1/3} R_*^{-1/2} = 0.048$$

$$p_0(t_c) = 1.5 F_c/\pi a_c^2 = 3540 \text{ N mm}^{-2} \qquad \delta_c/R_* = (a_c/R_*)^2 = 2.34 \times 10^{-3}$$

The duration of the contact period is less than 100 μs:

$$2t_c = 2\left(1.47 \delta_c/v_0 \right) = 69 \ \mu\text{s}.$$

For mild steel the quasistatic yield stress is $Y = 1000$ N mm^{-2}; i.e., this very modest drop height develops a maximum pressure p_0 that is substantially in excess of the uniaxial yield stress.

In the above example it is interesting to note that Eq. (6.31) gives a contact period that increases in proportion to the radius of the sphere. Hertz estimated that for a low impact speed $v_0 = 1$ m s^{-1}, an elastic sphere with radius equal to that of the earth would have a contact period of slightly more than one day.

It is worth noting that if a body is dropped onto an anvil from height h , the change in potential energy during the drop equals the work done during compression; for a homogeneous solid sphere this work W_c can be expressed as $W_c = (4\pi/3)\rho g h R_*^3$. Hence the drop height for initial yield h_Y is given by $\rho g h_Y/Y = (9/5\pi^2)(3Y/E_*)^4$. The drop height required for initial yield of metals is astonishingly small; for example a steel sphere dropped onto a steel anvil has $h_Y \approx 3$ mm. Measurements of indentation reported by Tabor (1951) have verified these predictions.[6]

---

[6] Local impact damage to contact surfaces has been considered by Evans, Gulden and Rosenblatt (1978) and Engel (1976).

6.2.2 Eccentric Planar Impact of Rough Elastic–Plastic Bodies

Continuous changes in relative velocity across the small deforming region that surrounds the contact area can be obtained by supposing that the deforming region is an *infinitesimal deformable particle* located between the colliding rigid bodies at contact point C. This particle is assumed to have negligible tangential compliance. In contrast to the method developed in the previous section, this gives changes in velocity that are independent of any compliance relation for the contact region. Instead, the coefficient of restitution is used to relate work done by the normal component of impulse during the separate periods of restitution and compression. This relationship requires separation of the normal impulse into that acting during compression and that acting during restitution – a separation that can be made *a priori* only for rate-independent compliance relations.

Let $\mathbf{V}_C$ and $\mathbf{V}_{C'}$ be the velocities of the two bodies at C, and let the relative velocity $\mathbf{v} \equiv (v_1, v_3)$ across the deformable particle be defined as $\mathbf{v} \equiv \mathbf{V}_C - \mathbf{V}_{C'}$. This relative velocity is resolved into a component v_1 in the common tangent plane and a component v_3 normal to this plane; the relative velocity v_1 is termed *slip*. The coordinate system is oriented so that at incidence $v_1 \geq 0$ and $v_3 < 0$. An equation of motion for this system can be expressed in terms of components of impulse $\mathbf{p} \equiv (p_1, p_3)$ of the reaction force $\mathbf{F} \equiv (F_1, F_3)$ at the contact point C:

$$\begin{Bmatrix} dv_1 \\ dv_3 \end{Bmatrix} = m^{-1} \begin{bmatrix} \beta_1 & -\beta_2 \\ -\beta_2 & \beta_3 \end{bmatrix} \begin{Bmatrix} dp_1 \\ dp_3 \end{Bmatrix} \tag{6.33}$$

where the inertia parameters $\beta_1, \beta_2, \beta_3$ are a rearrangement of parameters defined by Wang and Mason (1992). The inertia parameters depend on the masses M, M' of the bodies, their radii of gyration $\hat{k}_r, \hat{k}_r'$ about their centers of mass, and the locations $\mathbf{r}, \mathbf{r}'$ of the centers of mass relative to the contact point C, where $\mathbf{r} \equiv (r_1, r_3)$:

$$\beta_1 = 1 + mr_3^2/M\hat{k}_r^2 + mr_3'^2/M'\hat{k}_r'^2$$

$$\beta_2 = mr_1 r_3/M\hat{k}_r^2 + mr_1' r_3'/M'\hat{k}_r'^2$$

$$\beta_3 = 1 + mr_1^2/M\hat{k}_r^2 + mr_1'^2/M'\hat{k}_r'^2.$$

The coordinate system for describing the configuration is shown in Fig. 3.1.

Equations of Relative Motion

With Coulomb's law of friction the equations of motion (6.33) can be expressed in terms of a monotonically increasing independent variable $p \equiv p_3$. The motion depends on whether the contact point is sliding or sticking. It is sliding if $v_1 \neq 0$ or $\mu < |\beta_2|/\beta_1$:

$$dv_1/dp = -(\hat{s}\mu\beta_1 + \beta_2)m^{-1}$$

$$dv_3/dp = (\beta_3 + \hat{s}\mu\beta_2)m^{-1}.$$

It is sticking if $v_1 = 0$ and $\mu \geq |\beta_2|/\beta_1$:

$$dv_3/dp = (\beta_3 - \beta_2^2/\beta_1)m^{-1}$$

where $\hat{s} = \text{sgn}(v_1)$ is the direction of sliding and μ is the coefficient of friction. (For simplicity the static and kinetic coefficients of friction are assumed to be equal.) These equations can be integrated to give the relative velocity at any impulse p during a period of unidirectional slip in direction s:

$$v_1(p) = v_1(0) - (\hat{s}\mu\beta_1 + \beta_2)p/m \qquad (6.34a)$$

$$v_3(p) = v_3(0) + (\beta_3 + \hat{s}\mu\beta_2)p/m. \qquad (6.34b)$$

Hence the normal compression impulse p_c can be obtained from the condition $v_3(p_c) = 0$:

$$p_c = -(\beta_3 + \hat{s}\mu\beta_2)^{-1}mv_3(0). \qquad (6.35)$$

Work and Indentation at Transition from Compression to Restitution

The contact force does work on the rigid bodies that is the negative of the work W_c done on the deforming region during the same period. For each separate period of unidirectional sliding this work can be calculated by a theorem (Stronge, 1992) that goes back to Kelvin. Here for example we consider the limiting case of unidirectional sliding. Analyses for sliding that halts before separation, however, have been described in Chapter 3. For continuous slip the work done by the normal force during compression is as follows:

$$\frac{W_c}{W_Y} = \frac{1}{\beta_3 + \hat{s}\mu\beta_2}\frac{v_3^2(0)}{v_Y^2} \qquad (6.36)$$

where again we employ the definition (6.32) for the normal incident speed at yield. For collinear collisions, $\beta_2 = 0$ and $\beta_3 = 1$, so that effects of normal and tangential force are decoupled. Thus irrespective of the initial slip velocity, at the end of compression $p_c = -mv_3(0)$ for a collinear collision,

$$\frac{W_c}{W_Y} = \frac{v_3^2(0)}{v_Y^2}. \qquad (6.37)$$

Expression (6.36) or (6.37) can be equated with (6.19) or (6.24) to obtain the *maximum indentation* δ_c/δ_Y for any particular geometric configuration and incident velocity. This maximum indentation is required to calculate the part of the normal energy of relative motion that is recovered during restitution, W_r.

Table 6.1 lists the smallest normal impact speed that initiates yield, v_Y, for several different metals. This was calculated from measurements of the energy loss in collisions of

Table 6.1. *Material Properties from Indentation and Impact Tests*

Material	Density $\rho(g\ cm^{-3})$	Young's Modulus E_i (GPa)	Static Yield Y_s(MPa)	Dynamic Yield Y_d(MPa)	Impact Speed to Initiate Yield v_Y(m s^{-1})	Source
Mild steel:						
As received	7.8	210	600	583–780	.049–.101	Author ($\kappa_Y = 2.8$)
Work-hardened	7.8	210	650	1160	.055	Tabor (1951)
Brass (drawn)	8.5	100	200	250	.007	Tabor (1948)
Aluminum:						
1180-H14	2.7	69	110	130	.004	
2014-T6	2.8	69	410	410	.076	

small spheres at incident speeds of $0.1 < |v_3(0)| < 3$ m s^{-1}. The dynamic yield stress Y_d that corresponds to each v_Y has been calculated according to Eq. (6.32). For rate-dependent materials, Y_d is larger than the yield stress Y_s obtained from quasistatic indentation tests.

6.3 Coefficient of Restitution for Elastic–Plastic Solids

The energetic coefficient of restitution is the ratio of work done on the small deforming region during compression to work done by this region on the surrounding rigid bodies during restitution; i.e., it is a measure of that part of the kinetic energy of normal relative motion (the energy transformed to internal energy of deformation during compression) which is recoverable during restitution. Section 3.4.3 gave the following definition: The square of the coefficient of restitution, e_*^2, is minus the ratio of the elastic strain energy released at the contact point during restitution to the energy absorbed by internal deformation during compression. This energy ratio can be calculated from the work done by the normal component of contact force if tangential compliance is negligible (Stronge, 1995):

$$e_*^2 = -\frac{W_r}{W_c} = \frac{W_Y}{W_c}\left(\frac{8}{5}\frac{W_c}{W_Y} - \frac{3}{5}\right)^{3/4} \tag{6.38}$$

This final expression combines Eq. (6.27) with the energy transformed in compression, W_c/W_Y, obtained from Eq. (6.19) or (6.24). Together with (6.36), it relates e_* to the *damage number* $\rho v_0^2/Y$ defined by W. Johnson (1972). The result is similar to that obtained by Adams and Tran (1993), but here expression (6.38) explicitly incorporates the effect of friction in various possible slip processes. For elastic–perfectly plastic solids with a convex spherical contact surface, the relation between the coefficient of restitution and the normal impact speed is shown in Fig. 6.6.

In the limit of $W_c/W_Y \gg 1$ Eq. (6.38) indicates that $e_* \approx [v_2(0)/v_Y]^{1/4}$, where v_Y depends on effective mass, contact curvature and material properties but is *independent*

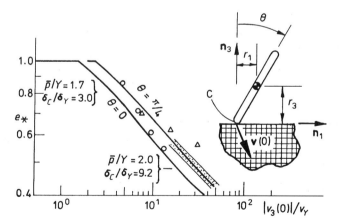

Figure 6.6. The coefficient of restitution for impact of elastic–perfectly plastic solids depends on the eccentricity of the impact configuration as well as the normal impact speed. For a rigid rod inclined at $\theta = \pi/4$ or 0 the lines compare the analytical expressions with two sets of experimental data ($\theta = 0$). For the eccentric configuration the narrow band at large speeds indicates the range of values for opposing directions of gross slip if the coefficient of friction $\mu = 0.5$.

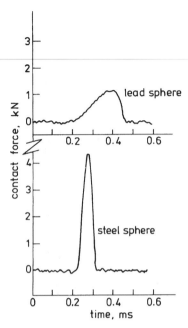

Figure 6.7. Force from coaxial impact of steel and lead spheres on end of steel rod at 2.3 m s^{-1}. Steel and lead spheres have masses of 64.2 and 73.3 g respectively.

of impact configuration. This functional relation between coefficient of restitution and normal impact speed agrees with measurements on a wide range of metals reported by Goldsmith (1960). The present expression neglects strain hardening and effects of high strain rates; for some materials these effects are important if the impact speed is moderately large (Mok and Duffy, 1965; Davies, 1949).

Figure 6.7 shows the stress pulses resulting from coaxial impact of a sphere against the end of a strain-gauged steel rod. The contact force as a function of time is shown for both steel and lead spheres striking the bar at 2.3 m s^{-1}. At this impact speed both spheres suffer plastic deformation, but since the lead sphere has a smaller yield stress, this reduces the maximum force and prolongs the contact period in comparison with the steel sphere. Table 6.2 provides a comparison between the measured values and Hertz elastic analysis for these collisions.

6.4   Partition of Internal Energy in Collision between Dissimilar Bodies

6.4.1   Composite Coefficient of Restitution for Colliding Bodies with Dissimilar Hardness

Where colliding bodies are dissimilar, the loss of kinetic energy due to irreversible internal deformation can be divided into the losses in the two bodies by considering separately the work done on each body by contact forces. This separation associates these energy losses with properties in the contact region – it is independent of the partitioned loss of kinetic energy obtained in Sect. 2.7. If each body has a coefficient of restitution $\tilde{e}_*$ or $\tilde{e}'_*$ obtained for a collision at the same impact speed against a body

Table 6.2. *Measurements from Coaxial Collision of Spheres on a 24.5 mm Dia. Steel Bar*

Sphere	Impact Speed v_0 (m s^{-1})	Rebound Speed v_f (m s^{-1})	Compression Impulse p_c (N s)		Restitution Impulse p_r (N s)		Maximum Force F_c (kN)		Compression Period t_c (μs)		Expt. Contact Period t_f (μs)	C.O.R. $e_* = \dfrac{v_f}{v_0}$	Expt. Impulse Ratio $e_0 = \dfrac{p_r}{p_c}$
			Calc.	Expt.	Calc.	Expt.	Hertz	Expt.	Hertz	Expt.			
Steel:													
$D = 25$ mm	0.91	0.68	0.058	0.051	0.043	0.046	2.40	1.94	45	52	100	0.74	0.9
$M = 64.2$ g	1.15	0.82	0.074	0.073	0.052	0.073	3.18	3.18	43	46	91	0.71	1.0
$E = 210$	1.41	0.91	0.091	0.075	0.058	0.075	4.05	3.70	41	41	81	0.65	1.0
kN mm^{-2}	1.69	1.05	0.108	0.092	0.068	0.082	5.02	4.03	40	46	86	0.62	0.89
	2.0	1.20	0.128	0.126	0.077	0.100	6.16	4.58	38	41	79	0.60	0.79
Lead:													
$D = 23$ mm	0.91	0.11	0.067	0.065	0.008	0.024	1.29	0.80	95	120	168	0.12	0.37
$M = 73.3$ g	1.15	0.13	0.085	0.092	0.010	0.028	1.71	0.85	91	179	232	0.12	0.30
$E = 14.7$	1.41	0.16	0.103	0.102	0.011	0.039	2.18	0.92	87	181	259	0.11	0.38
kN mm^{-2}	1.69	0.18	0.123	0.135	0.013	0.061	2.70	0.90	84	235	301	0.11	0.45
	2.0	0.20	0.146	0.130	0.015	0.062	3.31	1.16	81	193	273	0.10	0.48

composed of the same material and geometrically similar to itself, then these coefficients can be combined to calculate an *effective coefficient of restitution* e_* for collision between dissimilar bodies. In order to achieve this amalgamation, we consider for each body the ratio of the work $\tilde{W}_r$ done by the normal contact force during recovery to the work $\tilde{W}_c$ done during compression, and note that this ratio is equivalent to the coefficient of restitution:

$$\tilde{e}_*^2 = -\tilde{W}_r / \tilde{W}_c, \quad \tilde{e}_*'^2 = -\tilde{W}_r' / \tilde{W}_c', \qquad \text{while} \quad e_*^2 = -W_r / W_c. \tag{6.39}$$

During collision the normal component of contact force does work on each separate body; this work is in proportion to their respective indentations δ_i, $i = $ B, B$'$, because equal but opposed forces act on the colliding bodies. Hence

$$\tilde{e}_*^2 = -\tilde{\delta}_r / \tilde{\delta}_c, \quad \tilde{e}_*'^2 = -\tilde{\delta}_r' / \tilde{\delta}_c' \quad \text{and} \quad e_*^2 = -\delta_r / \delta_c$$

so if, for example, the bodies remain elastic during compression, Eq. (6.4) gives the individual indentations as $\delta_i = \pi \bar{a} p_0 (1 - \nu_i^2) / E_i$. Since the part of the total indentation for each body is approximately equal to the part of the deformation (or strain) energy, the internal deformation energy is distributed between two colliding bodies in inverse proportion to the ratio of their elastic moduli $E_i / (1 - \nu_i^2)$. This energy distribution is *independent of the relative curvature of the contact surfaces or the size of the bodies.* For either frictionless or collinear collisions this implies that the composite coefficient of restitution e_* is obtained from

$$\frac{e_*^2}{E_*} = \frac{\tilde{e}_*^2(1 - \tilde{\nu}^2)}{\tilde{E}} + \frac{\tilde{e}_*'^2(1 - \tilde{\nu}'^2)}{\tilde{E}'} \tag{6.40}$$

Hence *colliding bodies composed of the same material absorb equal parts of the kinetic energy of normal relative motion irrespective of any difference in size.* This is a consequence of the deforming region being a negligibly small part of the mass of either body. In Fig. 6.8 this theory is compared with experimental measurements of the coefficient of restitution obtained for collisions between pairs of spheres that are identical and other collisions between pairs of spheres composed of dissimilar materials.

6.4.2   Loss of Internal Energy to Elastic Waves

During any collision between nonconforming bodies, the bodies come together with some relative speed, and it is the local deformation of the bodies that generates the contact pressures that act to prevent mutual interference (or overlapping) between the bodies. The stresses generated by local deformation cause the stress waves that radiate away from the contact region. These stress waves transmit changes in velocity from the contact area and cause the momentum of the bulk of each colliding body to change. Unless the surfaces of the colliding bodies have a very particular shape, elastic waves are reflected from different parts of the surface at different times and in different directions, so that there is not a coherent reflected wave that returns to the contact region to relieve the contact pressure. Rather, the duration of contact is controlled by the time required for the contact pressure to accelerate the two bodies until they separate. Because the region with significant deformation is ordinarily very small in comparison with both the cross-section and the depth normal to the contact surface, the contact duration depends on the effective mass $m = M_B M_{B'} / (M_B + M_{B'})$. That is, unless the bodies are bounded by a surface with a focal point in the contact region, the contact duration for a collision is determined by

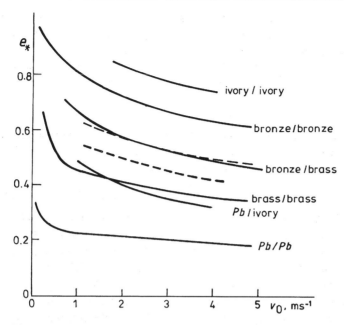

Figure 6.8. Coefficient of restitution e_* for direct impact of spheres of equal size. The solid lines are a best fit to experimental data collected by Goldsmith (1960), while the dashed lines for dissimilar materials are from Eq. (6.40).

quasistatic compliance and the effective mass m of the colliding bodies rather than wave propagation considerations.

Elastic stress waves play a part in energy dissipation during collision only if the relative size of the bodies, $R_B/R_{B'}$, is quite different from unity. In this case, there may be time for elastic waves to redistribute stresses and velocities in the smaller body in accord with rigid body dynamics while the larger body continues to suffer dispersal of energy from the contact region by elastic waves. In the smaller body the energy transmitted by elastic waves is not lost – this energy is redistributed by the waves until it is approximately equal to the energy distribution required by rigid body dynamics. For the larger body, however, there may be insufficient time for some parts of the radiating stress wave to reflect from any boundary, so that the distributions of kinetic energy and strain energy during contact are not even approximately equivalent to those required by rigid body dynamics.

Hunter (1956) examined the *vibration energy* in an elastic half space struck by an elastic sphere. He showed that the work done by the contact force in driving the half space is less than 1% of the energy absorbed, if the elastic wave speed c_1 is much larger than the normal component of relative velocity at impact $[(c_1/v_0)^{3/5} \gg 1]$; i.e., most of the absorbed energy is represented by elastic strain energy in material near the contact point C. This view was corroborated by a dynamic analysis of elastic impact by Tsai (1971). Early in the compression period the radial stress near the surface is larger than expected from a quasistatic analysis; nevertheless, Tsai concluded that the Hertz theory is a good approximation for the contact force during impact. Essentially, very large stresses are concentrated in a small contact region by geometric effects. A spherical elastic wave expands as it radiates away from the small contact region; a part of the expanding stress wave is continuously being reflected back towards the source due to stiffness that increases as the surface area of the wavefront increases. Thus at locations which are far from the contact region only very modest stresses are required to accelerate the bulk of the mass from the initial to the separation speed. The jump in stress $[\sigma]$ across an elastic wave provides an estimate of the stress required: $[\sigma]/E \approx v_0/c_1$. Consequently, compact bodies with a small contact region have most of the impact energy absorbed by deformation of material near the contact point. For collisions involving 3D deformation fields (spherical contact), vibrational energy can only be significant if at least one of the bodies is slender, i.e. if it has a dynamic response that is similar to that of a beam, plate or shell.

Hunter (1956) obtained his estimate of energy loss in an elastic half space by considering the steady state solution for an oscillating normal force acting in a circular region on the surface of the half space and calculating the work done by this force during half a period of loading. That is, he approximated the contact force as being sinusoidal and obtained the work done by this force during the contact period $t_f = \pi/\omega_0 \approx 1.068v_0/\delta_c$ where for bodies composed of rate-independent materials the maximum indentation δ_c occurs simultaneously with the maximum force F_c. In an elastic half space with Young's modulus E and Poisson's ratio ν the work W done during the contact period was calculated by Hunter (1956) as

$$W = \frac{(1+\nu)\beta}{\rho c_0^3}\left(\frac{1-\nu^2}{1-2\nu}\right)^{1/2} F_c^2\omega_0 = \frac{\beta}{\rho c_1^3}\left(\frac{1-\nu}{1-2\nu}\right)^2 F_c^2\omega_0 \qquad (6.41)$$

where the parameter $\beta = \beta(\nu)$ is a function of Poisson's ratio:

$$\beta(0.25) = 0.537, \qquad \beta(0.33) = 0.415.$$

The expression (6.30b) can be used to obtain the ratio between the work done during contact (or energy loss due to elastic waves) and the incident kinetic energy of normal relative motion:

$$\frac{W}{mv_0^2/2} = 4.74 \times \beta(1+\nu)\left(\frac{1-\nu^2}{1-2\nu}\right)^{1/2}\left(\frac{\rho R_*^3}{m}\right)^{1/2}\left(\frac{mv_0^2/2}{ER_*^3}\right)^{3/10}$$

Here the penultimate term is a geometric factor that depends on the curvature of the contact region R^{-1} and the shape of the colliding body, e.g. whether it is spherical, ellipsoidal, or a slender rod. For a solid colliding body that is spherical, the energy loss to elastic waves can be expressed also as

$$\frac{W}{mv_0^2/2} = 3.85 \times \beta(1+\nu)\left(\frac{1-\nu^2}{1-2\nu}\right)^{1/2}\left(\frac{v_0}{c_0}\right)^{3/5} \qquad (6.42a)$$

$$= 3.85 \times \beta\left(\frac{1-\nu}{1-2\nu}\right)^2\left(\frac{c_0}{c_1}\right)^{12/5}\left(\frac{v_0}{c_1}\right)^{3/5} \qquad (6.42b)$$

Here it is clear that the energy losses in elastic waves are a negligibly small part of the incident kinetic energy of relative motion if a small body collides against an elastic half space. Hunter (1956) stated that for a steel half space struck by a high-strength steel sphere the loss ratio is $2W/mv_0^2 = 1.04(v_0/c_0)^{3/5}$, while for a half space made of glass it is $2W/mv_0^2 = 1.27(v_0/c_0)^{3/5}$

6.5   Applicability of the Quasistatic Approximation

The previous example demonstrates that elastic collisions between slightly deformable bodies have small contact areas and very brief collision periods; consequently very large contact pressures develop during the collision. This raises the question of applicability of the quasistatic analysis.

Almost all of the mass in colliding bodies is decelerated by stresses that are transmitted from the contact region by elastic stress waves; the predominant mode of energy transmission is by dilatational waves travelling at speed $c_1 = [E_i(1 - \nu_i)/\rho_i(1 - \nu_i - 2\nu_i^2)]^{1/2}$. This speed depends on the bulk modulus of the material – for metals it is typically of the order of 4 or 5×10^3 m s^{-1} (see Table 7.1).

The contact period for elastic collisions $2t_c$ can be expressed in terms of the wave speed c_1. For a collision between two identical spheres this gives $m = m_B/2$, $R = R_B/2$ and $E = E_B/2(1 - \nu_B^2)$; thus the elastic contact period is given by

$$2t_c = 5.07 \frac{R_B}{c_1} \left[\frac{(1 - \nu_B)^4 c_1}{(1 - 2\nu_B)^2 v_0} \right]^{1/5} \tag{6.43}$$

The time t_B for an elastic wave to transit each sphere is $t_B = 2R_B/c_1$ so that the number of wave transits n during the contact period can be expressed as

$$n = \frac{2t_c}{t_B} = 2.56 \left[\frac{(1 - \nu_B)^4}{(1 - 2\nu_B)^2} \frac{c_1}{v_0} \right]^{1/5} \tag{6.44}$$

Thus collisions of compact or stocky bodies at low impact speeds (up to a few meters per second) result in contact periods that are long enough for several but not a very large number of transits by elastic waves.

Love (1906) proposed that the quasistatic Hertz theory applies only if the number of elastic wave transits is very large, i.e. $(c_1/v_0)^{1/5} \gg 1$. If this condition is not satisfied, Love presumed that a significant part of the impact energy remains in the bodies after contact ceases – remaining in the form of elastic vibration energy. By considering collisions where one body is very large in comparison with the other it can be shown that this notion is not correct.

Rather, the quasistatic contact theory gives accurate results for collisions if the region of significant deformation or internal energy density remains small in comparison with all dimensions of the colliding bodies; i.e., applicability depends on the geometry of the colliding bodies and how this affects diffusion of energy in 2D or 3D elastic waves emanating from the contact region.

The Hertz quasistatic analysis neglects the effect of stresses distributed throughout the body. By considering the quasistatic analysis of stress distribution in a heavy sphere, Villaggio (1996) investigated the error introduced by assuming that all compliance is lumped in a small region of negligible mass. This approximation causes a small reduction in the contact period without any significant change in the load–deflection relation.

6.6   Transverse Impact of Rough Elastic–Plastic Cylinders – Applicability of Energetic Coefficient of Restitution

Impact of hard metal bodies at modest speeds results in plastic deformation around the point of initial contact, and this has been shown to be a major source of energy

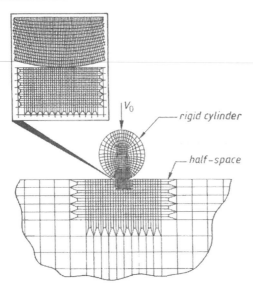

Figure 6.9. Finite element mesh for analyzing transverse impact of cylinder on elastic–plastic half space.

dissipation during impact of these bodies. For oblique impact between bodies with rough surfaces, both normal and friction forces contribute to the local stress field. The energetic coefficient of restitution, as defined in Chapter 3, depends on the normal component of force only. Consequently, for those collisions where the incident velocity is large enough to cause plastic deformation, it is important to understand whether the coefficient of restitution is sensitive to frictional impulse, i.e. the extent to which normal forces during elastic–plastic impact are affected by friction.

To assess the effect of friction on the coefficient of restitution for oblique impact of rough elastic–plastic bodies, 2D plane strain analysis of a cylinder colliding transversely against an elastic–plastic half space has been performed using the finite element code DYNA2D (see Lim and Stronge, 1998a). The finite element mesh is illustrated in Fig. 6.9. This continuum analysis of impact on a deformable body is compared with a hybrid quasistatic analysis where normal forces derived for elastic–plastic local deformation are combined with elastic tangential compliance and Coulomb friction. The analytical solution for the normal force and indentation follows closely the pattern presented in Sect. 6.1 for a colliding body with a spherical contact surface.

Consider two parallel cylinders with radii R_B and $R_{B'}$ that are composed of materials with Young's moduli E_B and $E_{B'}$. Here we again employ the definitions of an effective radius R_* and an effective elastic modulus E_*:

$$R_* = \left(R_B^{-1} + R_{B'}^{-1}\right)^{-1}$$

$$E_* = \left[\left(1 - \nu_B^2\right)E_B^{-1} + \left(1 - \nu_{B'}^2\right)E_{B'}^{-1}\right]^{-1}$$

6.6.1   Elastic Normal Compliance

For elastic deformations, an elliptic pressure distribution over the contact radius a gives a normal contact force F per unit depth and mean contact pressure $\bar{p}$ that are related to the contact area by

$$\frac{\bar{p}}{E_*} = \frac{F}{2aE_*} = \frac{\pi a}{8R_*}$$

and a normal indentation δ that depends on the normal force according to

$$\frac{\delta}{R_*} = \frac{F}{\pi E_* R_*}\left(2\ln\bar{\xi} - \frac{v}{1-v}\right)$$

where $\bar{\xi} = 2d/a$ is a characteristic depth for the deformation field. This relation is insensitive to the characteristic depth for $\bar{\xi} \gg 1$, so that in the elastic range, indentation depth is proportional to normal contact force:

$$\frac{F}{F_Y} = \frac{\delta}{\delta_Y}.$$

6.6.2   Yield for Plane Strain Deformation

For 2D deformation associated with indentation by a long cylinder, yield initiates at a mean pressure $\bar{p}/Y = 1.5 \equiv \bar{p}_Y/Y$, i.e. at a pressure somewhat larger than that required for yield under a spherical indenter. Once again the normal force and indentation at yield are material properties,

$$\frac{F_Y}{YR_*} = \frac{36}{\pi}\frac{Y}{E_*}$$

$$\frac{\delta_Y}{R_*} = \left(\frac{6Y}{\pi E_*}\right)^2 \frac{1}{1-v}\left[2\ln\frac{\pi E_*}{6Y} + 2\ln\bar{\xi} - \frac{v}{1-v}\right],$$

$$\frac{a_Y}{R_*} = \frac{12}{\pi}\frac{Y}{E_*}. \tag{6.45}$$

Equating the work done during compression and the incident kinetic energy of normal relative motion for a cylinder of mass m and length L, the incident normal velocity at yield v_Y is obtained as

$$v_Y^2 = \frac{3YR_*^2L}{(1-v^2)m}\left(\frac{6Y}{\pi E_*}\right)^3\left[4\ln\bar{\xi} + 4\ln\frac{\pi E_*}{6Y} - \frac{1+v}{1-v}\right]. \tag{6.46}$$

6.6.3   Elastic–Plastic Indentation

Employing Johnson's model for constrained plastic deformation, the mean pressure and indentation in the elastic–plastic range of deformation are expressed as

$$\frac{\bar{p}}{\bar{p}_Y} = 1 + \frac{2\sqrt{3}}{9}\ln\left(\frac{a}{a_Y}\right), \qquad \frac{\delta}{\delta_Y} = 1 + \frac{a_Y^2}{R_*\delta_Y}\left[\left(\frac{a}{a_Y}\right)^2 - 1\right] \tag{6.47}$$

This pressure gives the normal contact force as a function of deflection,

$$\frac{F}{F_Y} = \left[1 + \frac{R_*\delta_Y}{a_Y^2}\left(\frac{\delta}{\delta_Y} - 1\right)\right]^{1/2}\left\{1 + \frac{\sqrt{3}}{9}\ln\left[1 + \frac{R_*\delta_Y}{a_Y^2}\left(\frac{\delta}{\delta_Y} - 1\right)\right]\right\}$$

Elastic–plastic deformation terminates and uncontained plastic deformation begins when the mean contact pressure equals $\bar{p}/Y = 2.4$.

Table 6.3. *Incident Normal Velocities for Yield and Uncontained Plastic Deformation in Plane Strain*

	v_Y (m s^{-1})		v_P (m s^{-1})	
	Analytical	DYNA2D	Analytical	DYNA2D
Aluminum alloy, 2014-T6	1.62	1.55	18.12	17.50
Stainless steel, 302 cold-rolled	0.36	0.37	8.39	7.90

6.6.4   Fully Plastic Indentation

In the range of uncontained plastic deformation, additional indentation occurs with no increase in contact pressure; the contact area merely spreads outward over a radius

$$a/a_Y = 0.625 F/F_Y.$$

The indentation is related to contact area by the same relation (6.47), which gives no normal displacement at the periphery of the contact strip. The normal contact force that is developed by the uniform contact pressure equals

$$\frac{F}{F_Y} = 1.6 \left[1 + \frac{1}{4} \left(\frac{\pi E_*}{6Y} \right)^2 \frac{\delta_Y}{R_*} \left(\frac{\delta}{\delta_Y} - 1 \right) \right]^{1/2}$$

The incident impact speeds for yield v_Y and uncontained plastic deformation v_P are listed in Table 6.3 for an aluminum alloy and a stainless steel. The normal velocity at yield for 2D plane strain deformation is an order of magnitude larger than the corresponding value for spherical contact.

6.6.5   Analyses of Contact Forces for Oblique Impact of Rough Cylinders

For oblique impact there are tangential forces in the contact region in addition to the normal force analyzed in the previous section. The tangential forces are assumed to be due to dry friction that can be represented by Coulomb's law with a single coefficient that represents both static and dynamic friction.[7]

In the present analysis the energy losses are assumed to be entirely due to friction plus the work absorbed due to the hysteresis of the normal component of force. The latter factor is represented by the energetic coefficient of restitution. Here, the coefficient of restitution includes the effect of energy lost to elastic waves; for transverse impact of a cylinder on a half space at a normal speed equal to v_Y this gives a coefficient of restitution $e_* \approx 0.8$. Thus for elastic impact $[|v(0)| \leq v_Y]$, as shown in Fig. 6.10a, the normal component of relative velocity is somewhat smaller at separation than it was at incidence. The finite element calculations by DYNA2D show these same energy losses to elastic waves irrespective of the angle of incidence. At an impact speed that is large enough to

---

[7] While numerical simulations that incorporate separate values for static and dynamic coefficients of friction can achieve greater accuracy, for purposes of interpreting phenomenological effects this distinction merely muddies the water.

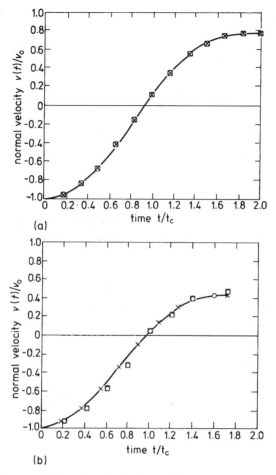

Figure 6.10. Normal velocity as a function of time during oblique impact on a rough half space with $\mu = 0.3$: (a) incident velocity in elastic range, $|v(0)/v_Y| = 0.67$, and (b) incident velocity in range of uncontained plastic indentation, $|v(0)/v_Y| = 33.3$.

cause uncontained plastic deformation $[|v(0)/v_Y| = 33.3]$, as shown in Fig. 6.10b, the energy lost to plastic work causes the separation velocity to be substantially smaller than that for elastic impact; nevertheless, for a coefficient of friction $\mu = 0.3$ these losses remain insensitive to friction.

The friction forces that result from oblique impact at a range of different incident velocities are shown in Fig. 6.11; this figure compares analytical results which incorporate elastic tangential compliance with DYNA2D finite element calculations. At small angles of incidence, shown in Fig. 6.11a, the contact initially sticks and does not begin sliding until late in the contact period. For incident normal speeds that give elastic behavior $[|v(0)/v_Y| < 1]$ the collision terminates when the contact separates at a time that is roughly double the compression period. At larger incident speeds, which result in more hysteresis, the period of restitution is smaller than the period of compression for these rate-independent materials.

For intermediate angles of incidence the contact initially slips, but slip vanishes during the contact period. During a subsequent period of stick the direction of slip reverses;

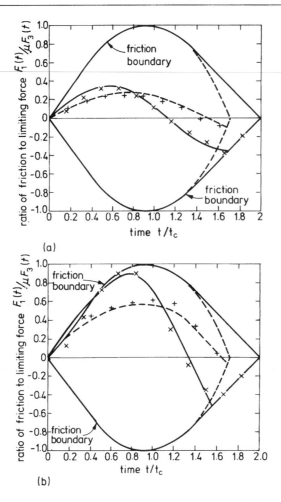

Figure 6.11. Variation of ratio between tangential and normal force during contact period resulting from oblique impact with $\mu = 0.3$ – a comparison of predictions of the linear tangential compliance model with finite element calculations: (a) small angle of incidence $[v_1(0)/v_3(0) = 0.176 < \mu\eta^2]$, exhibiting stick–slip behavior ($\eta^2 = 1.25$ for cylindrical contact in plane strain), and (b) intermediate angle of incidence $[v_1(0)/v_3(0) = 0.58]$, exhibiting slip–stick–slip behavior.

finally, however, as the normal contact force decreases, slip resumes. Figure 6.11b compares analytical and finite element calculations.

Other calculations at a large angle of incidence where there is gross slip in the initial direction similarly indicate that for normal impact speeds in the range $|v(0)/v_Y| < 33$ and coefficients of friction $\mu < 0.3$ the effect of tangential force on the normal compliance is negligible. Hence *the energetic coefficient of restitution is unaffected by friction for normal impact speeds that are below the limit for uncontained plastic deformation.*

6.6.6 Loss of Internal Energy to Elastic Waves for Planar (2D) Collisions

Calculations similar to those in Sect. 6.4.2 have been performed by Lim (1996) and Lim and Stronge (1998a), for transverse impact of a cylinder on an elastic half space.

Table 6.4. *DYNA2D Calculations of Energy Loss to Elastic Waves for Dissimilar Size Elastic Cylinders That Collide at Incident Speed $v_0 = 1$ m s^{-1}*

	Irrecoverable Part of Initial Kinetic Energy, $(T_0 - T_f)/T_0$					
Material	$R_B/R_{B'} = 1$	$R_B/R_{B'} = 2$	$R_B/R_{B'} = 4$	$R_B/R_{B'} = 8$	$R_B/R_{B'} = 16$	$R_B/R_{B'} \to \infty$
Aluminum	0.00	0.02	0.16	0.24	0.27	0.31
Steel	0.00	0.03	0.13	0.23	0.25	0.28

Transverse impact of parallel cylinders gives planar (2D) deformations[8]; these collision configurations result in a much larger part of the initial kinetic energy of relative motion being removed by elastic waves than occurs for impact between spheres:

$$W = 8.6 F_c^2 \omega_0 \frac{1+\nu}{\rho c_0^3} \left(\frac{1-\nu^2}{1-2\nu}\right)^{1/2} = \frac{8.6}{\rho c_1^3} F_c^2 \omega_0 \left(\frac{1-\nu}{1-2\nu}\right)^2 \qquad (6.48)$$

For two bodies of similar size and mass the loss of energy to elastic waves is negligible, as listed in Table 6.4.[9] For bodies that are very dissimilar in size, however, the irrecoverable energy due to elastic waves in the large body can be significant. In this case the size of the contact region and the strain energy stored in this region can be almost the same for each body (see Sect. 6.4.1), so that only the mass of the small body is important for determining the contact duration. For the large and massive body, the waves transmitting energy away from the contact region have little effect on the momentum in the normal direction. It is the energy in these waves which is lost in a collision if the masses of the colliding bodies are quite different.

6.7 Synopsis for Spherical Elastic–Plastic Indentation

Below is a synopsis of the maximum normal force F_c, maximum indentation depth δ_c and work done by the normal component of contact force at the termination of compression W_c for spherical indentation of elastic–perfectly plastic solids. The work W_c deforms the bodies and decreases the kinetic energy of relative motion. Also shown are the work done during restitution, W_r, that restores some of the kinetic energy of relative motion, and finally the coefficient of restitution e_*. All of these variables are expressed as functions of the nondimensional indentation δ/δ_Y.

Elastic:

$$\bar{p} < 1.1Y, \qquad \delta/\delta_Y < 1, \qquad a^2/a_Y^2 = \delta/\delta_Y$$

$$\frac{\bar{p}}{\bar{p}_Y} = \frac{a}{a_Y} = \left(\frac{\delta}{\delta_Y}\right)^{1/2}, \qquad \frac{F}{F_Y} = \left(\frac{\delta}{\delta_Y}\right)^{3/2}, \qquad \frac{W}{W_Y} = \left(\frac{\delta}{\delta_Y}\right)^{5/2}$$

$$e_* = 1$$

---

[8] For transverse compression of parallel cylinders the deformation field is 2D at points far from the ends. Near an end there is also an axial component of displacement, and this varies within the cross-section.

[9] Analyses of energy losses in elastic cylinders with distinct sizes were performed with ABAQUS/Explicit by Sunib Seah, National University of Singapore.

Elastic–plastic:

$$1.1Y < \bar{p} < 2.8Y, \qquad 1 < \delta/\delta_Y < 84, \qquad a^2/a_Y^2 = 2\delta/\delta_Y - 1$$

$$\frac{\bar{p}}{\bar{p}_Y} = 0.95 + 0.30 \ln\left(\frac{2\delta}{\delta_Y} - 1\right),$$

$$\frac{F}{F_Y} = \left(\frac{2\delta}{\delta_Y} - 1\right)\left[0.95 + 0.30 \ln\left(\frac{2\delta}{\delta_Y} - 1\right)\right]$$

$$\frac{W}{W_Y} = 1 + 0.531\left[\left(\frac{2\delta}{\delta_Y} - 1\right)^2 - 1\right] + 0.188\left(\frac{2\delta}{\delta_Y} - 1\right)^2 \ln\left(\frac{2\delta}{\delta_Y} - 1\right)$$

$$\frac{W_r}{W_Y} = -\left(\frac{\delta_r}{\delta_y}\right)^3 = -\left(\frac{2\delta_c}{\delta_Y} - 1\right)^{3/2}$$

$$e_* = \left\{\frac{\left(\frac{2\delta_c}{\delta_Y} - 1\right)^{3/2}}{1 + 0.531\left[\left(\frac{2\delta_c}{\delta_Y} - 1\right)^2 - 1\right] + 0.188\left(\frac{2\delta_c}{\delta_Y} - 1\right)^2 \ln\left(\frac{2\delta_c}{\delta_Y} - 1\right)}\right\}^{1/2}$$

Fully plastic:

$$\bar{p} = 2.8Y, \qquad \delta/\delta_Y > 84, \qquad a^2/a_Y^2 = 2\delta/\delta_Y - 1$$

$$\frac{\bar{p}}{\bar{p}_Y} = 2.55, \qquad \frac{F}{F_Y} = 2.55\left(\frac{2\delta}{\delta_Y} - 1\right)$$

$$\frac{W}{W_Y} = -3090 + 1.60\left(\frac{2\delta}{\delta_Y} - 1\right)^2,$$

$$\frac{W_r}{W_Y} = -\left(\frac{\delta_r}{\delta_y}\right)^3 = -\left(\frac{2\delta_c}{\delta_Y} - 1\right)^{3/2} \tag{6.49}$$

$$e_* = \left\{\frac{(2\delta_c/\delta_Y - 1)^{3/2}}{-3090 + 1.60(2\delta_c/\delta_Y - 1)^2}\right\}^{1/2} \tag{6.50}$$

From either elastic–plastic or fully plastic states the final or residual indentation δ_f is given by

$$\frac{\delta_f}{\delta_Y} = \frac{\delta_c}{\delta_Y} - \frac{\delta_r}{\delta_Y} = \frac{\delta_c}{\delta_Y} - \left(\frac{2\delta_c}{\delta_Y} - 1\right)^{1/2} \tag{6.51}$$

For colliding bodies with an effective mass m and a nominal initial radius of contacting surfaces R_*, the kinetic energy $\hat{T}_0$ of normal relative motion is transformed to deformation energy during compression, so that $W_c = \hat{T}_0 = mv_0^2/2$. Irrespective of the impact speed, the specific energy absorbed during compression,

$$\frac{W_c}{W_Y} = \frac{5\pi}{12}\left(\frac{E_*}{3Y}\right)^4 \frac{mv_0^2}{2YR_*^3} \tag{6.52}$$

can be used to calculate the minimum nondimensional contact radius $\bar{a}_c = a_c/a_Y$. Equation (6.52) applies to impacts of bodies with a normal effective mass m and an effective radius of curvature R_* in the contact region. For impact of an elastic sphere on an elastic half space,

$$\frac{W_c}{W_Y} = 2.74 \left(\frac{E_*}{3Y}\right)^4 \frac{\rho v_0^2}{Y}$$

PROBLEMS

6.1 Derive expression (2.8) from the sum of the kinetic energies of normal relative motion that are dissipated during compression. For an impact speed of 3 m s^{-1} and data from collisions between ivory and ivory and between lead and lead spheres, calculate the coefficient of restitution e_* for a collision between a lead and an ivory sphere of the same size. What assumption does this calculation make regarding the relative duration of the periods of contact for collisions between different pairs of identical spheres?

6.2 A ball peen hammer of mass $M = 1$ kg that is made of steel ($E = 210$ GPa, $Y = 600$ MPa, $\rho = 7.8$ g cm^{-3}) strikes a soft aluminum bar ($E = 70$ GPa, $Y = 120$ MPa, $\rho = 2.7$ g cm^{-3}); the bar has thickness $h = 6$ mm, while the impact surface of the hammer has a radius of 10 mm.

(a) Calculate the minimum impact speed V_Y for yield. For impact at this speed describe the location where yield initiates.

(b) Calculate an estimate of the coefficient of restitution e_* for an impact speed $V_0 = 0.1$ m s^{-1} (a light tap).

(c) Describe the final surface and subsurface deformation that you expect from the impact specified in (b).

CHAPTER 7

Axial Impact on Slender Deformable Bodies

Alice laughed. 'There's no use trying,' she said, 'one *can't* believe
impossible things.' 'I daresay you haven't had much practice,' said
the Queen. 'When I was your age, I always did it for half-an-hour a
day. Why, sometimes I've believed as many as six impossible things
before breakfast.'

Lewis Carroll, *Through the Looking Glass* (1872)

Axial impact on a deformable body results in a disturbance which initially
propagates away from the impact site at a specific speed. This disturbance is a pulse
or wave of particle displacement (and consequent stress). Wave propagation relates to
propagation of a coherent pulse of stress and particle displacement through a medium
at a finite speed. Familiar manifestations of this phenomenon are the transmission of
sound through air, water waves across the surface of the sea and seismic tremors through
the earth; thus, waves exist in gases, liquids and solids. Sources of excitation may be
either concentrated or distributed spatially, and brief or extended functions of time. The
unifying characteristic of waves is *propagation of a disturbance through a medium.*
Properties of the medium that result in waves and determine the speeds of propagation
are the density ρ and moduli of deformability (Young's modulus E, shear modulus G, bulk
modulus K, etc.).

7.1 Longitudinal Wave in Uniform Elastic Bar

Consider a uniform slender elastic bar of cross-sectional area A, elastic modulus
E and density ρ; the bar contains a region with axial stress $\sigma(x, t)$ that is propagating
in the positive x direction as shown in Fig. 7.1. For a differential element of the bar
located at spatial coordinate x let $u(x, t)$ be the axial displacement at any time t. The
stress pulse results in a difference in force $(A \, \partial\sigma/\partial x) \, dx$ across the differential element;
this difference gives an axial equation of motion for the element,

$$\rho A \frac{\partial^2 u}{\partial t^2} = A \frac{\partial \sigma}{\partial x}.$$

For a linear elastic material and a one dimensional state of stress, Hooke's law gives a
stress–strain relation,

$$\sigma = E \, \partial u/\partial x. \tag{7.1}$$

When E, A and ρ are independent of x the constitutive relation (7.1) results in a linear

146

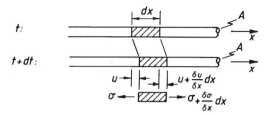

Figure 7.1. Differential element in bar of cross-sectional area A at successive times t and $t + dt$; the element is being traversed by a stress pulse traveling in direction x at speed c_0. Particle displacement $u(x, t)$ and axial stress $\sigma(x, t)$ vary across the element.

wave equation,

$$\frac{\partial^2 u}{\partial t^2} = c_0^2 \frac{\partial^2 u}{\partial x^2}, \qquad c_0^2 = \frac{E}{\rho} \tag{7.2}$$

where the wave propagation speed c_0 is known as the *bar velocity*. Equation (7.2) is a homogeneous, linear, hyperbolic partial differential equation (PDE). It is linear because the dependent variable $u(x, t)$ and its derivatives occur in the first degree only. Consequently, if u_1 and u_2 are two independent solutions, the sum $c_1 u_1 + c_2 u_2$ will also be a solution when c_1 and c_2 are arbitrary constants; i.e., the *principle of superposition* applies.

In contrast to many problems in wave propagation, there is a general solution to the homogenous wave equation,

$$u(x, t) = \tfrac{1}{2}[f(x - c_0 t) + g(x + c_0 t)] = \tfrac{1}{2}[f(\eta) + g(\xi)] \tag{7.3}$$

where f and g are arbitrary functions of their respective arguments $\eta = x - c_0 t$ and $\xi = x + c_0 t$. By differentiating, one can show that if f and g possess second derivatives with respect to their arguments at all but a finite number of points, they satisfy Eq. (7.2).[1] The functions f and g will be determined by *initial conditions* on the problem.

The form of Eq. (7.3) suggests the term *wave*. The function f represents a disturbance traveling in the direction of increasing x with speed c_0. This spatial variation in stress propagates in direction x without change in shape – no force is required to maintain the waveform. Similarly the function g represents a disturbance traveling in the direction of decreasing x without change in shape.

7.1.1 Initial Conditions

An initial displacement field $u(x, 0)$ results from a compressed or stretched segment of the bar. Similarly there can be an initial velocity field $\dot{u}(x, 0)$. Denoting the derivative of any function $f(\eta)$ with respect to its argument by $f' \equiv df/d\eta$, the initial

---

[1] The solution (7.3) to the homogenous wave equation (7.2) can be obtained by introducing a transformation of variables $\xi = x + ct$, $\eta = x - ct$:

$$c^2 \frac{\partial^2 u}{\partial x^2} - \frac{\partial^2 u}{\partial t^2} = 4c^2 \frac{\partial^2 u}{\partial \xi \, \partial \eta} = 0.$$

This implies that
$$\partial u / \partial \xi = g'(\xi)$$

$$u = g(\xi) + f(\eta), \qquad \text{where} \quad g = \int g' d\xi.$$

Note that the first order PDE $c \, \partial u / \partial x + \partial u / \partial t = 0$ also has this wave type solution.

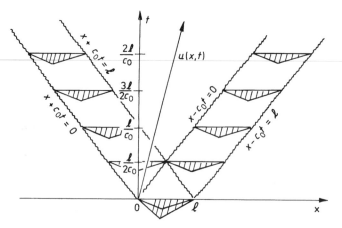

Figure 7.2. Distribution of displacement amplitude $u(x, t)$ at successive times following sudden release at time $t = 0$ of an initial compressed region of length l.

conditions are generally expressed as

$$u(x, 0) = \frac{1}{2}[f(x) + g(x)]$$

$$\dot{u}(x, 0) = \frac{c_0}{2}[-f'(x) + g'(x)], \quad \text{where} \quad \frac{\partial f(x)}{\partial t} = \frac{\partial f}{\partial \eta}\frac{\partial \eta}{\partial t} \equiv -c_0 f', \quad \eta \equiv x - c_0 t.$$

Prescribed Displacement $u(x, 0) = h(x)$ and Zero Velocity $\dot{u}(x, 0) = 0$

$$u(x, 0) = \frac{1}{2}[f(x) + g(x)] = h(x)$$

$$\dot{u}(x, 0) = \frac{c}{2}[-f'(x) + g'(x)] = 0.$$

The second condition is used to obtain the relationship between the functions f and g, viz. $g(x) = f(x)$. Hence, $f(x) = g(x) = h(x)$, so that $f(\eta) = h(\eta)$, $g(\xi) = h(\xi)$ and

$$u(x, t) = \frac{1}{2}[h(x - c_0 t) + h(x + c_0 t)]. \tag{7.4}$$

Hence, waves having the initial deformed shape $h(x)$ propagate in both directions away from their initial positions. The speed of propagation is $c_0 = \sqrt{E/\rho}$. The amplitude of each wave is $1/2$ the amplitude of the initial disturbance. Figure 7.2 illustrates this split of the solution into equal parts traveling in opposite directions away from the site of an initial disturbance.

Zero Displacement and Prescribed Velocity $\dot{u}(x, 0) = h(x)$

$$u(x, 0) = 0 \quad \Rightarrow \quad f(x) = -g(x) \quad \Rightarrow \quad f'(x) = -g'(x)$$

$$\dot{u}(x, 0) = h(x) \quad \Rightarrow \quad h(x) = -\frac{c_0}{2}[f'(x) + f'(x)]$$

$$\therefore \quad f(x) = \frac{-1}{c_0}\int_0^x h(\bar{x})\,d\bar{x}$$

$$u(x, t) = -\frac{1}{2c_0}\left[\int_0^{x-c_0 t} h(\bar{x})\,d\bar{x} - \int_0^{x+c_0 t} h(\bar{x})\,d\bar{x}\right] = \frac{1}{2c_0}\int_{x-c_0 t}^{x+c_0 t} h(\bar{x})\,d\bar{x} \tag{7.5}$$

Table 7.1. *Elastic Wave Speeds for Several Materials*

Material	ρ (kg/m^3)	E (kN/mm^2)	ν	c_0 (m s^{-1})	c_1 (m s^{-1})	c_2 (m s^{-1})	c_R (m s^{-1})
Aluminum alloy	2,700	70	.34	5092	6460	3100	2970
Brass	8,300	95	.35	3383	4300	2050	
Copper	8,500	114		3662		2300	
Lead	11,300	17.5	.45	1244	2200	700	
Steel	7,800	210	.31	5189	6060	3260	3040
Glass	1,870	55		5300	6800	3250	
Granite	2,700		.22	3120	2900		
Limestone	2,600		.33	4920			
Perspex			.40	2260	2600	1090	

$$c_0 = \left(\frac{E}{\rho}\right)^{1/2}, \quad c_1 = \left[\frac{E(1-\nu)/\rho}{(1+\nu)(1-2\nu)}\right]^{1/2}, \quad c_2 = \left(\frac{G}{\rho}\right)^{1/2}$$

With an imposed initial velocity field $h(x)$, the disturbance spreads from the initial disturbance with speed c_0.

In Table 7.1 the bar wave speed c_0 for uniaxial stress is shown for a variety of materials. Also shown are the wave speed c_1 for uniaxial displacement (plane strain), the shear wave speed c_2 and the Rayleigh wave speed c_R (surface waves). In hard materials (e.g. metals) all of these speeds are of the order of kilometers per second. The shear and Rayleigh wave speeds are slower than the dilatational (i.e. compressive) wave speeds.

Force F_0 Suddenly Applied to End of Finite Length Bar

A tension F_0 is suddenly applied to the end of a bar $0 \le x \le \infty$. Since the disturbance is outgoing from the source, $g(\xi) = 0$ and

$$u(x, t) = f(x - c_0 t) \quad \text{for} \quad x < c_0 t.$$

This gives $\partial u/\partial x = f'$ and $\partial u/\partial t = -c_0 f'$. Consequently, for a wave traveling in the positive x direction

$$\partial u/\partial x = -c_0^{-1} \partial u/\partial t \quad \text{for} \quad x < c_0 t. \tag{7.6}$$

Discontinuities in any dependent variable are indicated by a bracket [ ]. Equation (7.6) indicates that across a wavefront propagating in the positive x direction, discontinuities in stress and particle velocity are related by

$$[\sigma] = E[\partial u/\partial x] = -\rho c_0 [\dot{u}]. \tag{7.7}$$

Similarly, if F_0 is applied to the end of a bar where $-\infty < x \le 0$, the general solution has $f = 0$ (to satisfy the radiation condition) and $u(x, t) = g(x + c_0 t)$. Thus for a wave traveling in the negative x direction, discontinuities in stress and particle velocity satisfy

$$[\sigma] = E[\partial u/\partial x] = +\rho c_0 [\dot{u}]. \tag{7.8}$$

Thus Eq. (7.7) suffices for both directions of travel if the wave speed c_0 is given the sign of the direction of propagation. Note that for a longitudinal wave, a tensile discontinuity in stress $[\sigma_x] > 0$ results in a jump in particle velocity $[\dot{u}]$ that is opposite in sign from the wave speed c_0. A compressive stress wave $[\sigma_x] < 0$ has a jump in particle velocity in the same direction as the motion of the wavefront.

Example 7.1   A uniform elastic rod with Young's modulus E, density ρ and cross-sectional area A has constant pressure p_0 applied to the end $x = 0$ for a period of time τ. Obtain expressions for strain energy $U(\tau)$ and kinetic energy $T(\tau)$ in the bar at time $t = \tau$. Compare these parts of the total mechanical energy with work $W(\tau)$ done by the applied force that acts on the end of the rod.

   Solution   Boundary condition:

$$\sigma(0, t) = \begin{cases} -p_0, & t < \tau \\ 0, & \tau < t. \end{cases}$$

Solution for stress distribution:

$$\sigma(x, t) = \begin{cases} 0, & c_0 t < x \\ -p_0, & c_0(t - \tau) < x < c_0 t. \\ 0, & x < c_0(t - \tau) \end{cases}$$

Strain energy:

$$U(\tau) = \frac{1}{2E} \int_0^{c_0 \tau} \sigma^2(x, \tau) \, A \, dx = \frac{p_0^2 A c_0 \tau}{2E}$$

Kinetic energy:

$$T(\tau) = \frac{\rho}{2} \int_0^{c_0 \tau} \dot{u}^2(x, \tau) \, A \, dx = \frac{p_0^2 A c_0 \tau}{2E}$$

Work of external force:

$$W(\tau) = \int_0^{\tau} [-A\sigma(0, t)]\dot{u}(0, t) \, dt = \frac{p_0^2 A c_0 \tau}{E}$$

Partition of internal energy:

$$U/W = T/W = 1/2.$$

7.1.2   Reflection of Stress Wave from Free End

   If a constant tension F_0 is suddenly applied to one end of a free elastic bar, a stress wave of magnitude $F_0/A \equiv \sigma_0$ moves away from the loaded end with speed c_0; behind the wavefront (i.e. for $x < c_0 t$) the stress and the particle velocity $\dot{u}(x, t) = -\sigma_0/\rho c_0$ are constant. If the bar has length L then the wavefront approaches the free end at time $t = (L/c_0)-$. After incidence of the wave on the free end, Eq. (7.7) gives a stress σ_1 and particle velocity $\dot{u}_1$ behind the reflected wavefront,

$$[\sigma_1 - \sigma_0] = -\rho(-c_0)[u_1 - (-\sigma_0/\rho c_0)], \qquad L/c_0 < t < 2L/c_0. \tag{7.9}$$

At the free end there is a boundary condition $\sigma(L, t) = 0$ so that $\sigma_1 = 0$; thus for times

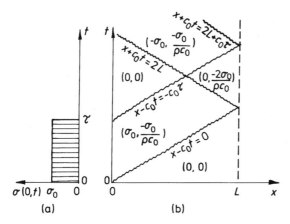

Figure 7.3. Characteristic diagram for a rod with a free end at $x = L$ that is subject to tensile stress σ_0 at end $x = 0$ during $0 < t < \tau$. In each region between wavefronts the stress and particle velocity $(\sigma, \dot{u})$ are indicated.

$L/c_0 < t < 2L/c_0$,

$$\sigma(x, t) = \sigma_0 \quad \text{and} \quad \dot{u}(x, t) = -\sigma_0/\rho c_0, \qquad\qquad x < 2L - c_0 t$$

$$\sigma(x, t) = 0 \quad \text{and} \quad \dot{u}(x, t) = \dot{u}_1 = -2\sigma_0/\rho c_0, \qquad x > 2L - c_0 t.$$

Behind the reflected wave propagating from the free end, the stress is zero to satisfy the boundary condition (BC) and the particle velocity is twice that of the incident wave.

Figure 7.3 is a characteristic diagram for stress waves in a rod of length L. The diagram illustrates the location of wavefronts emanating from the origin $x = 0$ at time $t = 0$ when a tensile stress σ_0 is suddenly applied and time $t = \tau$ when σ_0 is suddenly removed. The effects of subsequent reflections from the free end $x = L$ are also shown. In each region between wavefronts the stress and particle velocity $(\sigma, \dot{u})$ are indicated.

An arbitrary stress wave distribution $f(x - c_0 t)$ emanating from $x = 0$ has the following general solution during the period of time $t < L/c_0$ before the wavefront reaches the free end:

$$u(x, t) = \begin{cases} f(x - c_0 t), & x < c_0 t \\ 0, & c_0 t < x < L. \end{cases} \tag{7.10}$$

If this stress pulse is reflected from the free end, then after incidence during the period $L/c_0 < t < 2L/c_0$ the solution will be

$$u(x, t) = \begin{cases} f(x - c_0 t), & x < 2L - c_0 t \\ f(x - c_0 t) + g(x + c_0 t), & 2L - c_0 t < x < L. \end{cases} \tag{7.11}$$

The reflected wave component g is determined by the BC at the free end, viz., the strain $\partial u(x, t)/\partial x = f'(x - c_0 t) + g'(x + c_0 t)$ at $x = L$ and at any time t must be zero:

$$0 = \frac{\partial u(L, t)}{\partial x} = f'(L - c_0 t) + g'(L + c_0 t).$$

Let $\xi' = L + c_0 t$; hence the BC above gives $g'(\xi') = -f'(2L - \xi')$. Consequently by extension (i.e. substitution of $\xi' = x + c_0 t$) we have $g'(x + c_0 t) = -f'(2L - x - c_0 t)$,

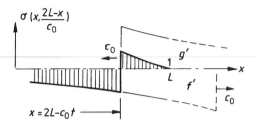

Figure 7.4. Stress distribution $\sigma(x, t)$ in a rod at time $t > L/c_0$ after reflection from free end.

so that behind the reflected wavefront

$$\sigma(x, t) = E \frac{\partial u(x, t)}{\partial x} = E\{f'(x - c_0 t) - f'(2L - x - c_0 t)\}. \tag{7.12}$$

The reflected stress wave has opposite sign to the incident stress wave and apparently is identical to a pulse with the same distribution but opposite sign that initiates at time $t = 0$ from $x = 2L$ in a virtual extension of the rod (see Fig. 7.4).

The reflected *particle velocity* wave has the same sign as the incident particle velocity wave. The displacement field at times $t > L/c_0$ is then

$$u(x, t) = f(x - c_0 t) + f(2L - x - c_0 t), \quad 2L - c_0 t < x < L \text{ and } t > L/c_0. \tag{7.13}$$

The second term above is the reflected wave from the free end.

7.1.3 Reflection from a Fixed End

Suppose a tension F_0 is suddenly applied to one end of an elastic bar of length L while the other end is fixed. Again, a stress wave of magnitude $\sigma = \sigma_0$ is generated at the point of loading and propagates toward the fixed end at speed c_0. At the fixed end we require the particle velocity to be zero. Applying Eq. (7.7) after incidence of the wave with the free end again gives Eq. (7.9). Together with the BC

$$\dot{u}(L, t) = 0, \qquad t > 0.$$

Eq. (7.9) gives stresses and particle velocity distributions during $L/c_0 < t < 2L/c_0$,

$$\sigma(x, t) = \sigma_0 \quad \text{and} \quad \dot{u}(x, t) = -\sigma_0/\rho c_0, \qquad x < 2L - c_0 t$$

$$\sigma(x, t) = 2\sigma_0 \quad \text{and} \quad \dot{u}(x, t) = \dot{u}_1 = 0, \qquad x > 2L - c_0 t.$$

At the fixed end the stress magnitude behind the reflected wave is twice the stress in the incident wave, while the particle velocity behind the reflected wave is zero.

7.1.4 Reflection and Transmission at Interface – Normal Incidence

When a wave crosses an interface between two materials, part of the wave is reflected and part transmitted. Because the wave equation is linear, after the wave crosses the interface the stress (and particle velocity) in the first material is the sum of those of the incident and reflected waves, $f' + g'$. The relative magnitude of reflected and transmitted components of the wave will depend on the density ρ_j and elastic modulus E_j of the two

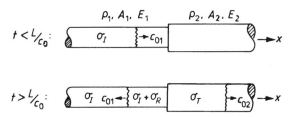

Figure 7.5. Stresses in a bar at times preceding and following the wavefront crossing a discontinuity in material of cross-sectional area A_j. The incident, reflected and transmitted components of stress are denoted by subscripts I, R and T respectively.

materials and the cross-sectional areas $A_j (j = 1, 2)$ of the rods. The relative magnitudes are determined by the ratio of impedance ρc_0 of the materials.

Figure 7.5 is an illustration of a wave with stress σ_I and particle velocity $\dot{u}_I$ approaching an interface in a slender composite bar. At the interface in a slender bar, contact force $A_j \sigma_j$ and particle velocity $\dot{u}_j$ are continuous.[2] When the bars are unstressed before the arrival of a wave, Eq. (7.7) gives the reflection and transmission coefficients γ_R, γ_R for stress and particle velocity:

$$\frac{\sigma_R}{\sigma_I} = \frac{\Gamma - 1}{\Gamma + 1} \equiv \gamma_R, \qquad \text{where} \quad \Gamma = \frac{A_2 \rho_2 c_2}{A_1 \rho_1 c_1} \tag{7.14a}$$

$$\frac{\sigma_T}{\sigma_I} = \frac{2\Gamma (A_1 / A_2)}{\Gamma + 1} \equiv \gamma_T \tag{7.14b}$$

$$\frac{\dot{u}_R}{\dot{u}_I} = \frac{1 - \beta}{1 + \beta} = -\gamma_R \tag{7.14c}$$

$$\frac{\dot{u}_T}{\dot{u}_I} = \frac{2}{\beta + 1} = \frac{A_2}{A_1 \Gamma} \gamma_T. \tag{7.14d}$$

The reflection and transmission coefficients for stresses are shown in Fig. 7.6 for bars of equal area $A_2 = A_1$ and varying impedance ratio Γ.

When a transient wave passes through a layered material (e.g. a composite), the succession of reflections and transmissions from interfaces can be analyzed using this one dimensional model. Interlaminar tensile fractures can be formed near a surface where compressive forces are applied for a short duration. Tensile stresses are large if the acoustic impedance of adjacent layers are quite different and the impedance of the surface layer is relatively small (Achenbach, Hemann and Ziegler, 1968).

7.1.5 Spall Fracture Due to Reflection of Stress Waves

At a free surface or an interface where the acoustic impedance increases in the direction of propagation, compressive incident waves are reflected as tensile waves. When behind the wavefront the stress magnitude decreases, a net tensile stress occurs in the bar

---

[2] Discontinuities in cross-sectional area are handled by similar means in a 1D theory. At the interface, the force acting on the end of the second bar is equal in magnitude and opposite in direction to the force acting on the first bar. Consequently, the boundary condition on stress is replaced by $(\sigma_I + \sigma_R) A_1 = \sigma_T A_2$.

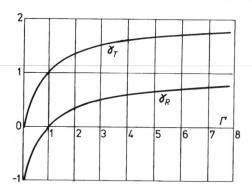

Figure 7.6. Reflection and transmission coefficients as a function of impedance ratio $\Gamma = \rho_2 c_2 A_2 / \rho_1 c_1 A_1$ for a one-dimensional stress wave in a slender composite bar having equal areas $A_2 = A_1$.

after reflection. A tensile fracture, or *spall*, results where the stress equals an ultimate stress.[3]

Consider a free bar of length L with an exponentially decreasing compressive stress suddenly applied to one end: $\sigma(0, t) = -\sigma_0 \exp(-t/\tau)$.[4] The BC at the origin is used to evaluate the function $f(x - c_0 t)$ for a radiating wave:

$$\sigma(0, t) = -\sigma_0 \exp(-t/\tau) = Ef'(0 - c_0 t), \qquad t > 0.$$

Substituting $\xi' = -c_0 t$, we obtain the function $f(\xi')$ and subsequently by extension $f(x - c_0 t) = -(\sigma_0 c_0 \tau / E) \exp[(x - c_0 t)/c_0 \tau]$. Consequently,

$$\sigma(x, t) = Ef'(x - c_0 t) = \begin{cases} -\sigma_0 \exp[(x - c_0 t)/c_0 \tau], & x < c_0 t \\ 0, & x > c_0 t \end{cases}$$

$$\dot{u}(x, t) = -c_0 f'(x - c_0 t) = \begin{cases} (\sigma_0/\rho c_0) \exp[(x - c_0 t)/c_0 \tau], & x < c_0 t \\ 0, & x > c_0 t. \end{cases}$$

The bar has a free end at $x = L$, where the incident stress wave is reflected with a change in sign. At the free end the boundary condition is

$$0 = \sigma(L, t) = E\{f'(L - c_0 t) + g'(L + c_0 t)\}.$$

Hence $g'(\xi') = -f'(2L - \xi')$, so that

$$g'(x + c_0 t) = \frac{\sigma_0}{E} \exp\left(\frac{2L - x - c_0 t}{c_0 \tau}\right)$$

---

[3] In ductile metals, fracture occurs by tearing between voids that have opened from fracture initiation sites. The density of sites is stress-dependent. Consequently, when an ultimate plastic shear strain is required for tearing between voids, ductile fracture is time-dependent. The apparent ultimate stress for failure caused by short duration impact loads is increased by this effect (Meyers and Murr, 1980).

[4] This theory assumes a uniform distribution of stress and displacement across the section, i.e. one dimensional (1D) deformation. It will be a poor approximation for higher frequency components of a wave which have wavelength shorter than the bar diameter.

Consequently, during the period $L/c_0 < t < 2L/c_0$ after the first reflection,

$$\sigma(x, t) = -\sigma_0 \begin{cases} \exp[(x - c_0 t)/c_0 \tau] & x < 2L - c_0 t \\ \exp[(x - c_0 t)/c_0 \tau] - \exp[(2L - x - c_0 t)/c_0 \tau] & x > 2L - c_0 t \end{cases}$$

Figure 7.4 shows the distribution of stress at a time in the period $L/c_0 < t < 3L/c_0$ after reflection of an initially compressive stress wave. If the stress wave decreases exponentially behind the wavefront, during this period the largest tensile stress will be at the reflected wavefront where $2L - x = ct$. This stress increases with distance between the wavefront and the free end; at any location x the maximum stress σ_{max} is given by

$$\sigma_{max}\left(x, \frac{2L - x}{c}\right) = \sigma_0\{1 - e^{2(x-L)/c\tau}\}. \tag{7.15}$$

The spall thickness is determined by the ratio of maximum applied pressure $-\sigma_0$ to fracture stress σ_f and by the rate of decay τ in pressure amplitude behind the front.

Example 7.2 Consider coaxial impact of two long slender bars, each of length L and cross-sectional area A, as shown if Fig. 7.7. The impact occurs at time $t = 0$, and the impact interface is designated as $x = 0$. Bar 2 is butted against a rigid barrier at end $x = L$, and initially it is stationary while bar 1 is moving at initial velocity V_0. Let the impedance of bar 1 be $\rho_1 c_{01}$ and that of bar 2 be $\rho_2 c_{02}$, where $c_{02}/c_{01} > 1$. Obtain the distribution of stress and particle velocity in the rod as a function of time.

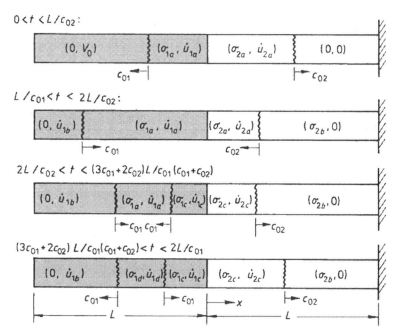

Figure 7.7. Elastic waves from collinear impact of traveling bar against stationary bar shown at successive times after impact. The wavefronts are indicated, and between them there is a constant state of stress and particle velocity $(\sigma, \dot{u})$.

Solution   Equations (7.7) and (7.8) for the velocity change across a wavefront are used together with boundary and interface conditions to obtain the stress and particle velocity distributions at successive times:

$$0 < t < L/c_{02}: \qquad \sigma_{1a} = \sigma_{2a} = -(1 + \Gamma)^{-1}\rho_2 c_{02} V_0, \qquad \Gamma = \rho_2 c_{02}/\rho_1 c_{01}$$

$$\dot{u}_{1a} = \dot{u}_{2a} = (1 + \Gamma)^{-1} V_0$$

$$L/c_{01} < t: \qquad \sigma_{1b} = 0, \quad \sigma_{2b} = -2(1 + \Gamma)^{-1}\rho_2 c_{02} V_0$$

$$\dot{u}_{1b} = (1 - \Gamma)(1 + \Gamma)^{-1} V_0, \quad \dot{u}_{2b} = 0.$$

At time $2L/c_{02}$ the faster wave returns to the interface $x = 0$, where it is partly reflected and partly transmitted:

$$2L/c_{02} < t < 4L/(c_{01} + c_{02}): \qquad \sigma_{2c} = -(3 + \Gamma)(1 + \Gamma)^{-2}\rho_2 c_{02} V_0$$

$$\dot{u}_{1c} = \dot{u}_{2c} = (1 - \Gamma)(1 + \Gamma)^{-2} V_0.$$

At time $(3c_{01} + 2c_{02})L/(c_{01}^2 + c_{01}c_{02})$ the two waves cross in bar 1:

$$4L/(c_{01} + c_{02}) < t < 2L/c_{01}: \qquad \sigma_{1d} = -2(1 + \Gamma)^{-2}\rho_2 c_{02} V_0$$

$$\dot{u}_{1d} = (1 - 2\Gamma - \Gamma^2)(1 + \Gamma)^{-2} V_0.$$

Notice that at the interface for time $t < 2L/c_{01}$ the stress is compressive ($\sigma_{2c} < 0$) irrespective of the impedance ratio Γ. At larger times the stress at the interface must be checked to ensure that it remains negative – when this condition is no longer satisfied, the interface condition changes so that the interface is free of stress.

7.2   Planar Impact of Rigid Mass against End of Elastic Bar

Let the end $x = L$ of a uniform, straight elastic bar be struck by a rigid body of mass M that is moving with initial speed V_0 while the end $x = 0$ is restrained by a viscoelastic damper as shown in Fig. 7.8. Here we assume that changes in stress are rapid so that they are transmitted by waves propagating at an elastic wave speed c_0. For a wave traveling in the direction of decreasing (increasing) x, stress and particle velocity are related by conservation of momentum across a wavefront [Eq. (7.7)]:

$$[\sigma] = \rho c_0 [\dot{u}], \qquad c_0 < 0 \tag{7.7a}$$

$$[\sigma] = -\rho c_0 [\dot{u}], \qquad c_0 > 0 \tag{7.7b}$$

where $\dot{u}(x, t)$ is the particle velocity at x.

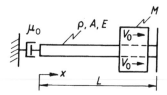

Figure 7.8. Impact of heavy mass against end of elastic bar. The distal end of the bar is restrained by a viscous dashpot generating a force proportional to the rate of extension $\dot{u}(0, t)$.

7.2.1 Boundary Condition at Impact End

At any instant the stress (and particle velocity) at each end of the bar is a sum of an incident (or incoming) and a reflected (or outgoing) wave with magnitudes σ_I and σ_R respectively. Suppose mass M collides against end $x = L$ with incident speed V_0. Subsequently the mass has velocity $V(t)$. While contact persists, this velocity is the same as the particle velocity at the impact end $\dot{u}(L, t)$. The BC is

$$V(t) = \dot{u}_I(L, t) + \dot{u}_R(L, t).$$

The accelerations of the colliding mass depends on the force at the end of the bar:

$$M \, dV/dt = -[\sigma_I(L, t) + \sigma_R(L, t)]A = \rho c_0 A[\dot{u}_I(L, t) - \dot{u}_R(L, t)]$$

$$= \rho c_0 A[2\dot{u}_I(L, t) - V(t)]$$

where the last equality follows from the BC above. Denote the ratio of masses of colliding missile and bar by $\alpha = M/\rho AL$ and solve the ordinary differential equation for the velocity $V(t)$ of the colliding mass,

$$V(t) = \left(\frac{2c_0}{\alpha L}\right) e^{-c_0 t/\alpha L} \left\{ C + \int_0^t \dot{u}_I(L, t') e^{-c_0 t'/\alpha L} \, dt' \right\} \tag{7.16}$$

Initially $V(0) \equiv V_0$ and $\dot{u}_I(L, t) = 0$, so that $C = (\alpha L/2c_0)V_0$ and

$$\frac{V(t)}{V_0} = \frac{\dot{u}(L, t)}{V_0} = e^{-c_0 t/\alpha L}, \qquad c_0 t < 2L. \tag{7.17}$$

During the first transit period, the outgoing wave from the impact end is obtained by extending this BC (i.e. replacing $c_0 t$ by $c_0 t + x - L$):

$$\frac{\dot{u}_R(x, t)}{V_0} = \frac{\sigma_R(x, t)}{\rho c V_0} = e^{-(c_0 t + x - L)/\alpha L}, \qquad c_0 t < 2L. \tag{7.18}$$

7.2.2 Boundary Condition at Dashpot End

The dashpot force is proportional to the velocity at the end $\dot{u}_I(0, t) + \dot{u}_R(0, t)$. If the constant of proportionality is μ_0, equilibrium of forces gives the boundary condition at the restrained end $x = 0$,

$$0 = [\sigma_I(0, t) + \sigma_R(0, t)]A - \mu_0[\dot{u}_I(0, t) + \dot{u}_R(0, t)]$$

$$= \rho c_0 A[\dot{u}_I(0, t) - \dot{u}_R(0, t)] - \mu_0[\dot{u}_I(0, t) + \dot{u}_R(0, t)]. \tag{7.19}$$

Let $\bar{\Gamma} = \mu_0/\rho c_0 A$, and rearrange to obtain the reflection coefficient γ_R:

$$\gamma_R \equiv \frac{\sigma_R(0, t)}{\sigma_I(0, t)} = -\frac{\dot{u}_R(0, t)}{\dot{u}_I(0, t)} = -\frac{\bar{\Gamma} - 1}{\bar{\Gamma} + 1}. \tag{7.20}$$

During the first transit period $L/c_0 < t < 3L/c_0$, the reflected (outgoing) wave at the dashpot end ($x = 0$) is determined from an incident wave (7.19) and the reflection coefficient (7.20),

$$\dot{u}_R(0, t) = -\gamma_R \dot{u}_I(0, t) = -V_0 e^{-(c_0 t - L)/\alpha(L)}, \qquad L/c_0 < t < 3L/c_0. \tag{7.21}$$

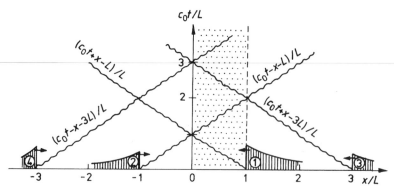

Figure 7.9. Characteristic diagram for wavefront components in extended virtual bar. Location of bar is the grey band $0 < x/L < 1$.

The outgoing wave is obtained by extending this BC (i.e. replacing $c_0 t - L$ by $c_0 t - x - L$):

$$\dot{u}_R(x, t)/V_0 = -\gamma_R e^{-(c_0 t - x - L)/\alpha L}, \qquad x < c_0 t - L \text{ and } L/c_0 < t < 3L/c_0.$$

$$(7.22)$$

7.2.3 Distribution of Stress and Particle Velocity

The stress wave components obtained from continuation of the solution by successive use of boundary conditions are superposed to obtain the impact response. At any time the solution for a region $0 \le x \le L$ is the sum of those components that are generated by the initial impact and subsequent reflections. An additional component enters at alternate ends of region $0 \le x \le L$ after each time period $c_0 t/L$ as illustrated in the characteristic diagram, Fig. 7.9. For each successive wave entering the bar at the impact end, the constant of integration in Eq. (7.16) is determined by continuity of $V(t)$:

$$\frac{\sigma_1}{\rho c_0 V_0} = \frac{\dot{u}_1}{V_0} = \exp\left[\frac{-1}{\alpha}\left(\frac{c_0 t}{L} + \frac{x}{L} - 1\right)\right], \qquad x > L - c_0 t$$

$$\frac{\sigma_2}{\rho c_0 V_0} = \frac{-\dot{u}_2}{V_0} = \gamma_R \exp\left[\frac{-1}{\alpha}\left(\frac{c_0 t}{L} - \frac{x}{L} - 1\right)\right], \qquad x < -L + c_0 t$$

$$\frac{\sigma_3}{\rho c_0 V_0} = \frac{\dot{u}_3}{V_0} = \gamma_R\left\{1 - \frac{2}{\alpha}\left(\frac{c_0 t}{L} + \frac{x}{L} - 3\right)\right\} \exp\left[\frac{-1}{\alpha}\left(\frac{c_0 t}{L} + \frac{x}{L} - 3\right)\right],$$

$$x > 3L - c_0 t$$

$$\frac{\sigma_4}{\rho c_0 V_0} = \frac{-\dot{u}_4}{V_0} = \gamma_R^2\left\{1 - \frac{2}{\alpha}\left(\frac{c_0 t}{L} - \frac{x}{L} - 3\right)\right\} \exp\left[\frac{-1}{\alpha}\left(\frac{c_0 t}{L} - \frac{x}{L} - 3\right)\right],$$

$$x < 3L + c_0 t.$$

The collision terminates at some time $t_f \ge 2L/c_0$ when $\sigma(L, t_f) = 0$. Thereafter a reflected wave at the free end just cancels $\sigma_I(L, t)$.

Figure 7.10 shows the variation of stress with time at the impact end of the bar for two values of the colliding mass ratio, $\alpha = 1$ and 4 and a fixed distal end $\bar{\Gamma} \gg 1$. After initial

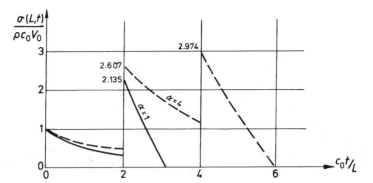

Figure 7.10. Stress at impact end as a function of time for two mass ratios, $\alpha = 1$ and 4.

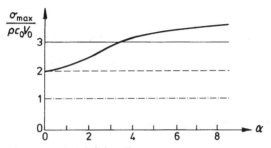

Figure 7.11. Maximum stress at fixed distal end ($\bar{\Gamma} \gg 1$) as function of mass ratio α: elastic wave solution (solid line), dynamic solution assuming uniform strain (dashed line) and quasistatic solution assuming uniform strain (dash-dot line).

incidence the light mass rapidly decelerates and the bodies separate at time $c_0 t/L \approx 3$, while with the heavier mass, separation occurs at $c_0 t/L \approx 6$. At the impact end of the bar there is a jump in stress each time the initial wavefront returns.

At the distal end of the bar the effect of mass ratio α on maximum stress is shown in Fig. 7.11. The maximum stress calculated from the elastic wave solution is larger than the maximum stress obtained from a strength-of-materials analysis where at the instant the colliding mass is brought to rest the initial kinetic energy of the colliding body is distributed uniformly along the bar.

7.2.4 Experiments

John Hopkinson (1872) and Bertram Hopkinson (1913) performed experiments with steel spheres striking the end of an iron wire. The wire was threaded through a hole in the sphere, and the sphere was dropped from a height h. Within a broad range of colliding mass, the minimum height h required to break the wire was independent of the mass M of the sphere. Moreover, the wire broke at the upper (fixed) end and not at the impact end. If the drop height was increased above a limiting value h_{cr}, fracture occurred at the impact end if $h > h_{cr}$.

These results were explained in terms of stress waves in an elastic wire. The stress at the impact end is

$$\sigma(0, t) = \rho c_0 V_0 e^{-c_0 t/\alpha L}$$

while the stress at the clamped end ($\bar{\Gamma} \gg 1$) is twice this value during $L < ct < 3L$. Subsequent reflections increase the stress further. G. I. Taylor (1946) showed that without fracture, the maximum stress occurring at the fixed end is approximately

$$\sigma_{\max}(L, t) = \rho c_0 V_0 \left\{1 + \sqrt{\alpha + 2/3}\right\}.$$

7.3   Impact, Local Indentation and Resultant Stress Wave

When an elastic body strikes the end of an elastic rod, both bodies suffer local indentation as described in Chapter 6. The axial force in the contact region is the source that generates a stress wave in the rod – a wave that propagates away from the impact site during an initial period of motion.

Suppose an elastic sphere B_2 traveling in direction x with an initial velocity V_0 suddenly strikes the domed end of a stationary elastic bar as shown in Fig. 7.12. Let the sphere have radius R_2 and be composed of material with density ρ_2 and elastic modulus E_2, while the domed end of the bar has radius R_1. The bar is slender and is composed of material with density ρ_1 and elastic modulus E_1; the cross-section has area A_1. With these material properties, the mass of the solid sphere is $M_2 = \frac{4}{3}\pi\rho_2 R_2^3$, while the wave speed in the bar is $c_0 = \sqrt{E_1/\rho_1}$.

The stress wave propagating away from the domed end of the bar results in an axial displacement $u(x, t) = f(t - x/c_0)$. Consequently sections of the bar have particle velocity $\dot{u}(x, t)$ and strain $\varepsilon(x, t) \equiv \partial u/\partial x$ given by

$$\dot{u}(x, t) = f'(t - x/c_0), \qquad \varepsilon(x, t) = -c_0^{-1} f'(t - x/c_0).$$

Following incidence there is local indentation $\delta(t)$ that develops at the end of the rod. The rate of indentation is given by

$$\dot{\delta} = V(t) - \dot{u}(0, t) = V(t) - f'(t). \tag{7.23}$$

At the contact surface $x = 0$ the positive axial contact force $F(0, t)$ is related to both the strain and the local indentation:

$$F(0, t) = \frac{A_1 E_1}{c_0} f'(t) = \kappa_s \delta^{3/2} \tag{7.24}$$

where $E_* \equiv E_1 E_2/(E_1 + E_2)$, $R_* \equiv R_1 R_2/(R_1 + R_2)$, and the indentation stiffness is $\kappa_s = \frac{4}{3} E_* R_*^{1/2}$ in accord with Eq. (6.8). It is this contact force which propagates away from the contact surface as a stress wave. After substituting the expression for f' into

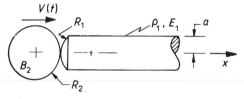

Figure 7.12. Coaxial impact of elastic sphere of mass M_2 striking domed end of slender elastic rod. Axial speed of sphere $V(t)$ is retarded during collision.

(7.22) and noting that $F(0, t) = -M_2 \, dV/dt$, the equation of motion is obtained as

$$0 = \frac{d^2\delta}{dt^2} + \frac{\kappa_s c_0}{A_1 E_1} \frac{d\delta^{3/2}}{dt} + \frac{\kappa_s}{M_2} \delta^{3/2}$$

Making the change of variables $y \equiv \delta/R_*$, $\tau \equiv t\sqrt{\kappa_s R_*^{1/2}/M_2}$ and denoting the ratio of axial rod to local indentation compliance by $2\zeta = (3c_0/2A_1 E_1)\sqrt{M_2\kappa_s R_*^{1/2}}$, we obtain the equation of motion in terms of nondimensional variables:

$$0 = \ddot{y} + 2\zeta y^{1/2}\dot{y} + y^{3/2} \tag{7.25}$$

where $\dot{y} \equiv dy/d\tau$.

To interpret the nondimensional results obtained from this analysis it is useful to note that for a bar with circular cross-section of diameter $2a$ the parameters can be expressed as

$$\zeta^2 = \frac{E_* \rho_2}{\pi E_1 \rho_1} \left(\frac{R_2}{a}\right)^3 \left(\frac{R_*}{a}\right) \quad \text{and} \quad \tau = \frac{t}{R_2}\sqrt{\frac{E_* R_*}{\pi \rho_2 R_2}}$$

while the initial conditions are

$$y(0) = 0, \qquad dy(0)/d\tau = V_0(\pi\rho_2/E_*)^{1/2}(R_2/R_*)^{5/4}$$

Notice that the analysis above requires that the contact radius be small in comparison with the radius of the bar so that the Hertz stress field is representative of contact conditions. Consequently, the asymptotic limit as $R_*/a \to \infty$ of the analysis above does not approach the solution for planar axial impact of a rigid body (Sect. 7.2).

Example 7.3 A sphere of radius $R_2 = a$ collides against the flat end of a bar made of an identical material. Find the contact force $F(t)$ for impact speeds $V_0 = 1$ and 10 m s^{-1}.

Solution For a solid sphere that collides against the flat end of a rod made of the same material,

$$\zeta^2 = (2\pi)^{-1}(R_2/a)^4, \qquad \tau = (2\pi)^{-1}(a/R_2)c_0 t/a, \qquad \dot{y}(0) = (2\pi)^{1/2}V_0/c_0.$$

With these relations, Eq. (7.24) has been used to calculate the interface force as a function of nondimensional time τ for impact speeds of $V_0 = 1$ and 10 m s^{-1}. Figure 7.13 shows that the contact duration decreases slowly with increasing impact speed. In Fig. 7.14 the maximum force is plotted as a function of the nondimensional incident velocity V_0/c_0. Also shown is the 1D stress wave solution for coaxial impact of a rigid mass on the end of a slender elastic bar. For impact between elastic bodies, the local compliance in the contact region reduces somewhat the maximum force at the interface; there is a corresponding increase in the duration of contact. In these elastic collisions against an infinitely long bar, the part of the kinetic energy of relative motion that is lost during collision equals the sum of the strain plus the kinetic energy in the bar.[5]

---

[5] Saint-Venant (1867) mentioned that for a collision at the end of an elastic bar, the stress wave propagating away from the end transports energy away from the contact region; consequently, this energy is not available during restitution to restore the initial kinetic energy of the colliding body.

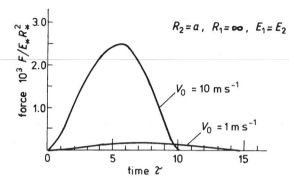

Figure 7.13. Contact force $F/E_* R_*^2$ during impact of steel sphere on flat end of steel bar where $R_2 = a$. Impact speeds of $V_0 = 1, 10 \, \text{m s}^{-1}$ give initial conditions $dy(0)/d\tau = 0.48 \times 10^{-3}, 4.8 \times 10^{-3}$.

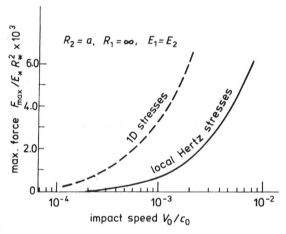

Figure 7.14. Maximum contact force $F_{\max}/E_* R_*^2$ for elastic impact of steel sphere on flat end of steel bar where $R_2 = a$. The dashed line shows the 1D stress wave solution for coaxial impact of a rigid body against the end of an elastic bar; this gives a maximum force $F/E R_2^2 = \pi V_0/c_0$.

7.4   Wave Propagation in Dispersive Systems

The wave equation has a solution which represents a pulse which propagates without distortion. There are variants of the wave equation which also represent disturbances that initially propagate away from a point of initiation, but whose solutions involve distortion of the pulse shape as it propagates. This continuous variation of the pulse shape is known as *dispersion*.

Consider a longitudinal wave propagating in a bar that has continuous elastic support as shown in Fig. 7.15; the bar has cross-sectional area A, Young's modulus E and density ρ. At any point the support provides a distributed force per unit length $-\kappa u(x, t)$ that is proportional to the displacement $u(x, t)$. The equation of motion for this system is a linear Klein–Gordon equation,

$$\frac{\partial^2 u}{\partial x^2} - \frac{\kappa}{E} u = \frac{1}{c_0^2} \frac{\partial^2 u}{\partial t^2}, \qquad c_0^2 = \frac{E}{\rho} \tag{7.26}$$

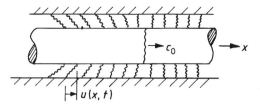

Figure 7.15. Axial wave motion in a bar embedded in an elastic medium.

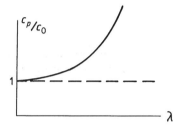

Figure 7.16. Variation of phase velocity c_p with wavelength λ in elastically embedded bar.

If the solution is both spatially and temporally harmonic, we can write $u(x, t) = u_0 \exp$ $[i(kx - \omega t)]$, where u_0 is the amplitude, $i \equiv \sqrt{-1}$, $k = 2\pi/\lambda$ is the wave number (inversely proportional to the wavelength λ), and at any location ω is the frequency of oscillation. Substituting this doubly harmonic form of solution into the equation of motion gives a *dispersion relation* between frequency ω and wave number k,

$$\omega^2/c_0^2 = k^2 + \kappa/E. \tag{7.27}$$

At any time the harmonic disturbance has a wavelength λ. This disturbance propagates with the bar wave speed c_0, resulting in a frequency ω for an observer at any spatial location x; i.e. $kx - \omega t = k(x - c_p t)$. Consequently $\omega = kc_p$, and the *phase velocity* c_p varies with the wavelength of the disturbance:

$$\frac{c_p^2}{c_0^2} = 1 + \frac{\kappa}{E}\left(\frac{\lambda}{2\pi}\right)^2. \tag{7.28}$$

The phase velocity is the speed of propagation of a component of the stress pulse that has wavelength λ. The phase velocity increases with the wavelength as shown in Fig. 7.16.

If one considers propagating waves with wavelength that grows without bound ($\lambda \to \infty$), the corresponding frequency decreases to a cutoff frequency ω_c. Excitations at frequencies smaller that ω_c do not propagate – they are dissipated:

$$u = u_0 e^{-ikx} e^{-i\omega t}, \quad \text{for} \quad \omega < \omega_c. \tag{7.29}$$

This oscillating displacement field which decays exponentially with distance from a source is known as an *evanescent wave*.

7.4.1   Group Velocity

A stone dropped into a pond (Fig. 7.17) causes ripples that propagate outward on the water surface. Rayleigh observed these wave groups and the individual waves within a group. He wrote that "when a group of waves advances into still water, the velocity of

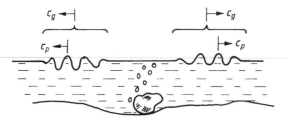

Figure 7.17. Phase and group velocities c_p and c_g for ripples spreading on surface of shallow water.

the group is less than that of the individual waves of which it is composed; the waves appear to advance through the group, dying away as they approach its anterior limit."

This behavior can be understood by examining the superposition of two simple harmonic waves of equal amplitude and slightly different wavelengths:

$$u = u_0 \left[e^{i(kx - \omega t)} + e^{i(k_0 x - \omega_0 t)} \right]$$

$$= 2u_0 e^{i(\bar{k}x - \bar{\omega}t)} \cos(\Delta k \, x - \Delta \omega \, t), \qquad \bar{k} = \frac{k + k_0}{2}, \quad \bar{\omega} = \frac{\omega + \omega_0}{2}$$

Letting $2\,\Delta k = k - k_0$, $2\,\Delta \omega = \omega - \omega_0$ yields the phase velocity c_p and the group velocity c_g:

$$c_p = \bar{\omega}/\bar{k}, \qquad c_g = \Delta \omega / \Delta k \approx d\omega/dk. \tag{7.30}$$

Noting that $\omega = kc_p$ gives $c_g = c_p + k \, dc_p/dk = c_p - \lambda \, dc_p/d\lambda$. In effect there is a carrier frequency $\bar{\omega}$ propagating with the phase velocity and an amplitude modulation term that travels with the group velocity. Three conditions can occur:

(a) $c_p > c_g$: waves appear at the back of a group, travel to the front and disappear. This is normal dispersion.
(b) $c_p = c_g$: no dispersion and no change in pulse shape.
(c) $c_p < c_g$: waves appear at the front of a group and travel to the back.

For the previous example of longitudinal waves in a bar embedded in an elastic medium, $c_g/c_0 = c_0/c_p = (1 + \kappa/Ek^2)^{-1/2}$, which is case (a).

The wave group between two nodes has a certain energy. *In a dispersive medium, energy propagates with the group velocity.* Thus the flexural response of a beam to transverse impact is a wave traveling away from the impact site; this wave has a leading element, or wavefront, that travels at a speed equal to the group velocity c_g.

7.5   Transverse Wave in a Beam

7.5.1   Euler–Bernoulli Beam Equation

Flexural waves in a beam involve translation perpendicular to the axis. From Fig. 7.18 the equation of motion can be written

$$\rho A \, dx \, \frac{\partial^2 w}{\partial t^2} = \frac{\partial^2 M}{\partial x^2} \, dx + q \, dx$$

where $w(x, t)$ is the transverse displacement, $S(x, t)$ is the shear force, $M(x, t)$ is the bending moment, $S = \partial M/\partial x$, and q is a distributed force per unit length. Ignoring any

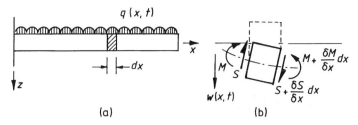

Figure 7.18. (a) Beam loaded by downward pressure $q(x, t)$, and (b) stress resultants acting on deformed element of beam.

effects of rotational inertia of the element dx, the shear force and bending moment are related by $S = \partial M / \partial x$. With the moment–curvature relation $M = -EI\,\partial^2 w / \partial x^2$ (which assumes that plane sections remain plane) one obtains the Euler–Bernoulli equation for beam elements,

$$\frac{\partial^2 w}{\partial t^2} = -c_0^2 k_r^2 \frac{\partial^4 w}{\partial x^4} + \frac{q(x,t)}{\rho A}, \qquad c_0^2 = \frac{E}{\rho} \tag{7.31}$$

and the radius of gyration k_r is related by $k_r^2 = I/A$ to the second moment of area, I, of the cross-section about the neutral axis.

For a free wave solution ($q = 0$), substitute $w = w_0 \exp i(kx - \omega t)$ into the homogeneous Euler–Bernoulli equation. This gives a dispersion relation,

$$\omega = \pm c_0 k_r k^2$$

and a phase speed and group velocity,

$$c_p/c_0 = k_r k, \qquad c_g/c_0 = 2 k_r k.$$

Although this system exhibits normal dispersion, it is unrealistic in that both phase speed and group velocity become large without bound as the wavelength $\lambda \to 0$ (i.e. $k \to \infty$). Experimentally it has been observed that the transverse motions of a beam have a finite maximum wave speed irrespective of the wavelength.

7.5.2 Rayleigh Beam Equation

The Euler–Bernoulli equation is based on the assumption that rotary inertia is negligible, an assumption that is valid for long wavelengths only. To consider short wavelengths the rotary inertia of sections must be included. Rayleigh first developed the equation of motion incorporating rotary inertia for transverse waves in a beam. Considering the equation of rotational acceleration due to moments acting on a differential element of length dx gives

$$S\,dx - \frac{\partial M}{\partial x}\,dx = \rho I\,dx\,\frac{\partial^2 \theta}{\partial t^2}$$

where the (small) rotation of the element is $\theta \approx dw/dx$. Therefore

$$\frac{\partial S}{\partial x} = \frac{\partial^2 M}{\partial x^2} + \frac{\partial}{\partial x}\left(\rho I \frac{\partial^3 w}{\partial x\,\partial t^2}\right) = -\frac{\partial^2}{\partial x^2}\left(EI \frac{\partial^2 w}{\partial x^2}\right) + \frac{\partial}{\partial x}\left(\rho I \frac{\partial^3 w}{\partial x\,\partial t^2}\right).$$

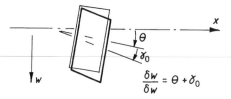

$$\frac{\delta w}{\delta w} = \theta + \delta_0$$

Figure 7.19. Element rotation $\theta + \gamma_0$ caused by combination of flexure and shear warping.

By substitution into (7.31) Rayleigh obtained the following equation for a uniform beam:

$$\frac{\partial^2 w}{\partial t^2} = -c_0^2 k_r^2 \frac{\partial^4 w}{\partial x^4} + k_r^2 \frac{\partial^4 w}{\partial x^2 \partial t^2} + \frac{q(x,t)}{\rho A}. \tag{7.32}$$

This equation has a homogeneous solution $w = w_0 \exp i(kx - \omega t)$ with a dispersion relation

$$\omega = \pm c_0 k_r k^2 \left(1 + k_r^2 k^2\right)^{-1/2}$$

This gives a phase speed and group velocity,

$$c_p/c_0 = k_r k \left(1 + k_r^2 k^2\right)^{-1/2}, \qquad c_g/c_0 = k_r k \left(2 + k_r^2 k^2\right)\left(1 + k_r^2 k^2\right)^{-3/2}$$

While this analytical model provides a bounded phase speed and group velocity for short wavelengths, the asymptotic limits are too large. The assumptions of this analysis are an improvement over the Euler–Bernoulli theory but still are not representative of transients that include short wavelengths.

7.5.3   Timoshenko Beam Equation

Short wavelength deformations ($k_r/\lambda > 0.1$) require more accurate representation of the deformation field than occurs with the Rayleigh beam equation. This equation considers the rotation θ due to flexure $\partial\theta/\partial x = -M/EI$, but it neglects warping of initially plane sections – warping caused by the shear stress distribution over the cross-section. The change of inclination caused by shear varies from zero on the top and bottom surfaces to a maximum γ_0 at the neutral axis. In the deformed configuration, lines that initially were perpendicular to the neutral axis have a maximum rotation $\partial w/\partial x = \theta + \gamma_0$ at the neutral axis, as shown in Fig. 7.19.

Section rotation caused by warping is overestimated by the warping rotation of the neutral axis γ_0. An effective rotation from warping is $\bar{\gamma} = \Xi\gamma_0$, where the constant Ξ is obtained on the basis of average shear stress (or strain) across the section, $0 < \Xi < 1$. Alternatively the coefficient $\Xi = \bar{\gamma}/\gamma_0$ can be obtained by equating the work per unit length done by the shear force $(S\gamma_0/2)$ to the shear strain energy per unit length. These alternative definitions for the *Timoshenko beam coefficient* Ξ result in somewhat different values, as shown in Table 7.2.

The effective shear stress across the section $G\bar{\gamma}$ is related to the shear force S by

$$S = GA\Xi\gamma_0 = GA\Xi(\partial w/\partial x - \theta)$$

while the bending moment is related to flexural rotation of the neutral axis by

$$M = -EI\,\partial\theta/\partial x.$$

Table 7.2. *Timoshenko Beam Coefficient* Ξ *of Various Cross-Sections*

Cross-Section	Ξ	
	From Av. Shear Stress	From Shear Strain Energy
Rectangle	0.67	0.88
Circle	0.75	0.9
Isosceles triangle	0.75	

Hence for a uniform beam with a distributed load $q(x, t)$, Eq. (7.31) for rotation of a differential element results in

$$GA\Xi\left(\frac{\partial w}{\partial x} - \theta\right) + EI\frac{\partial^2\theta}{\partial x^2} = \rho I\frac{\partial^2\theta}{\partial t^2}. \tag{7.33}$$

An independent equation representing transverse motion of a beam can be obtained as

$$GA\Xi\left(\frac{\partial^2 w}{\partial x^2} - \frac{\partial\theta}{\partial x}\right) + q(x, t) = \rho A\frac{\partial^2 w}{\partial t^2}. \tag{7.34}$$

Noting that $dS/dx = GA\,\Xi(\partial^2 w/\partial x^2 - \partial\theta/\partial x) = \rho A\,\partial^2 w/\partial t^2$, Eqs. (7.33) and (7.34) can be combined to give the Timoshenko beam equation;

$$\frac{EI}{\rho A}\frac{\partial^4 w}{\partial x^4} - \frac{I}{A}\left(1 + \frac{E}{G\Xi}\right)\frac{\partial^4 w}{\partial x^2\,\partial t^2} + \frac{\partial^2 w}{\partial t^2} + \frac{\rho I}{GA\Xi}\frac{\partial^4 w}{\partial t^4}$$

$$= \left\{\frac{q(x, t)}{\rho A} + \frac{k_r^2}{\Xi}\frac{\partial^2 q}{\partial t^2} - \frac{EI}{\rho A\Xi}\frac{\partial^2 q}{\partial x^2}\right\}$$

Solutions to the Timoshenko beam equation are most clearly expressed from solutions to the coupled equations for rotational displacement (7.33) and transverse displacement (7.34). Harmonic solutions representing propagating waves $[q(x, t) = 0]$ will be of the form

$$w = w_0 \exp i(kx - \omega t), \qquad \theta = \theta_0 \exp i(kx - \omega t)$$

where the wave number $k = 2\pi/\lambda$. Solutions to (7.33) and (7.34) of this form satisfy

$$0 = iGA\Xi k w_0 - (GA\Xi + EIk^2 - \rho I\omega^2)\theta_0 \tag{7.35a}$$

$$0 = (GA\Xi k^2 - \rho A\omega^2)w_0 + iGA\Xi k\theta_0. \tag{7.35b}$$

These equations provide an amplitude ratio θ_0/w_0 and a characteristic equation for the frequency ω:

$$\frac{\theta_0}{w_0} = \frac{i(GA\Xi k^2 - \rho A\omega^2)}{GA\Xi k} = \frac{iGA\Xi k}{GA\Xi + EIk^2 - \rho I\omega^2} \tag{7.36a}$$

$$0 = \frac{\rho I}{GA\Xi}\omega^4 - \left(\frac{1}{k^2} + \frac{I}{A} + \frac{EI}{GA\Xi}\right)k^2\omega^2 + \frac{EI}{\rho A}k^4 \tag{7.36b}$$

Recalling that $\omega = kc_p$ and $k_r^2 = I/A$, we have the dispersion relation

$$0 = c_p^4 - \left(k_r^2 k^2 + \frac{G\Xi}{E}(1 + k_r^2 k^2)\right)\frac{c_0^2 c_p^2}{k_r^2 k^2} + \frac{G\Xi}{E}c_0^4. \tag{7.37}$$

Limiting cases for short and long wavelengths can be examined after defining a nondimensional phase speed $\bar{c}_p \equiv c_p/c_0$:

(a) Short wavelengths ($k \to \infty$):

$$0 = \bar{c}_p^4 - \left(1 + \frac{G\Xi}{E}\right)\bar{c}_p^2 + \frac{G\Xi}{E}, \qquad \therefore \quad \bar{c}_p^2 = \frac{G\Xi}{E}, 1 + \frac{G\Xi}{E}. \tag{7.38}$$

These limits represent the longitudinal bar and effective shear wave speeds, respectively.

(b) Long wavelengths ($k \to 0$):

$$0 = -\omega^4 + \left(\frac{G\Xi}{E}\right)\frac{c_0^2 \omega^2}{k_r^2}, \qquad \therefore \quad \omega_c^2 = 0, \frac{G\Xi}{E}\frac{c_0^2}{k_r^2}. \tag{7.39}$$

These limits on the admissible range of frequencies represent cutoff frequencies ω_c for predominately flexural and predominately shear deformation, respectively.

The mode of deformation corresponding to the cutoff frequency $\omega_c = (c_0/k_r)\sqrt{G\Xi/E}$ can be obtained by substituting this limiting frequency into the characteristic equations (7.35). Letting $o(\varepsilon)$ denote small terms as $k \to 0$,

$$0 = io(\varepsilon)w_0 - \left(GA\Xi - \rho I\omega_c^2\right)\theta_0$$

$$0 = -\rho A\omega_c^2 w_o + io(\varepsilon)\theta_0.$$

If the frequency approaches the cutoff frequency $\omega_c \to (c_0/k_r)\sqrt{G\Xi/E}$, then $w = 0$ and $\theta = \theta_0 \exp(-i\omega_c t)$, where for long wavelengths both w and θ are independent of the spatial coordinate x.

7.5.4 Comparison of Euler–Bernoulli, Rayleigh and Timoshenko Beam Dynamics

Figure 7.20 compares the phase speed c_p for flexural (and shear) waves that was obtained from the three models of elastic beam deformation in the range of long wavelengths $0 < h/\lambda < 2$. This comparison was calculated for a beam of rectangular cross-section with depth h and Poisson ratio $\nu = 0.3$. Notice that for short wavelengths ($\lambda \to 0$), the wave propagation speed has an upper limit in both the Rayleigh and Timoshenko theories. In the latter case, the group velocity approaches the Rayleigh surface wave speed $c_R/c_0 = 0.56$, which is slightly less than the shear wave speed $c_2/c_0 = 0.57$. Throughout this range of wavelengths the Timoshenko theory is within 1% of the exact solution given by a 2D elasticity solution. It should not be expected however, that the Timoshenko theory is accurate for wavelengths that are much smaller than the cross-section depth. Cremer and Heckl (1973) explained why higher order shear and through-thickness deformation modes need to be considered if the wavelength is very small. Calculations of the beam transient response to a suddenly applied transverse force were published by Boley (1955).

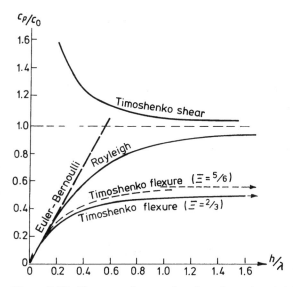

Figure 7.20. Phase speed c_p as a function of wavelength λ for Euler–Bernoulli, Rayleigh and Timoshenko beam theories.

PROBLEMS

7.1 A uniform elastic bar with Young's modulus E, density ρ and cross-sectional area A has pressure $p(t)$ applied to the end $x = 0$ for a period of time τ. The pressure $p(t)$ satisfies $p = 2p_0 t/\tau$ for $t/\tau < 1/2$, $p = 2p_0(1 - t/\tau)$ for $1/2 < t/\tau < 1$ and $p = 0$ for $1 < t/\tau$. Obtain expressions for the strain energy $U(\tau)$ and kinetic energy $T(\tau)$ in the bar at time $t = \tau$. Compare these parts of the total mechanical energy with the work $W(\tau)$ done by the applied force that acts on the end of the rod.

7.2 Two long uniform elastic bars are moving in the same direction before they collide end on. The first bar, of length $L/2$, is moving with initial speed V_1, and the second bar, of length L, is moving with initial speed V_2. The bars are identical in all other respects. Find the intensity of the stress due to longitudinal impact, and determine how far the interface between the bars moves between start and end of contact. On a characteristic diagram plot the position of the bar ends as a function of time.

7.3 For the previous problem show that at time $t = L/4c_0$ the sum of the kinetic and strain energies in the two bars is equal to the kinetic energy of the system before the collision. Also show that at this time the axial momentum of the system is equal to the momentum before collision. Identify the collision periods in which energy and/or momentum are conserved.

7.4 A thin aluminum alloy bar of length L has an axial velocity V_0 before impact with one end of an initially stationary, equal diameter, steel bar of length $3L$. The steel bar is suspended by long, flexible strings which keep the axes of the two bars aligned. Denote the elastic wave speed in the aluminum and steel bars by c_a and c_s respectively. Plot the position of the ends of each bar for time $0 < t < 4L/c_a$. Show that contact between the rods ceases at time $t = 2L/c_a$, and calculate the mean velocity of each rod at this time. Determine if momentum and energy are conserved in this impact, and account for any losses. What is the impact velocity if the largest stress equals the yield stress in one of the bars?

7.5 A pile driver consists of a steel block, the *monkey*, which is dropped onto the top of a vertical timber pile. The length of the pile is L, and the impact speed is V_0. Assuming that the steel block is rigid compared with the timber, show that the initial retardation of the monkey is $(c_0 V_0/\alpha L) - g$, where g is the acceleration due to gravity, M is the mass of the monkey, $\alpha = M/\rho AL$ is the ratio of the mass of the monkey to the mass of the pile, and c_0 is the elastic bar wave speed for the timber. Assuming the foot of the pile to be rigidly fixed, prove that the retardation of the monkey just after time $2L/c_0$ from impact is given by

$$\frac{2c_0 V_0}{\alpha L} + \left(\frac{c_0 V_0}{\alpha L} - g\right) \exp(-2/\alpha).$$

7.6 A stress wave is traveling away from the end $x = 0$ of an elastic bar of length L and cross-sectional area A. Beginning at time $t = 0$, the end of the rod is driven axially at a velocity $V(t) = V_0 e^{-\alpha t}$.

(a) Find an expression for the particle velocity in the bar.

(b) The end of the rod at $x = L$ has an axial damper that provides a damping force $F(t) = -\mu_0 \dot{u}(L, t)$. Show that the amplitude $\dot{u}_R$ of the wavefront reflected from the damped end can be expressed as

$$\frac{\dot{u}_R}{V_0} = \frac{1 - \mu_0/\rho c_0 A}{1 + \mu_0/\rho c_0 A}.$$

For what value of the damping constant μ_0 is the wave completely absorbed so there is no reflection?

(c) Obtain an expression for the particle distribution for times $t > L/c_1$. Sketch this particle velocity distribution at time $t = 3L/2c_1$. For any location x give an upper limit for the time t when this expression is applicable.

7.7 A uniform, slender elastic bar of length L, cross-sectional area A, mass M and elastic wave speed c_0 is moving initially in an axial direction with speed $2V_0$. At time $t = 0$ one end of the moving bar collides against the end of an identical bar. The second bar is initially stationary and the bars are oriented coaxially. A heavy particle of mass M/α rests against the free end of the stationary bar at $x = L$, where spatial coordinate x is measured from the interface and directed parallel to the initial velocity. Axial motions are unconstrained.

(a) Produce sketches of the distribution of particle velocity $\dot{u}(x, t)$ and stress $\sigma(x, t)$ in the bars at time $t = L/2c_0$. On this sketch indicate the amplitude of any discontinuities. The sketches should be labeled so that positive values are upward.

(b) Calculate the speed of the heavy particle, $V(t)$; show that this speed equals

$$\frac{V}{V_0} = \begin{cases} 0, & t < L/c \\ 2 - 2\exp[\alpha(1 - ct/L)], & L/c < t < 3L/c. \end{cases}$$

(c) Sketch the distribution of particle velocity and stress in the bars at time $t = 3L/2c$; on this sketch indicate the amplitudes of any discontinuities.

(d) Find the first time when stress vanishes at the interface between the rods.

7.8 Two coaxial uniform, slender elastic bars are joined by a heavy particle with mass M'. Each bar has length L, cross-sectional area A, mass density ρ and elastic wave speed c. At time $t = -L/c$ an axial stress σ_0 is suddenly applied at a free end of one bar, where $x = 0$; i.e., $\sigma(0, t) = \sigma_0, t > -L/c$. The interface between the bars can support tension, and axial motions are unconstrained.

(a) For the stress wave incoming to the interface, the incident stress σ_I is related to the incident particle velocity $\dot{u}_I$ by $\sigma_I = -\rho c \dot{u}_I$. Write similar equations that relate reflected and transmitted stresses σ_R, σ_T to the reflected and transmitted particle velocities $\dot{u}_R, \dot{u}_T$.

(b) For a wave incident on the interface, express the continuity conditions across the interface in terms of a mass ratio $\alpha = M'/\rho A L$. These equations relate incident, reflected and transmitted stresses and particle velocities across the interface.

(c) Obtain the reflection and transmission coefficients $\gamma_R = \sigma_R/\sigma_I$ and $\gamma_T = \sigma_T/\sigma_I$ for waves incident on the interface, and show that they can be expressed as functions of $ct/\alpha L$.

(d) For mass ratio $\alpha = 1$, sketch the stress distribution along the rod at time $t = 0.5(L/c)$, i.e. after the wavefront has passed the interface. Indicate on your sketch the amplitude of stresses at the interface.

7.9 A long freight train has n identical wagons (freight cars), each of mass M and length l across the buffers. The buffer springs between two wagons are of combined stiffness S. The couplings put some initial compression in the springs.

(a) By considering the forces acting on a wagon and/or by analogy with a uniform bar, show that waves of tension and compression travel down the train with a velocity

$$c \approx l\sqrt{S/M}.$$

(b) If the driver suddenly applies the brakes to give the locomotive a severe uniform deceleration $-\dot{V}$ for a time $nl/4c$, suggest when and where is there a possibility of a coupling breaking in tension. Take the locomotive to be massive compared with the train of wagons.

Dispersive Waves

7.10 (a) A 1D system has a dispersion relation $\omega = \omega(k)$. Two waves of equal amplitude but slightly different frequencies travel along the system, in the same direction. Show that the *envelope* of the resulting wave pattern travels (approximately) at the group velocity.

(b) Consider a finite section of the same system, of length L. Make a plausible guess for the approximate form of standing waves (vibration modes) in the system. For modes well up the modal series for this system, what is the approximate interval between the wave numbers of successive modes? Hence show that the spacing (in hertz) between adjacent mode frequencies is given approximately by $\Delta f \approx c_g/2L$, where c_g is the group velocity.

7.11 A simple model for the vertical vibration of a railway rail on its bed of ballast is the accompanying drawing. An Euler–Bernoulli beam of flexural rigidity EI and mass per unit length m is supported on an *elastic foundation*, which exerts a restoring force $Sy(x, t)$ per unit length when the displacement is $y(x, t)$. The equation of motion is

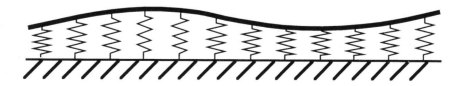

thus

$$EI\frac{\partial^4 y}{\partial x^4} + Sy = -m\frac{\partial^2 y}{\partial t^2}.$$

(a) Obtain the dispersion relation. For each (real) value of frequency ω, describe the possible propagating and/or evanescent waves on the rail.

(b) Calculate the phase velocity and the group velocity for the traveling waves, and plot both (on the same axes) as a function of frequency.

(c) A train wheel rolls along the rail and applies a broadband input force to the rail (arising from roughness of the rail surface and wheel tread). Describe (without calculations) the qualitative response pattern of the rail.

7.12 For axial displacement $u(x, t)$ in an elastic circular rod of radius a made of material with density ρ, Young's modulus E, and Poisson's ratio v, Love derived an approximate expression for the effect of lateral inertia on a longitudinal wave of axial displacement,

$$\frac{\partial^2 u}{\partial x^2} + \frac{\rho v^2 a^2}{2E}\frac{\partial^4 u}{\partial x^2 \partial t^2} = c_0^{-2}\frac{\partial^2 u}{\partial t^2}.$$

Find the dispersion relation by considering a solution of the form $u(x, t) = A\exp[i(\omega t - kx)]$. Calculate the phase velocity c_p and the group velocity c_g, and plot a graph of $\bar{c} = c_p/c_0$ as a function of the radius to wavelength ratio $\bar{k} = vak/\sqrt{2}$ in the range $0 < \bar{k} < 5$.

(a) Use your graph to delimit the circumstances where the elementary theory of longitudinal wave propagation in a bar is satisfactory.

(b) Compare your solution with the 2D Pochhammer–Chree solution for longitudinal waves in an elastic rod, and estimate the maximum value of $\bar{k}$ where Love's theory is satisfactory.

7.13 Two semi-infinite Euler–Bernoulli beams are connected together end to end by a pinned joint. Both beams have flexural rigidity EI and mass per unit length m. What are the appropriate boundary conditions for the displacement and its derivatives in the vicinity of the joint? Hence calculate the reflection and transmission coefficients for a flexural traveling wave that is incident on the joint from the left. Show also that the amplitudes of the evanescent waves on either side of the joint are equal; do this by finding reflection and transmission ratios, γ_{ER} and γ_{ET}, for evanescent waves.

CHAPTER 8

Impact on Assemblies of Rigid Elements

Physics is popularly deemed unnecessary for the astronomer, but
truly it is in the highest degree relevant to the purpose of this branch of
philosophy, and cannot indeed, be dispensed with by the astronomer.
For astronomers should not have absolute freedom to think up anything they please without reason; on the contrary, you should give
causas probabiles for your hypotheses which you propose as the
true cause of the appearances, and thus establish in advance the
principles of your astronomy in a higher science, namely physics or
metaphysics.

J. Kepler, *Epitome Astronomiae Copernicanae*, transl. N. Jardine,
The Birth of History and Philosophy of Science, CUP (1984)

Impact against a mechanism composed of nearly rigid bodies is a feature of many
practical machines. These systems may include mechanisms where the relative velocities
at joints are initially zero and finally must vanish or they can be another type of system
such as a gear train or an agglomerate of unconnected bodies where at each contact the
normal component of terminal relative velocity must be separating. These two classes of
multibody impact problems – mechanisms and separate bodies that are touching – are
distinguished from analyses of two-body impacts by the addition of constraint equations
that describe limitations on relative motion at each point of contact between bodies.
These constraint equations express the linkage between separate elements. Books on
dynamics typically analyze the impulsive response of systems composed of rigid bodies
linked by frictionless or nondissipative pinned joints. In these systems, pairs of bodies
are connected by a joint or hinge that imposes a *constraint* on the relative velocity at the
point of connection. A common assumption is that during impact the relative velocity at
joints remains negligible; this assumption applies if the compliance of the joints is small
in comparison with the compliance at the point of external impact. Ordinarily, however,
there is no reason for a point of external impact to be more compliant than all other points
of contact within the system.

Any connection between bodies can be represented by a constraint equation – an
equation that expresses the limitation on relative motion provided by the physical link. One
class of constraint equations represents pinned joints; these *bilateral* constraint equations
require that at the jth point of contact the relative velocity vanishes at initiation and
termination of impact: $\mathbf{v}_j(0) = \mathbf{v}_j(p_f) = \mathbf{0}$. A second class of constraint equations
represents contact where there is no physical connection; these *unilateral* constraint

equations require that at separation the normal component of relative velocity at each point of contact is positive. The latter form of constraint arises in systems where bodies are in initial contact but not physically linked.

For both classes of constraints, the action due to an external impact propagates away from the impact point. If the point of external impact is much more compliant than other contact points, it is satisfactory to assume that the reaction at each point of constraint acts simultaneously with the contact force at the point of external impact. For all other distributions of contact compliance, however, it is necessary to return to discrete modeling of the local compliance of each contact region in order to follow the chain reaction that transmits energy through the compliant elements of the system. Energy propagates outward from the point of external impact and travels through the connecting joints to adjacent bodies.

Irrespective of the form of constraint equations, convenient methods for analyzing the dynamics of multibody systems employ generalized coordinates which are associated with Lagrange's equations of motion. Lagrangian methods are advantageous for multibody systems because they eliminate any requirement to consider reaction forces arising at the constraints. This chapter employs Lagrangian dynamics to analyze the behavior of systems of compact colliding bodies – compact or stocky bodies that have small structural compliance in comparison with the compliance at points of connection between bodies in the system.

8.1   Impact on a System of Rigid Bodies Connected by Noncompliant Bilateral Constraints

Analysis of the dynamics of a mechanism composed of an assembly of rigid bodies that are connected by frictionless pinned joints (nondissipative, noncompliant bilateral constraints) is based on the following assumptions: (a) the time of contact is brief, so that there is no change in configuration during impact; and (b) the reactions at points of contact occur simultaneously. The latter assumption is appropriate if the compliance of each joint and body within the system is small in comparison with the contact compliance at any site of external impact.

8.1.1   Generalized Impulse and Equations of Motion

Let S be a set of n particles with the jth particle located at a position vector $\mathbf{r}_j$, $j = 1, \ldots, n$; each particle is subject to external impulse $\mathbf{p}_j$ applied at the instant of impact. The velocity $\mathbf{V}_j$ of the jth particle is the rate of change of the position vector, $\mathbf{V}_j(\mathbf{p}_j) \equiv d\mathbf{r}_j/dt$; the particle velocity is a function of the impulse $\mathbf{p}_j$. Suppose the particle velocities are subject to constraints that are represented by $3n - \bar{m}$ holonomic constraint equations; e.g., let there be a fixed distance between the jth and kth particles, so that $(\mathbf{V}_k - \mathbf{V}_j) \cdot (\mathbf{r}_k - \mathbf{r}_j) = 0$. Also there can be $\bar{m} - m$ nonintegrable (nonholonomic) constraint equations $\sum_{r=1}^{\bar{m}} a_{sr} \dot{q}_r + b_s$, $s = 1, \ldots, \bar{m} - m$. Then the particle velocities can be expressed in terms of generalized speeds $\dot{q}_r$, $r = 1, \ldots, m$, and time t; i.e., $\mathbf{V}_j = \mathbf{V}_j(\dot{q}_1, \dot{q}_2, \ldots, \dot{q}_m, t)$, where m is the minimum number of variables required to define the constrained motion of the system. The number m of generalized speeds $\dot{q}_r$ equals the number of degrees of freedom of the system.

Virtual displacements $\delta\mathbf{r}_j$ are a displacement field that is compatible with the displacement constraints. Similarly virtual velocities $\delta\mathbf{V}_j$ are compatible with the velocity constraints of the system. The virtual velocity of the jth particle can be expressed in terms of generalized speeds $\dot{q}_r$ as[1]

$$\delta \mathbf{V}_j = \sum_{r=1}^{\bar{m}} \frac{\partial \mathbf{V}_j}{\partial \dot{q}_r} \delta \dot{q}_r. \tag{8.1}$$

During impact on a system of rigid bodies the virtual differential of work $\delta(dW)$ done by external forces $d\mathbf{p}_j / dt$ is used to define a *differential of generalized impulse* $d\Pi_r$:

$$\delta(dW) = \sum_{j=1}^{n} \sum_{r=1}^{\bar{m}} d\mathbf{p}_j \cdot \frac{\partial \mathbf{V}_j}{\delta \dot{q}_r} \delta \dot{q}_r = \sum_{r=1}^{\bar{m}} (d\Pi_r) \delta \dot{q}_r \tag{8.2}$$

where $d\Pi_r \equiv \sum_{j=1}^{n} d\mathbf{p}_j \cdot (\partial \mathbf{V}_j / \partial \dot{q}_r)$.[2] It is important to recognize that generalized active forces $F_r = \sum_{j=1}^{n} (d\mathbf{p}_j / dt) \cdot (\partial \mathbf{V}_j / \partial \dot{q}_r)$ are the only forces that contribute to the differential of generalized impulse; other forces, which do not contribute, include any equal but opposite forces of interaction at rigid (i.e. noncompliant) constraints and external body forces or pressures which remain constant during impact.

A system of n particles connected by $3n - m$ velocity constraints has a kinetic energy T which is a scalar that varies with the applied impulse. This kinetic energy can be expressed either in a global coordinate system or as a function of the generalized speeds:

$$T = \frac{1}{2} \sum_{j=1}^{n} M_j (\mathbf{V}_j \cdot \mathbf{V}_j) = \frac{1}{2} \sum_{r=1}^{\bar{m}} \sum_{s=1}^{\bar{m}} m_{rs} \dot{q}_r \dot{q}_s \tag{8.3}$$

where M_j is the mass of the jth particle. The inertia matrix m_{rs} for the constrained system with generalized speeds $\dot{q}_r$ can be obtained from the expression for the kinetic energy of the constrained motion.

For a system subject to velocity constraints, the equations of motion in terms of generalized speeds $\dot{q}_r$ are obtained directly from the kinetic energy T and the differential of generalized impulse $d\Pi_r$.

For a system subject to a differential of generalized impulse $d\Pi_r$ and a set of $3n - m$ velocity constraints, the equations of motion in terms of m independent generalized speeds $\dot{q}_r$ are obtained as

$$d\frac{\partial T}{\partial \dot{q}_r} = d\Pi_r. \tag{8.4}$$

PROOF For external impulse $\mathbf{p}_j$ acting on a set of n particles, the differential form of Newton's second law is expressed as

$$d\left(\sum_{j=1}^{n} M_j \mathbf{V}_j \right) = d(M\hat{\mathbf{V}}) = \sum_{j=1}^{n} d\mathbf{p}_j$$

[1] If the velocity of the jth particle is expressed as $\mathbf{V}_j = \mathbf{V}_j(\dot{q}_1, \dot{q}_2, \ldots, \dot{q}_m, t)$, note the distinction between virtual velocity $\delta \mathbf{V}_j$ and the differential of velocity $d\mathbf{V}_j = \sum_{r=1}^{m} \{(\partial \mathbf{V}_j / \partial \dot{q}_r) \delta \dot{q}_r + (\partial \mathbf{V}_j / \partial t) dt\}$.

[2] This formulation requires that the differential of impulse at each compliant contact $d\mathbf{p}_j$ be included in calculating the generalized impulse in a mechanism consisting of rigid bodies linked by compliant constraints.

where the system of particles has mass $M = \sum_{j=1}^{n} M_j$ and the center of mass of the system has velocity $\hat{\mathbf{V}}$. Multiplying each side of the equation by the set of virtual velocity coefficients $\partial \mathbf{V}_j / \partial \dot{q}_r$, we obtain

$$\sum_{j=1}^{n} M_j \frac{\partial \mathbf{V}_j}{\partial \dot{q}_r} \cdot d\mathbf{V}_j = d\left(\frac{\partial T}{\partial \dot{q}_r}\right) - \sum_{j=1}^{n} M_j \mathbf{V}_j \cdot d\left(\frac{\partial \mathbf{V}_j}{\partial \dot{q}_r}\right) = d\Pi_r - \sum_{j=1}^{n} \mathbf{p}_j \cdot d\left(\frac{\partial \mathbf{V}_j}{\partial \dot{q}_r}\right).$$

Since $\partial \mathbf{V}_j / \partial \dot{q}_r$ depends solely on the configuration of the system and this does not change during the instant of impact, the differential of these virtual velocity coefficients must vanish $[d(\partial \mathbf{V}_j / \partial \dot{q}_r) = \mathbf{0}]$, giving

$$d\frac{\partial T}{\partial \dot{q}_r} = d\Pi_r. \qquad\qquad\qquad\qquad\qquad \text{q.e.d.}$$

The expression $d(\partial T / \partial \dot{q}_r)$ is termed a *differential of generalized momentum*[3]; it can be expressed in terms of generalized speeds as $d(\partial T / \partial \dot{q}_r) = m_{rs}\, d\dot{q}_s$.

Example 8.1   Six identical uniform rigid bars form a regular hexagon with frictionless pinned joints connecting adjacent bars at each corner. The hexagon lies on a smooth horizontal plane surface and is initially stationary. A transverse impulse $p\mathbf{n}$ acts at point A located at the center of one bar; the impulse acts in a direction normal to the bar and tangent to the plane. For this impulsive load, find the speed of the center of mass of the mechanism $\hat{V}$ and the speed V_A of the loaded side immediately after the impulse; thus show that $V_A / \hat{V} = 20/11$.

   Solution   Let each bar have mass $M_j = M$ and length $L_j = L$, and let $\dot{\theta}$ be the angular speed of each side bar (symmetry reduces the problem to one with two degrees of freedom, so there are only two independent generalized speeds, $\dot{q}_1 = \hat{V}$ and $\dot{q}_2 = L\dot{\theta}$). From (1.11) we obtain that the total kinetic energy T for this assembly of rigid bars is the sum of the kinetic energies of the bars: $T = \frac{1}{2}\sum(M_j \hat{\mathbf{V}}_j \cdot \hat{\mathbf{V}}_j + \boldsymbol{\omega}_j \cdot \hat{\mathbf{I}}_j \cdot \boldsymbol{\omega}_j)$; for symmetrical motion this hexagonal mechanism has a kinetic energy

$$T = \tfrac{1}{2} M\left(6\hat{V}^2 + \tfrac{11}{6}(L\dot{\theta})^2\right).$$

This symmetrical problem has two degrees of freedom. The impulse $p\mathbf{n}$ is applied at point A; this point has a velocity that can be expressed as $\mathbf{V}_A = (\hat{V} + L\dot{\theta}/2)\mathbf{n}$.
   Virtual velocity coefficients:

$$\frac{\partial \mathbf{V}_A}{\partial \hat{V}} = \mathbf{n}, \qquad \frac{\partial \mathbf{V}_A}{\partial (L\dot{\theta})} = \tfrac{1}{2}\mathbf{n}.$$

Generalized impulses:

$$\Pi_1 = p\mathbf{n} \cdot \mathbf{n} = p, \qquad \Pi_2 = p\mathbf{n} \cdot \frac{1}{2}\mathbf{n} = \frac{p}{2}.$$

---

[3] For an impulsive load it is not useful to define a Lagrangian function $L = T - U$ which incorporates the potential energy U of conservative active forces. These forces do no work during the instant that an impulse is applied, so they do not affect Eq. (8.4).

Equations of motion:

$$d\frac{\partial T}{\partial \dot{q}_1} = d\Pi_1 \quad \Rightarrow \quad \hat{V} = \frac{1}{6}M^{-1}p$$

$$d\frac{\partial T}{\partial \dot{q}_2} = d\Pi_2 \quad \Rightarrow \quad L\dot{\theta} = \frac{6}{11}M^{-1}p.$$

Speed of point A after application of impulse:

$$V_A = \hat{V} + \frac{L\dot{\theta}}{2} = \frac{20}{66}M^{-1}p.$$

$$\therefore \quad V_A/\hat{V} = 20/11.$$

8.1.2 Equations of Motion Transformed to Normal and Tangential Coordinates

In the preceding example a specified impulse was applied to the system. For two or more bodies that come into contact with a relative velocity at some contact point, additional relations are required to describe the properties of the contact region. In order to apply laws of impact and friction that are related to normal or tangential components of relative velocity at the contact point, the equations of motion must be transformed to components in the normal and tangential directions.

For a system of rigid bodies that are linked by nondissipative, noncompliant joints, let the velocity at contact point C be separated into a component $\mathbf{V_e}$ in the tangent plane and a component $V_3\mathbf{n}_3$ normal to the tangent plane; i.e., $\mathbf{V_C} = \{\mathbf{V_e}, V_3\}^T$. For a differential impulse $d\mathbf{p} = \{d\mathbf{p_e}, dp\}^T$ applied at C the equations of relative motion can be expressed as

$$\begin{Bmatrix} d\mathbf{V_e} \\ dV_3 \end{Bmatrix} = [N]\begin{Bmatrix} d\mathbf{p_e} \\ dp \end{Bmatrix}. \tag{8.5}$$

Batlle (1993, 1996) has shown that

$$[N] = \left\{\frac{\partial \mathbf{V_C}}{\partial \dot{q}_r} \cdot \mathbf{n}\right\}^T [m]\left\{\frac{\partial \mathbf{V_C}}{\partial \dot{q}_s} \cdot \mathbf{n}\right\} = \begin{bmatrix} [b] & \mathbf{h}^T \\ \mathbf{h} & a \end{bmatrix} \tag{8.6}$$

where $[m] = m_{rs}$ are generalized inertia coefficients and $\partial \mathbf{V_C}/\partial \dot{q}_r$ are virtual velocity coefficients for the system.

Impact is initiated when two systems come together with a relative velocity $\mathbf{v} = \mathbf{v_e} + v_3\mathbf{n}_3 \equiv \mathbf{V_C} - \mathbf{V_{C'}}$ between the contact point C and the coincident point C'. If there is slip at C (i.e., $\mathbf{n}_3 \times \mathbf{v} \neq 0$), then for dry friction with a coefficient of friction μ, Coulomb's law can be used to relate the tangential components of the contact force $d\mathbf{p_e}$ to the normal force dp; i.e., $d\mathbf{p_e} = -\mu \hat{\mathbf{s}} dp$, $\hat{\mathbf{s}} \equiv [(\mathbf{n}_3 \times \mathbf{v}) \times \mathbf{n}_3]/|(\mathbf{n}_3 \times \mathbf{v}) \times \mathbf{n}_3|$. In this case there is a set of $n + n'$ generalized speeds for the combined systems. Hence during slip, the equations of relative motion at a point of external impact can be expressed as

$$\begin{Bmatrix} d\mathbf{v_e}/dp \\ dv_3/dp \end{Bmatrix} = \begin{Bmatrix} \mathbf{h} - \mu[b]\hat{\mathbf{s}} \\ a - \mu[\mathbf{h}]^T\hat{\mathbf{s}} \end{Bmatrix}. \tag{8.7}$$

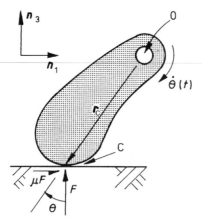

Figure 8.1. Rigid compound pendulum at angle of inclination θ when it strikes against an inelastic half space where the energetic coefficient of restitution is e_* and the coefficient of friction is μ.

Here the terms on the right side that come from $[N]$ are inertia properties of the system for a point of external impact C.

The next two examples consider systems that collide against a massive half space; i.e., the body B' has an indefinitely large mass. In this case $M'^{-1} \to 0$, so that the mass M' does not appear in the equations of relative motion for the contact point C.

Example 8.2 A compound pendulum with mass M pivots around a frictionless pin O, and the tip strikes a rough half space at contact point C. The position of C relative to O is $\mathbf{r}_C = -r_1\mathbf{n}_1 - r_3\mathbf{n}_3$, where the unit vector $\mathbf{n}_1$ is parallel to the tangent plane and $\mathbf{n}_3$ is the common normal direction as shown in Fig. 8.1. The pendulum has radius of gyration k_r for O and at the contact point the energetic coefficient of restitution is e_*. Find the ratio of terminal to incident angular velocities $\dot\theta_f / \dot\theta_0$ assuming that tangential compliance is negligible and that friction at C is represented by Coulomb's law with a coefficient of friction μ.

Solution

Kinetic energy:

$$T = \frac{1}{2}Mk_r^2\dot\theta^2 = \frac{1}{2}M\dot q_1^2, \qquad \dot q_1 \equiv k_r\dot\theta.$$

Velocity of C:

$$\mathbf{V}_C = \mathbf{V}_O + \boldsymbol\omega \times \mathbf{r}_C = \frac{r_3\dot q_1}{k_r}\mathbf{n}_1 - \frac{r_1\dot q_1}{k_r}\mathbf{n}_3.$$

The differential of impulse $d\mathbf{p} = dp_1\,\mathbf{n}_1 + dp_3\,\mathbf{n}_3$ at contact point C satisfies the Amontons–Coulomb law:

$$\left.\begin{array}{r}dp_1 = \mu\,dp \\ -\mu\,dp < dp_1 < \mu\,dp \\ dp_1 = -\mu\,dp\end{array}\right\} \quad \text{for} \quad \left\{\begin{array}{l}\mathbf{V}_C \cdot \mathbf{n}_1 < 0 \\ \mathbf{V}_C \cdot \mathbf{n}_1 = 0 \\ \mathbf{V}_C \cdot \mathbf{n}_1 > 0\end{array}\right\}, \qquad \text{where} \quad dp \equiv dp_3.$$

The differential of the generalized impulse is

$$d\Pi_1 = \frac{r_3 dp_1}{k_r} - \frac{r_1 dp_3}{k_r}$$

$$= \begin{cases} -k_r^{-1}(r_1 + \mu r_3)\,dp, & \mathbf{V}_C \cdot \mathbf{n}_1 < 0 \\ -k_r^{-1}(r_1 - \mu r_3)\,dp, & \mathbf{V}_C \cdot \mathbf{n}_1 > 0 \end{cases}.$$

The equations of motion are obtained from (8.4) after recognizing that the slip reverses in direction at impulse p_c simultaneously with the transition from compression to restitution. After integration, one obtains the following generalized speed as a function of normal impulse p:

$$\dot{q}_1(p) = \begin{cases} \dot{q}_1(0) - (k_r M)^{-1}(r_1 + \mu r_3)p, & p < p_c \\ -(k_r M)^{-1}(r_1 - \mu r_3)p, & p_c < p < p_f \end{cases}.$$

Notice that slip reversal requires $\mu < \bar{\mu} \equiv r_1/r_3$; otherwise the pendulum sticks in the compressed configuration.

Compression impulse p_c from $\mathbf{n}_3 \cdot \mathbf{V}_C(p_c) = 0$:

$$p_c = k_r(r_1 + \mu r_3)^{-1} M \dot{q}_1(0).$$

Work of normal impulse during compression $W_3(p_c)$:

$$W_3(p_c) = \int_0^{p_c} -\frac{r_1}{k_r}\left[\dot{q}_1(0) - \frac{r_1 + \mu r_3}{k_r M}p\right]dp = -\frac{r_1^2 + \mu r_1 r_3}{2k_r^2 M}p_c^2.$$

Work of normal impulse during restitution $W_3(p_f) - W_3(p_c)$:

$$W_3(p_f) - W_3(p_c) = \int_{p_c}^{p_f} \frac{r_1}{k_r}\left[\frac{r_1 - \mu r_3}{k_r M}p\right]dp = \frac{r_1^2 - \mu r_1 r_3}{2k_r^2 M}(p_f^2 - p_c^2).$$

Energetic coefficient of restitution:

$$e_*^2 = -\frac{W_3(p_f) - W_3(p_c)}{W_3(p_c)} = \frac{r_1^2 - \mu r_1 r_3}{r_1^2 + \mu r_1 r_3}\left[\frac{p_f^2}{p_c^2} - 1\right].$$

Terminal impulse p_f as function of angle of inclination $\theta = \cot^{-1}(r_3/r_1)$:

$$\frac{p_f}{p_c} = 1 + e_*\sqrt{\frac{r_1^2 + \mu r_1 r_3}{r_1^2 - \mu r_1 r_3}} = 1 + e_*\sqrt{\frac{1 + \mu \cot \theta}{1 - \mu \cot \theta}}.$$

Ratio of final to initial angular speed:

$$\frac{\dot{\theta}_f}{\dot{\theta}_0} = \frac{\dot{q}_1(p_f)}{\dot{q}_1(0)} = -e_*\sqrt{\frac{1 - \cot \theta}{1 + \cot \theta}}.$$

Figure 8.2 illustrates the ratio of angular speeds as a function of the eccentricity angle θ of the pendulum at impact and the energetic coefficient of restitution e_*. The result shows the effect of energy dissipated by friction even if the bodies are elastic. At small angles of eccentricity θ the work done by friction is large in comparison with the work done

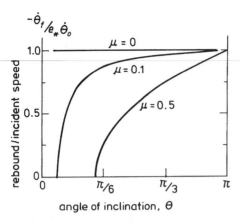

Figure 8.2. Rebound of compound pendulum as a function of configuration angle θ for coefficients of friction $\mu = 0.1$ and 0.5.

by the normal component of contact force. Also for small θ there is no rebound (i.e., the contact sticks) if friction is sufficiently large ($\mu \geq \bar{\mu} = \tan \theta$) (Stronge, 1991).

A more complex problem of impact at the tip of a double compound pendulum was proposed by Kane and Levinson (1985). The double pendulum is swinging when the tip strikes against a rough inelastic half space. This problem has generated renewed interest in analytical methods for representing impact with friction.

Example 8.3 Two identical uniform rods OB and BC are joined at ends B by a frictionless joint in order to form a double pendulum; the other end of OB is suspended from a frictionless hinge at O as shown in Fig. 8.3. When the free end C of rod BC strikes against a rough half space, the rods have angles of inclination from vertical denoted by θ_1 and θ_2 and angular speeds $\dot{\theta}_1$ and $\dot{\theta}_2$ respectively. Denote the coefficient of friction between C and the half space by μ, and the energetic coefficient of restitution at the same location by e_*. Assume the motion is planar.

    (a) Find an expression for the critical coefficient of friction $\bar{\mu}$ that prevents slip reversal.

    (b) For $\mu < \bar{\mu}$ find expressions for the angular speeds at separation.

    Solution   After defining generalized speeds $\dot{q}_1 = L\dot{\theta}_1$, $\dot{q}_2 = L\dot{\theta}_2/2$, the kinetic energy of the system can be expressed as

$$T = \frac{ML^2}{6}\left[4\dot{\theta}_1^2 + \dot{\theta}_2^2 + 3\dot{\theta}_1\dot{\theta}_2 \cos(\theta_2 - \theta_1)\right]$$
$$= \frac{M}{3}\left[2\dot{q}_1^2 + 2\dot{q}_2^2 + 3\dot{q}_1\dot{q}_2 \cos(\theta_2 - \theta_1)\right]$$

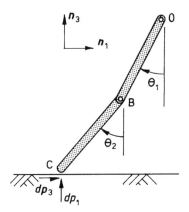

Figure 8.3. Double pendulum colliding against an inelastic half space where the energetic coefficient of restitution is e_* and the coefficient of friction is μ.

so that the generalized momenta are

$$\frac{\partial T}{\partial \dot{q}_1} = \frac{M}{3}[4\dot{q}_1 + 3\dot{q}_2 \cos(\theta_2 - \theta_1)]$$

$$\frac{\partial T}{\partial \dot{q}_2} = \frac{M}{3}[4\dot{q}_2 + 3\dot{q}_1 \cos(\theta_2 - \theta_1)]. \tag{a}$$

The velocity of the contact point $\mathbf{V}_C$ can be written, with $s\theta_i \equiv \sin\theta_i$ and $c\theta_i \equiv \cos\theta_i$,

$$\mathbf{V}_C = \dot{\theta}_1(-Lc\theta_1\mathbf{n}_1 + Ls\theta_1\mathbf{n}_3) + \dot{\theta}_2(-Lc\theta_2\mathbf{n}_1 + Ls\theta_2\mathbf{n}_3)$$

$$= -(\dot{q}_1 c\theta_1 + 2\dot{q}_2 c\theta_2)\mathbf{n}_1 + (\dot{q}_1 s\theta_1 + 2\dot{q}_2 s\theta_2)\mathbf{n}_3. \tag{b}$$

The differential of generalized impulse $d\Pi_i$ for increment of impulse $d\mathbf{p} = dp_1 \mathbf{n}_1 + dp_3 \mathbf{n}_3$ is

$$d\Pi_1 = -dp_1 c\theta_1 + dp_3 s\theta_1, \qquad d\Pi_2 = -2\,dp_1\, c\theta_2 + 2\,dp_3\, s\theta_2.$$

Initially the slip is in direction $\mathbf{n}_1$, so Coulomb's law gives $dp_1 = -\mu\,dp_3 \equiv -\mu\,dp$ and (8.4) results in the equations of motion

$$\begin{bmatrix} 4 & 3c(\theta_2 - \theta_1) \\ 3c(\theta_2 - \theta_1) & 4 \end{bmatrix} \begin{Bmatrix} d\dot{q}_1/dp \\ d\dot{q}_2/dp \end{Bmatrix} = \frac{3}{M} \begin{Bmatrix} \mu c\theta_1 + s\theta_1 \\ 2\mu c\theta_2 + 2s\theta_2 \end{Bmatrix} \equiv \begin{Bmatrix} b_1 \\ b_2 \end{Bmatrix}$$

After solving for the differentials and then integrating with initial conditions $\dot{q}_i(0)$, we obtain

$$\begin{Bmatrix} \dot{q}_1 \\ \dot{q}_2 \end{Bmatrix} = \begin{Bmatrix} \dot{q}_1(0) \\ \dot{q}_2(0) \end{Bmatrix} + \frac{1}{\Delta} \begin{Bmatrix} 4b_1 - 3b_2 \cos(\theta_2 - \theta_1) \\ 4b_2 - 3b_1 \cos(\theta_2 - \theta_1) \end{Bmatrix} p, \qquad p < p_s \tag{c}$$

$$\Delta = 16 - 9\cos^2(\theta_2 - \theta_1)$$

where at impulse

$$p_s = \frac{-[\dot{q}_1(0)c\theta_1 + 2\dot{q}_2(0)c\theta_2]\Delta}{[4b_1 - 3b_2 c(\theta_2 - \theta_1)]c\theta_1 + 2[4b_2 - 3b_1 c(\theta_2 - \theta_1)]c\theta_2} \tag{d}$$

the tangential speed vanishes ($\mathbf{V}_C \cdot \mathbf{n}_1 = \dot{q}_1 c\theta_1 + 2\dot{q}_2 c\theta_2 = 0$), while at impulse

$$p_c = \frac{-[\dot{q}_1(0)s\theta_1 + 2\dot{q}_2(0)s\theta_2]\Delta}{[4b_1 - 3b_2 c(\theta_2 - \theta_1)]s\theta_1 + 2[4b_2 - 3b_1 c(\theta_2 - \theta_1)]s\theta_2} \tag{e}$$

Table 8.1. *Results for Double Pendulum Striking Rough Half Space*

Coeff. of Friction, μ	Initial Velocity (rad s^{-1}) $\dot\theta_1(0)$ $\dot\theta_2(0)$		Rel. Impulse (slip = 0), p_s/p_c	Rel. Impulse (separation), p_f/p_c	Final Velocity (rad s^{-1}) $\dot\theta_1(p_f)$ $\dot\theta_2(p_f)$		Final Dir. of Slip, $(+/-)$	Final Normal Velocity, $V_3(p_f)/V_3(0)$	Final Kin. Energy, T_f/T_0
0.0	−0.1	−0.2	1.32	1.50	−0.230	+0.292	−	−0.50	0.707
0.2	−0.1	−0.2	1.30	1.52	−0.214	+0.259	−	−0.42	0.621
0.5	−0.1	−0.2	1.29	1.56	−0.199	+0.223	−	−0.33	0.552
0.7	−0.1	−0.2	1.28	1.58	−0.193	+0.210	Stick	−0.28	0.527

Configuration gives a coefficient of friction for stick of $\bar\mu = 0.62$. Coefficient of restitution, $e_* = 0.5$

the normal component of relative velocity vanishes ($\mathbf{V}_C \cdot \mathbf{n}_3 = \dot q_1 s\theta_1 + 2\dot q_2 s\theta_2 = 0$). If the contact point C slides in the positive direction, during compression the normal component of impulse does work $W_3(p_c)$ equal to

$$W_3(p_c) = \{\dot q_1(0)s\theta_1 + 2\dot q_2(0)s\theta_2\}p_c$$
$$+\{[4b_1 - 3b_2c(\theta_2 - \theta_1)]s\theta_1 + 2[4b_2 - 3b_1c(\theta_2 - \theta_1)]s\theta_2\}\frac{p_c^2}{2\Delta}. \quad (f)$$

After initial sliding is brought to a halt, if sliding resumes, it occurs in direction $-\mathbf{n}_1$, so that coefficients b_1 and b_2 transform to $\bar b_1$ and $\bar b_2$ where

$$\begin{Bmatrix} \bar b_1 \\ \bar b_2 \end{Bmatrix} \equiv \frac{3}{M} \begin{Bmatrix} -\mu c\theta_1 + s\theta_1 \\ -2\mu c\theta_2 + 2s\theta_2 \end{Bmatrix}.$$

For impulse applied after slip is halted ($p_s < p < p_f$) the critical coefficient of friction for stick $\bar\mu$ is obtained from $\mathbf{V}_C \cdot \mathbf{n}_1 = \dot q_1 c\theta_1 + 2\dot q_2 c\theta_2 = 0$ with $\dot q_i(\bar b_1, \bar b_2)$; i.e.,

$$\bar\mu = \frac{8s\theta_2 c\theta_2 + 2s\theta_1 c\theta_1 - 3s(\theta_2 + \theta_1)c(\theta_2 - \theta_1)}{8c^2\theta_2 + 2c^2\theta_1 - 6c\theta_1 c\theta_2 c(\theta_2 - \theta_1)}. \quad (g)$$

For some specific initial values,[4] $\theta_1 = \pi/9$, $\theta_2 = \pi/6$, $\dot\theta_1 = -0.1$ rad s^{-1}, $\dot\theta_2 = -0.2$ rad s^{-1}, $e_* = 0.5$, Table 8.1 contains results for this double pendulum obtained with an energetic coefficient of restitution at C.

8.2 Impact on a System of Rigid Bodies Connected by Compliant Constraints

Unilateral constraints require that at separation each pair of contact points have a normal component of relative velocity that is nonnegative, i.e. that finally the contact points move apart or remain in contact – they cannot interpenetrate. This constraint is appropriate to problems of granular flow where a colliding body may strike against a

---

[4] For this case of a noncollinear configuration with friction, alternative definitions of the coefficient of restitution result in calculated values of $T_f/T_0 > 1.0$.

cluster of separate bodies or else against a body that is at rest against a wall. In each case the bodies have no specified relative velocity during contact; at each contact point there is a constraint on the relative velocity at separation only.

8.2.1 Comparison of Results from Alternative Analytical Approximations for Multibody Systems with Unilateral Constraints

Impulse–momentum relations have been a direct and effective route for analyzing the process of impact between two rigid bodies (Keller, 1986; Stronge, 1990). For multibody systems, however, use of impulse–momentum relations implicitly incorporates one of two assumptions – either (a) impulsive reactions occur simultaneously or (b) they occur sequentially. Alternative formulations based on sequential impacts have been proposed by Johnson (1976), Han and Gilmore (1993) and Adams (1997); on simultaneous impacts, by Glockner and Pfeiffer (1995) and by Pereira and Nikravesh (1996). Here we will demonstrate that neither the simultaneous nor the sequential approach gives results in agreement with a simple experiment on a multibody system where elastic bodies are linked by compliant unilateral constraints.

When two bodies collide, contact pressures develop in a small area around the point of initial contact. These pressures are compatible with the local deformations of the bodies, and they are just sufficient to prevent interpenetration of the bodies. If the bodies are composed of sufficiently hard, rate-independent materials, the resultant force $\mathbf{F}(t)$ that acts on each body has a component normal to the common tangent plane $F(t) = \mathbf{F} \cdot \mathbf{n}$ that increases while the colliding bodies outside the local contact region have a relative velocity of approach during a period of compression. During a subsequent period of restitution the colliding bodies are driven apart by strain energy stored in the contact region during compression. For these materials the transition from compression to restitution occurs when the normal relative velocity across the contact region vanishes and the normal component of contact force is a maximum (Fig. 2.2).

Hard bodies have deformations that remain localized around a small contact area, so that the contact compliance is very small and the period of impact is extremely brief; these conditions are necessary for changes in inertia properties during impact to be negligible. Despite the brief period of contact, it is important to recognize that during impact, changes in relative velocity across each contact region are a continuous function of either time or impulse. The continuous nature of changes in relative velocity across the contact regions can be modeled most effectively by introducing an infinitesimal deformable particle between each pair of contact points; in multicontact problems, although the contact compliances are very small, they are not negligible. In particular, they are not negligible in comparison with the compliance at other connections to adjacent bodies. With the artifice of an infinitesimal deformable particle separating each pair of contact points, bodies B and B′ have coincident contact points C and C′ with normal components of velocity $V_C(t)$ and $V_C'(t)$ respectively. These velocities vary continuously during impact as a function either of the time t or of the normal component of impulse, $p(t) \equiv \int_0^t \mathbf{n} \cdot \mathbf{F}(t) \, dt$.

The following comparisons of different approximations for impact relations have been selected to bring out the most important features representing contact interactions during multibody impact. In that spirit we analyze the dynamic response to axial impact on a periodic series of collinear bodies that are initially at rest and touching.

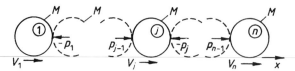

Figure 8.4. Direct impact on a uniform collinear system of n identical rigid bodies with unilateral constraints at $n - 1$ contact points.

Simultaneous Reactions at Multiple, Noncompliant Points of Contact

Consider direct impact on a set of n identical bodies in a collinear configuration with a unique common normal direction x as shown in Fig. 8.4. Newton's second law of motion provides the relation for differential changes of velocity for body j in terms of the impulse acting at each contact point. Let the normal component of velocity of body j be V_j, and denote the contact point between body j and body $j + 1$ as C_j, consistent with the multibody system shown in Fig. 8.4. After recognizing that the two bodies touching at unilateral constraint C_j have equal but opposite impulses p_j, and denoting the normal component of relative velocity at contact point C_j by $v_j \equiv V_{j+1} - V_j = \mathbf{n} \cdot (\mathbf{V}_{j+1} - \mathbf{V}_j)$, the differential equations for the relative velocities across this series of contact points can be written as (see Wittenburg, 1977)

$$dv_j = M^{-1} \Phi_{jk} \, dp_k, \qquad j = 1, 2, \ldots, n - 1 \tag{8.8}$$

where the inverse of the inertia matrix, Φ_{jk}, generally depends on the directions of common normals as well as the inertia properties of each body. For the present case of parallel collinear collisions between identical bodies one obtains

$$\Phi_{jk} = \begin{bmatrix} 2 & -1 & 0 & \cdots & 0 \\ -1 & 2 & -1 & \cdots & 0 \\ 0 & -1 & 2 & \cdots & 0 \\ \vdots & \vdots & \vdots & \ddots & \vdots \\ 0 & 0 & 0 & \cdots & 2 \end{bmatrix} \tag{8.9}$$

The narrow bandwidth of this matrix results from each body being subject to a pair of collinear impulses only; in systems containing branches or loops the bandwidth is much larger.

Equation (8.8) can be integrated to obtain the changes in relative velocity as a function of the set of impulses p_k:

$$v_j(p) = v_j(0) + M^{-1} \Phi_{jk} p_k. \tag{8.10}$$

To analyze collisions between two bodies, a *characteristic normal impulse for compression* p_c was defined as the impulse at the transition from compression to restitution, so that $v(p_c) = 0$. For multicontact point collisions the equivalent idea requires that each contact point have a *simultaneous time of transition* from compression to restitution. This gives

$$p_j(t_c) = -M \Phi_{jk}^{-1} v_k(0) \tag{8.11}$$

where the inertia matrix

$$
\Phi_{jk}^{-1} = \frac{1}{n}
\begin{bmatrix}
n-1 & n-2 & n-3 & \cdots & 1 \\
n-2 & 2(n-1) & & \cdots & 2 \\
n-3 & & & & \\
\vdots & \vdots & & \ddots & \vdots \\
1 & 2 & & \cdots & n-1
\end{bmatrix}
$$

The ratio between the terminal impulse $p_j(t_f)$ and the normal impulse $p_j(t_c)$ for compression at the jth contact point is given by the local *coefficient of restitution*

$$
e_j = \frac{p_j(t_f) - p_j(t_c)}{p_j(t_c)}. \tag{8.12}
$$

At different contacts this coefficient can differ, since it depends on material properties, the impact configuration, incident relative velocities, etc.

With incident relative velocities $v_j(0)$ Eq. (8.10), (8.11) and (8.12) give

$$
v_j(p) = v_j(0) - (1 + e_k)\Phi_{jk}\Phi_{kl}^{-1}v_l(0) \tag{8.13}
$$

e.g., the initial conditions $v_j(0) = \{-v_0, 0, 0, \cdots, 0\}^T$ give terminal relative velocities

$$
v_1(p_f)/v_0 = -1 + 2n^{-1}(n-1)(1+e_1) - n^{-1}(n-2)(1+e_2)
$$
$$
v_2(p_f)/v_0 = 0 - n^{-1}(n-1)(1+e_1) + 2n^{-1}(n-2)(1+e_2) - \left(\tfrac{n-3}{n}\right)(1+e_3).
$$
$$
\vdots
$$

For example, if all collisions are elastic ($e_j = 1$), these initial conditions yield a distribution of terminal velocity $V_j(p_f)/V_0 = 2n^{-1}\{1 - n/2, 1, 1, \ldots, 1\}^T$.

Simple experiments on a periodic collinear system of identical elastic spheres in initial contact (*Newton's cradle*) suggest that this solution is not representative of the dynamics of multibody collisions (Johnson, 1976).[5]

Sequential Reactions at Multiple, Noncompliant Points of Contact
In multibody systems with compliant unilateral constraints an alternative method of relating the reaction impulses is to assume that they act sequentially, usually in order of increasing distance from the site of an external impact. This assumption is unequivocal as long as the series of bodies is linked solely by single points of contact, but for any body in contact with three or more bodies (i.e. at a branch point), some other *ad hoc* assumption regarding the order of impulses is required.

For the simple collinear system of identical bodies with equal masses $M_i = M$ described in Fig. 8.4, each body in turn is accelerated by impulse from the preceding moving body while the speed of the striking body is retarded. If all contacts are elastic and the collision has incident velocity $V_j(0) = \{V_0, 0, 0, \ldots, 0\}^T$, this interaction hypothesis yields a set of terminal velocities $V_j(p_f)/V_0 = \{0, 0, 0, \ldots, 1\}^T$. This result agrees with crude observations of Newton's cradle. The question is whether the sequential method is generally applicable.

---

[5] The name "Newton's cradle" seems to have evolved from Kerwin's (1972) description of a "Newton momentum-conservation apparatus". Analyses of impact against collinear series of spheres, however, began much earlier (for references see Chapman, 1960).

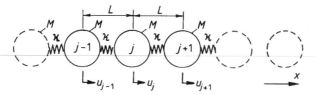

Figure 8.5. Typical elements in a uniform collinear system with bilinear springs representing local elastic compliance at points of contact.

Note that both simultaneous and sequential solutions above satisfy conservation of momentum and conservation of energy; i.e., *for multibody elastic collisions, conservation of momentum and energy are not sufficient to uniquely determine the solution.*

A third, more detailed approach employed by Cundall and Strack (1979) and by Stronge (1994b) abandons impulse–momentum relations and develops contact relations directly in terms of interaction forces. This approach obtains changes in relative velocity at contact points as continuous functions of time (or relative displacement), but has the disadvantage of requiring additional information about the contact geometry and material properties in order to represent the compliance of each local contact region.

Force–Acceleration Relations for Multibody Systems with Compliant Points of Contact
The compliance of hard bodies with small areas of contact results from local deformation around the contact area. Irrespective of whether the bodies are elastic or inelastic, the magnitude of contact stresses rapidly decreases with increasing distance from the contact region; thus strain energy is localized in a small neighborhood around contact area. The small size of the region of large strain causes the compliance of this region to be small and consequently the contact period to be very brief.

This small local compliance at each contact point can be modeled as either an elastic or an inelastic spring of infinitesimal length oriented normal to the surface. Outside the contact region the bodies are assumed to be rigid. For this periodic collinear system, let the element spacing be L, the normal contact stiffness at each interface be κ and the infinitesimal displacement of the jth element be u_j, as shown in Fig. 8.5. A typical element has an equation of motion

$$\ddot{u}_j + 2\omega_0^2 u_j = \omega_0^2(u_{j-1} + u_{j+1}) \quad \text{for} \quad u_j \leq u_{j-1}, \qquad \omega_0^2 = \kappa/M_j = \kappa/M. \quad (8.14)$$

For a wave of slowly varying amplitude the displacement varies according to

$$u_j = Ae^{i(kx - \omega t)} = Ae^{i(kjL - \omega t)}$$

where the wave number k is related to the wavelength λ by $k = 2\pi/\lambda$, and $i \equiv \sqrt{-1}$. Substitution into (8.14) gives an expression for the dispersion relation [see (7.27)]

$$\omega/\omega_0 = \pm 2 \sin(kL/2), \qquad \omega/\omega_0 < 2. \qquad (8.15)$$

Individual wavelength components of the wave propagate with phase velocity

$$c_p = \frac{\omega}{k} = \pm\omega_0 \frac{\sin(kL/2)}{kL/2} = \pm\frac{\omega_0\lambda}{\pi} \sin\left(\frac{\pi L}{\lambda}\right) \qquad (8.16)$$

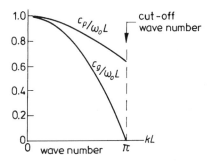

Figure 8.6. Phase velocity c_p and group velocity c_g as functions of wave number k for a periodic system of equal masses separated by bilinear compliant elements.

while energy propagates at the group velocity

$$c_g = \frac{d\omega}{dk} = \pm\omega_0 L \cos\left(\frac{kL}{2}\right) \tag{8.17}$$

Figure 8.6 illustrates that this system has a cutoff wave number $k = \pi/L$, which gives a lower bound for wavelength λ of a propagating disturbance: propagation occurs only if $\lambda > 2L$. This minimum compares with a value of roughly $3L$ measured from photoelastic patterns in a collinear series of thin disks driven axially by a small explosive charge (Singh, Shukla and Zervas, 1996).

The limiting wave speed is obtained for waves with vanishingly small wave number $k = 0$; this gives a group velocity $c_g = \omega_0 L$ that can be calculated for spherical elastic elements with individual masses $M_j = (4\pi/3)\rho R^3$. If the linear spring constant κ represents contact stiffness, the work done during indentation to the yield limit δ_Y, Eqs. (6.8) and (6.10a) for Hertzian compliance give an equivalent linear contact stiffness κ that depends on the yield pressure $\vartheta_Y Y$, viz. $\kappa = (4\pi/5)\vartheta_Y Y R_*$. Thus in a collinear series of identical elastic spheres, the limiting wave speed is

$$c_g = \sqrt{\frac{6}{5}\frac{\vartheta_Y Y}{\rho}}.$$

For metal spheres, this speed is of the order of $1/10$ the elastic wave speed of the medium.

High Frequency Waves, $\omega > 2\omega_0$

Depending on size and constitution of the colliding bodies, the contact force $F(t)$ generated by the external impact may contain frequency components larger than the cutoff frequency. In this system these high frequency components cannot propagate. What happens to the energy contained in high frequency components of the collision force?

Consider a complex wave number $k = k_R + ik_I$, where $k_R = \pi/L$ gives continuity with lower frequencies at the cutoff frequency $\omega = 2\omega_0$. The wave solution becomes

$$u_j = Ce^{-k_I jL}e^{i(\pi j - \omega t)}, \qquad i = \sqrt{-1}, \quad j = \text{element number}.$$

This steady state solution is a standing wave (nonpropagating) that decays exponentially in amplitude with increasing distance from the end, $x = 0$. The solution is termed an *evanescent wave*; each element has an oscillatory motion with displacement that is out of phase by π with the motion of adjacent elements.

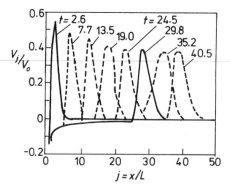

Figure 8.7. Velocity distribution at successive times in a collinear periodic system of rigid bodies separated by bilinear compliant elements; initially one end is struck by a single colliding element moving at speed V_0 in the coaxial direction.

8.2.2   Numerical Simulation and Discussion of Multibody Impact

Figure 8.7 illustrates the spatial velocity distribution at successive times in a periodic array of identical elements connected by springs which are linear in compression only; the bilinear springs provide no resistance to extension. As an initial disturbance propagates through the system, it slowly decreases in amplitude and the pulse width broadens; i.e., the wave is dispersive. Less obvious is the fact that each element near the impact end finally separates from its neighbor with a small negative velocity – the amplitude of these residual velocities decreases almost exponentially with increasing distance from the external impact point. In Fig. 8.7 the curve with negative velocity represents this residual velocity distribution.

In a more limited collinear system of six touching spheres, the temporal variation in velocity for each element illustrated in Fig. 8.8a clearly shows both dispersion and the exponentially decreasing residual negative velocity. If the bilinear springs of the system in Fig. 8.8a are replaced by nonlinear springs where the compressive force–deflection relation $F_j = \kappa_s \delta_j^{3/2}$ is obtained from the Hertz contact law, then the velocity distribution as a function of time is shown in Fig. 8.8b. In comparison with similar results for bilinear springs, the nonlinear compliance relation results in a smaller rate of decrease in peak amplitude and a smaller amplitude of residual negative velocity. Nevertheless, both the bilinear and the nonlinear compliance relations result in dispersion and evanescence (Stronge, 1999).

Figure 8.9 illustrates the effect of two colliding spheres striking a chain of four identical spheres with Hertzian compliance relations. Here waves of change in relative velocity propagate in both directions away from the off center point of impact. If the chain is struck coaxially by a pair of spheres with twice the modulus of the spheres in the chain, the final velocity distribution is more widely spread than that anticipated from experiments with a Newton cradle.

In systems of bodies connected by unilateral constraints, the forces are a reaction to relative displacements between adjacent bodies, and they develop from the relative displacements as a result of small local compression. Any analysis of the effect of these forces generally requires consideration of infinitesimally small relative displacements which develop across contact regions. In general, analyses of multibody impact need to consider local displacements at contact points although global displacements during the

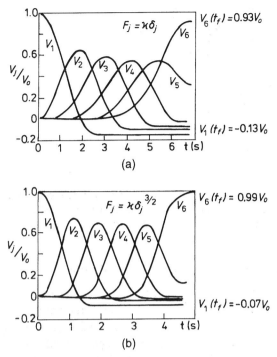

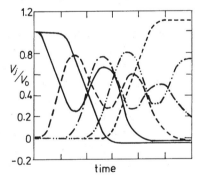

Figure 8.8. Variation in velocity for each ball in a collinear elastic system where a single ball strikes a set of five identical balls: (a) Linear interaction compliance and (b) Hertz nonlinear interaction compliance.

Figure 8.9. Variation in velocity for each ball in collinear set of four balls struck collinearly by two identical balls traveling at an initial speed V_0. Speeds calculated using Hertz nonlinear interaction compliance.

contact period may be sufficiently small so that the configuration can be considered to be time-invariant.

The present formulation of equations of motion is for systems of rigid bodies connected by compliant constraints; this analytical simulation uses two distinct scales for the effect of displacements. Infinitesimal displacements generate interaction forces at compliant constraints, so it is necessary that they be included in order to represent the interaction forces that prevent interference. These infinitesimal displacements, however, have no effect on the inertia or the kinetic energy T of the system. Thus the equations of motion (8.4) do not have terms arising from changes in the impact configuration during contact.

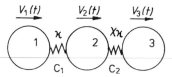

Figure 8.10. Set of three spheres in collinear configuration; spheres are separated by bilinear springs with differing contact stiffness.

8.2.3   Spatial Gradation of Normal Contact Stiffness $\kappa_j = \chi\kappa$

In an assembly of rigid bodies where there are multiple deformable points of contact,
the simultaneity of reaction forces depends on the distributions of inertia and contact
compliance. By considering the dynamic response of a small system of three collinear
spheres that are initially in contact at two contact points, the effect of gradation of contact
stiffness will be investigated.

Consider the set of three identical spheres numbered 1–3 with equal masses $M_j = M$
as shown in Fig. 8.10. Let the contact points be numbered 1, 2 successively, and at
the contact points denote the normal relative displacement between bodies (indentation
positive) by $\delta_j \equiv u_j - u_{j+1}$. For a bilinear contact stiffness, the normal contact force F_j at
the jth contact point is $F_j = -\kappa_j\delta_j$ if $\delta_j > 0$ and $F_j = 0$ if $\delta_j \leq 0$. Let $\kappa_j = \chi^{j-1}\kappa$, and
recall that equal but opposite reaction forces act on adjacent bodies to derive equations
of relative motion for indentation δ_j at contact points 1 and 2,

$$\begin{Bmatrix} \ddot{\delta}_1 \\ \ddot{\delta}_2 \end{Bmatrix} + \omega_0^2 \begin{bmatrix} 2 & -\chi \\ -1 & 2\chi \end{bmatrix} \begin{Bmatrix} \delta_1 \\ \delta_2 \end{Bmatrix} = \begin{Bmatrix} 0 \\ 0 \end{Bmatrix}, \qquad \omega_0^2 \equiv \kappa/M \tag{8.18}$$

during the period when $\delta_1 > 0$, $\delta_2 > 0$. These equations of relative motion give modal
frequencies ω_i for the first and second modes,

$$\omega_i/\omega_0 = \left[(1 + \chi) \pm (1 - \chi + \chi^2)^{1/2}\right]^{1/2} \tag{8.19}$$

Figure 8.11 shows these frequencies as a function of the stiffness gradient χ and illustrates
that the ratio of characteristic frequencies ω_1/ω_2 is a minimum near $\chi = 1$.

For an unstressed initial state $\delta_1(0) = \delta_2(0) = 0$ the modal solution for indentations
can be expressed as

$$\begin{Bmatrix} \delta_1(t) \\ \delta_2(t) \end{Bmatrix} = \begin{bmatrix} A_1 & A_2 \\ A_1\chi^{-1}\Omega_1 & A_2\chi^{-1}\Omega_2 \end{bmatrix} \begin{Bmatrix} \sin \omega_1 t \\ \sin \omega_2 t \end{Bmatrix}, \qquad \Omega_i = 2 - \omega_i^2/\omega_0^2 \tag{8.20}$$

where the ith eigenvector equals $\left\{1 \quad \chi^{-1}\Omega_i\right\}^T$. If the first body has initial speed V_0 when
it collides against two stationary balls, the initial conditions $\dot{\delta}_1(0) = V_0$, $\dot{\delta}_2(0) = 0$ give
the constants

$$A_1 = \left(1 - \frac{\Omega_1}{\Omega_2}\right)^{-1} \frac{V_0}{\omega_1}, \qquad A_2 = -\frac{\omega_1\Omega_1}{\omega_2\Omega_2} A_1. \tag{8.21}$$

This solution has been used to evaluate the times of maximum compression at contact
points 1 and 2 as shown in Fig. 8.12. For $\chi < 1$ (decreasing stiffness with increasing x)
the times of maximum compression differ by more than a factor of 2; thus decreasing

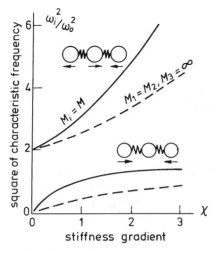

Figure 8.11. Characteristic frequencies for system of three masses as a function of stiffness gradient χ in a system with springs linear in compression.

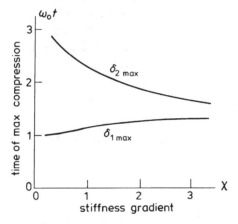

Figure 8.12. Time of maximum compression at contacts C_1 and C_2 as a function of stiffness gradient χ in a system of two spheres hit in the axial direction by a third sphere.

contact stiffness can be approximated as a sequence of individual collisions between pairs of bodies. On the other hand, for $\chi > 3$ (increasing stiffness with increasing x) the times of maximum compression asymptotically approach the same limit; this case is closely approximated by simultaneous collisions at all points of contact.

The final velocities of the three spheres as functions of the stiffness ratio χ are shown in Fig. 8.13. When χ is very small or very large these terminal velocity distributions approach the results of sequential or simultaneous collision approximations, respectively. For uniform stiffness ($\chi = 1$), however, the velocity distributions are between these two limits.

Example 8.4 A stack of two spherical balls has a common initial velocity $-V_0\mathbf{n}$ when contact point C_1 on ball B_1 strikes a massive barrier. The impact configuration is collinear

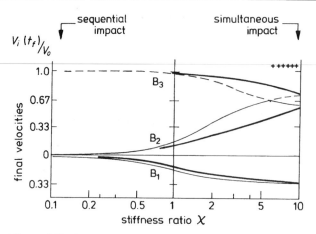

Figure 8.13. Terminal velocities after collinear impact of a sphere against two identical spheres. Heavy lines are for spring stiffness given by Hertz contact law, $F = \kappa_s \delta^{3/2}$. Light lines are for linear stiffness, $F = \kappa \delta$.

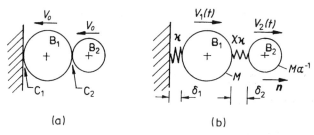

Figure 8.14. Normal collinear impact of a quiescent stack of two dissimilar balls against an immovable barrier: (a) initial velocities and (b) analytical model for compliant bodies with bilinear contact stiffness.

as shown in Fig. 8.14. The mass of ball B_1 is M, and that of ball B_2 is M/α. Assume that at contact point C_1 ball B_1 and the barrier are separated by an elastic element whose spring constant equals κ if the relative displacement is compressive ($\delta_1 < 0$); similarly, at contact C_2 between balls B_1 and B_2 assume that there is an elastic element whose spring constant equals $\chi\kappa$ if the relative displacement at C_2 is compressive ($\delta_2 < 0$). At both contact points the spring constant $\kappa = 0$ if the elastic element is extended ($\delta_j > 0$). The bodies are assumed to be rigid outside of these elastic elements. Find the ratios V_j^+/V_0 of final to initial velocities for each of the balls as functions of the mass ratio α for stiffness ratios $\chi = 1/6, 1/2, 1$.

Solution
Let the rates of change of relative displacements be defined as

$$\dot{\delta}_1 = V_1, \qquad \dot{\delta}_2 = V_2 - V_1.$$

The equations of relative motion for initial period where $\delta_1 < 0$, $\delta_2 < 0$ are

$$\begin{Bmatrix} 0 \\ 0 \end{Bmatrix} = \begin{Bmatrix} \ddot{\delta}_1 \\ \ddot{\delta}_2 \end{Bmatrix} + \omega_0^2 \begin{bmatrix} 1 & -\chi \\ 1 & (1+\alpha)\chi \end{bmatrix} \begin{Bmatrix} \delta_1 \\ \delta_2 \end{Bmatrix}, \qquad \omega_0^2 = \frac{\kappa}{M}.$$

The determinant of coefficients gives the natural (modal) frequencies,

$$\frac{\omega_1}{\omega_0} = \left\{ \frac{1}{2}(1 + \chi + \alpha\chi) \left[1 - \left(1 - \frac{4\alpha\chi}{(1 + \chi + \alpha\chi)^2}\right)^{1/2} \right] \right\}^{1/2}$$

$$\frac{\omega_2}{\omega_0} = \left\{ \frac{1}{2}(1 + \chi + \alpha\chi) \left[1 + \left(1 - \frac{4\alpha\chi}{(1 + \chi + \alpha\chi)^2}\right)^{1/2} \right] \right\}^{1/2}$$

The normal mode solution satisfying initial conditions $\delta_1(0) = \delta_2(0) = 0$ is

$$\delta_1(t) = A_1 \sin\omega_1 t + A_2 \sin\omega_2 t, \qquad \Omega_1 = \chi^{-1}\left(1 - \omega_1^2/\omega_0^2\right)$$

$$\delta_2(t) = A_1\Omega_1 \sin\omega_1 t + A_2\Omega_2 \sin\omega_2 t, \qquad \Omega_2 = \chi^{-1}\left(1 - \omega_2^2/\omega_0^2\right).$$

The initial conditions $\dot\delta_1(0) = -V_0$, $\dot\delta_2(0) = 0$ give

$$A_1 = \frac{-\Omega_2 V_0}{\omega_1(\Omega_2 - \Omega_1)}, \qquad A_2 = \frac{\Omega_1 V_0}{\omega_2(\Omega_2 - \Omega_1)}.$$

The relative displacements and velocities can be expressed as

$$\left\{ \begin{matrix} \omega_0\delta_1/V_0 \\ \omega_0\delta_2/\Omega_1 V_0 \end{matrix} \right\} = \frac{-\omega_0/\omega_1}{1 - \Omega_1/\Omega_2} \begin{bmatrix} 1 & -\omega_1\Omega_1/\omega_2\Omega_2 \\ 1 & -\omega_1/\omega_2 \end{bmatrix} \left\{ \begin{matrix} \sin\omega_1 t \\ \sin\omega_2 t \end{matrix} \right\}$$

$$\left\{ \begin{matrix} \omega_0\dot\delta_1/V_0 \\ \omega_0\dot\delta_2/\Omega_1 V_0 \end{matrix} \right\} = \frac{-1}{1 - \Omega_1/\Omega_2} \begin{bmatrix} 1 & -\Omega_1/\Omega_2 \\ 1 & -1 \end{bmatrix} \left\{ \begin{matrix} \cos\omega_1 t \\ \cos\omega_2 t \end{matrix} \right\}$$

Double contact ceases at time t_s when contact separates at either C_1 or C_2 ($\delta_1 \leq 0$ or $\delta_2 \leq 0$).

The equation of motion after initial separation at C_1, i.e. for $t > t_s$ and $\delta_1 \leq 0$, is

$$\ddot\delta_2 + \bar\omega^2\delta_2 = 0, \qquad \bar\omega^2 = (1 + \alpha)\chi\kappa/M$$

with transition conditions $\delta_{2s} \equiv \delta_2(t_s)$, $\dot\delta_{1s} \equiv \dot\delta_1(t_s)$ and $\dot\delta_{2s} \equiv \dot\delta_2(t_s)$, so that

$$\delta_2 = (\dot\delta_{2s}/\bar\omega)\sin\bar\omega(t - t_s) + \delta_{2s}\cos\bar\omega(t - t_s)$$

$$\dot\delta_2 = \dot\delta_{2s}\cos\bar\omega(t - t_s) - \bar\omega\delta_{2s}\sin\bar\omega(t - t_s)$$

giving velocities

$$V_1 = V_1(t_s) - (1 + \alpha)^{-1}(\dot\delta_2 - \dot\delta_{2s}), \qquad V_2 = V_2(t_s) + \alpha(1 + \alpha)^{-1}(\dot\delta_2 - \dot\delta_{2s}).$$

These equations apply unless there is a second hit at contact point C_1, i.e. $\delta_1(t) = 0$, $t > t_s$, where

$$\delta_1(t) = (t - t_s)\dot\delta_{1s} + (1 + \alpha)^{-1}\dot\delta_{2s}[t - t_s - \bar\omega^{-1}\sin\bar\omega(t - t_s)]$$

$$\quad -(1 + \alpha)^{-1}\delta_{2s}[1 - \cos\bar\omega(t - t_s)].$$

Alternatively, the equation of motion after initial separation at C_2, i.e. for $t > t_s$ and $\delta_2 < 0$, is

$$\ddot\delta_1 + \omega_0^2\delta_1 = 0$$

with transition conditions $\delta_{1s} \equiv \delta_1(t_s)$, $\dot\delta_{1s} \equiv \dot\delta_1(t_s)$ and $\dot\delta_{2s} \equiv \dot\delta_2(t_s)$, so that

$$\delta_1(t) = (\dot\delta_{1s}/\omega_0)\sin\omega_0(t - t_s) + \delta_{1s}\cos\omega_0(t - t_s)$$

$$\dot\delta_1(t) = \dot\delta_{1s}\cos\omega_0(t - t_s) - \omega_0\delta_{1s}\sin\omega_0(t - t_s)$$

giving velocities

$$V_1 = \dot{\delta}_{1s} \cos \omega_0(t - t_s) - \omega_0 \delta_{1s} \sin \omega_0(t - t_s), \qquad V_2 = V_2(t_s).$$

A second hit occurs at contact point C_2 if $\delta_2(t) = 0$, $t > t_s$, where

$$\delta_2 = (t - t_s)V_{2s} - \omega_0^{-1}\dot{\delta}_{1s} \sin \omega_0(t - t_s) + \delta_{1s}[1 - \cos \omega_0(t - t_s)].$$

Figure 8.15 illustrates these results for three values of the spring constant ratio χ. In each case the curves can be compared with the result of rigid body assumptions of (a) simultaneous impact $V_1^+ / V_0 = V_2^+ / V_0 = -1$ or (b) sequential impact. The final velocities resulting from an assumption of sequential rigid body impact are shown as the light lines on

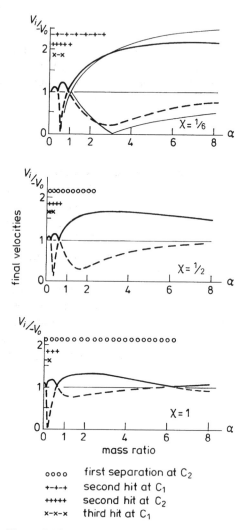

Figure 8.15. Final velocities of compliant spheres B_1 and B_2 with masses M and M/α respectively, as functions of mass ratio α, for three ratios of local elastic stiffness. In each case multiple hits occur between B_1 and B_2 if $\alpha < 1$. The graph for rapidly decreasing stiffness ($\chi = 1/6$) has light lines that show the result of assuming that at contact points between rigid bodies the collisions occur sequentially.

the graph for $\chi = 1/6$. If the mass ratio $\alpha > 1$ (i.e. $M_2 < M_1$), the assumption of simultaneous impact provides a rough approximation of the elastic result if $\chi \geq 1$ (see Harter, 1971). On the other hand, sequential rigid body impact is a good approximation if the stiffness decreases rapidly ($\chi \leq 1/6$) with increasing contact number starting from the impact site.

At small mass ratios $\alpha < 1$ (i.e. $M_1 < M_2$) after initial separation at contact C_2, body B_1 rebounds from second strikes against both the massive barrier and body B_2. As α decreases towards zero the number of impacts between the bodies increases without bound – in the limit as α approaches zero, body B_1 can be thought of as a small elastic particle oscillating between the surfaces of two immovable bodies.

Eccentric Impact with Compliance at Multiple Points of Contact

The previous examples considered multibody impact in a collinear impact configuration. The following problem considers the impact of a rigid compound pendulum against an elastic half space when there is compliance at the pivot as well as at the contact with the half space; in this case the configuration is eccentric. Effects of friction are assumed to be negligible.

Example 8.5   A compound pendulum pivots around a frictionless pivot C_1 before colliding with an elastic half space at contact point C_2. The pendulum has mass M and a radius of gyration $\hat{k}_r$ about the center of mass. Denote the position vectors from the center of mass G to C_1 and C_2 by $\mathbf{r}'$ and $\mathbf{r}$, respectively, as illustrated in Fig. 8.16. The pendulum has rotational velocity $\dot{\theta}\mathbf{n}_2$, while the translational velocity of the center of mass is denoted by $\hat{\mathbf{V}} = \hat{V}_1\mathbf{n}_1 + \hat{V}_3\mathbf{n}_3$. Contact C_1 is constrained to move parallel to the tangent plane. At C_1 and C_2 there are contact forces $\mathbf{F}'_1 = F'_1\mathbf{n}_1 + F'_3\mathbf{n}_3$ and $\mathbf{F}_2 = F_3\mathbf{n}_3$ if friction is negligible. These forces arise from linear compliance at each contact; let a representative spring constant at C_1 be denoted by κ, and at C_2 let the spring constant be $\chi\kappa$ in compression and 0 in extension. Find the ratio of incident to rebound speeds for the normal component of relative velocity at C_2 as a function of the stiffness ratio χ. Assume the displacements are small so that inertia properties do not change during contact.

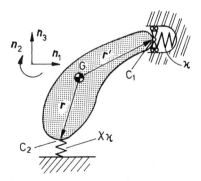

Figure 8.16. Collision of compound pendulum having linear elastic compliance at pivot C_1 as well as bilinear compliance at impact point C_2.

Solution

Velocity constraint on vertical motion at C_1:

$$\hat{V}_3 = r_1'\dot{\theta}.$$

Definition of generalized speeds that satisfy the constraint:

$$\dot{q}_1 = \hat{V}_1, \qquad \dot{q}_2 = \hat{k}_r\dot{\theta}.$$

Velocities at contact points C_1 and C_2:

$$\mathbf{V}_1 = (\hat{V}_1 + r_3'\dot{\theta})\mathbf{n}_1 = \dot{q}_1 + \frac{r_3'}{\hat{k}_r}\dot{q}_2.$$

$$\mathbf{V}_2 = (\hat{V}_1 + r_3\dot{\theta})\mathbf{n}_1 + (\hat{V}_3 - r_1\dot{\theta})\mathbf{n}_3 = \left(\dot{q}_1 + \frac{r_3}{\hat{k}_r}\dot{q}_2\right)\mathbf{n}_1 + \frac{r_1' - r_1}{\hat{k}_r}\dot{q}_2\mathbf{n}_3.$$

Differentials of generalized impulse, $d\Pi_j = (\partial\mathbf{V}_1/\partial\dot{q}_j)\cdot\mathbf{F}'\,dt + (\partial\mathbf{V}_2/\partial\dot{q}_j)\cdot\mathbf{F}\,dt$:

$$d\Pi_1 = F_1'\,dt, \qquad d\Pi_2 = \frac{r_3'}{\hat{k}_r}F_1'\,dt + \frac{r_1' - r_1}{\hat{k}_r}F_3\,dt. \tag{a}$$

Kinetic energy:

$$T = \frac{M}{2}(\hat{V}_1^2 + \hat{V}_3^2 + \hat{k}_r^2\dot{\theta}^2) = \frac{M}{2}\left[\dot{q}_1^2 + \left(1 + \frac{r_1'^2}{\hat{k}_r^2}\right)\dot{q}_2^2\right].$$

Generalized momenta:

$$\frac{\partial T}{\partial\dot{q}_1} = M\dot{q}_1, \qquad \frac{\partial T}{\partial\dot{q}_2} = M\left(1 + \frac{r_1'^2}{\hat{k}_r^2}\right)\dot{q}_2. \tag{b}$$

Relations (a) and (b) give

$$\begin{Bmatrix}\ddot{q}_1 \\ \ddot{q}_2\end{Bmatrix} = \frac{1}{M}\left(\frac{\hat{k}_r^2}{\hat{k}_r^2 + r_1'^2}\right)\begin{bmatrix}1 + r_1'^2/\hat{k}_r^2 & 0 \\ r_3'/\hat{k}_r & (r_1' - r_1)/\hat{k}_r\end{bmatrix}\begin{Bmatrix}F_1' \\ F_3\end{Bmatrix}. \tag{c}$$

At each contact a component of relative velocity is defined for each component of active force (positive for separation):

$$\dot{\delta}_1 \equiv -\mathbf{V}_1\cdot\mathbf{n}_1 = -\dot{q}_1 - \frac{r_3'}{\hat{k}_r}\dot{q}_2$$

$$\dot{\delta}_2 \equiv \mathbf{V}_2\cdot\mathbf{n}_3 = \frac{r_1' - r_1}{\hat{k}_r}\dot{q}_2. \tag{d}$$

The contact forces related to relative displacements are

$$F_1' = \kappa\delta_1, \qquad F_3 = \begin{cases}0, & \delta_2 \geq 0 \\ -\chi\kappa\delta_2, & \delta_2 < 0\end{cases}. \tag{e}$$

Substituting (d) and (e) into (c) results in

$$\begin{Bmatrix}\ddot{\delta}_1 \\ \ddot{\delta}_2\end{Bmatrix} + \omega_0^2\begin{bmatrix}b_{11} & -b_{12}\chi \\ -b_{12} & b_{22}\chi\end{bmatrix}\begin{Bmatrix}\delta_1 \\ \delta_2\end{Bmatrix} = \begin{Bmatrix}0 \\ 0\end{Bmatrix}, \qquad \omega_0^2 = \frac{\kappa}{M} \tag{f}$$

where b_{ij} are configuration parameters, $b_{11} = 1 + r_3'^2/(r_1'^2 + \hat{k}_r^2)$, $b_{12} = r_3'(r_1' - r_1)/(r_1'^2 + \hat{k}_r^2)$, $b_{22} = (r_1' - r_1)^2/(r_1'^2 + \hat{k}_r^2)$ Equation (f) has characteristic frequencies

$$\frac{\omega_i}{\omega_0} = \left\{ \frac{b_{11} + \chi b_{12}}{2} \left[1 \mp \left(1 - \frac{4\chi(b_{11}b_{22} - b_{12}^2)}{(b_{11} + \chi b_{12})^2} \right)^{1/2} \right] \right\}^{1/2}.$$

Modal solution from substitution into (f) results in

$$\delta_1 = A_1 \sin \omega_1 t + A_2 \sin \omega_2 t + A_3 \cos \omega_1 t + A_4 \cos \omega_2 t$$

$$\delta_2 = A_1 \Omega_1 \sin \omega_1 t + A_2 \Omega_2 \sin \omega_2 t + A_3 \Omega_1 \cos \omega_1 t + A_4 \Omega_2 \cos \omega_2 t$$

with $\Omega_1 \equiv (\chi b_{12})^{-1}(b_{11} - \omega_1^2/\omega_0^2)$, $\Omega_2 \equiv (\chi b_{12})^{-1}(b_{11} - \omega_2^2/\omega_0^2)$. When impact initiates at time t_s, the transition conditions[6] $\delta_1(t_s) \equiv \delta_{1s}, \delta_2(t_s) \equiv 0, \dot{\delta}_1(t_s) \equiv \dot{\delta}_{1s}, \dot{\delta}_2(t_s) \equiv \dot{\delta}_{2s}$ give relative displacements and relative velocities at C_1 and C_2,

$$\delta_1 = (\Omega_2 - \Omega_1)^{-1} \left\{ \omega_1^{-1}(\Omega_2 \dot{\delta}_{1s} - \dot{\delta}_{2s}) \sin \omega_1(t - t_s) - \omega_2^{-1}(\Omega_1 \dot{\delta}_{1s} - \dot{\delta}_{2s}) \right.$$
$$\left. \times \sin \omega_2(t - t_s) + \delta_{1s}[\Omega_2 \cos \omega_1(t - t_s) - \Omega_1 \cos \omega_2(t - t_s)] \right\}$$

$$\delta_2 = (\Omega_2 - \Omega_1)^{-1} \left\{ \omega_1^{-1} \Omega_1 (\Omega_2 \dot{\delta}_{1s} - \dot{\delta}_{2s}) \sin \omega_1(t - t_s) \right.$$
$$- \omega_2^{-1} \Omega_2 (\Omega_1 \dot{\delta}_{1s} - \dot{\delta}_{2s}) \sin \omega_2(t - t_s)$$
$$\left. + \Omega_1 \Omega_2 \delta_{1s}[\cos \omega_1(t - t_s) - \cos \omega_2(t - t_s)] \right\}$$

$$\dot{\delta}_1 = (\Omega_2 - \Omega_1)^{-1} \left\{ (\Omega_2 \dot{\delta}_{1s} - \dot{\delta}_{2s}) \cos \omega_1(t - t_s) \right.$$
$$- (\Omega_1 \dot{\delta}_{1s} - \dot{\delta}_{2s}) \cos \omega_2(t - t_s) - \delta_{1s}[\omega_1 \Omega_2 \sin \omega_1(t - t_s)$$
$$\left. - \omega_2 \Omega_1 \sin \omega_2(t - t_s)] \right\}$$

$$\dot{\delta}_2 = (\Omega_2 - \Omega_1)^{-1} \left\{ \Omega_1 (\Omega_2 \dot{\delta}_{1s} - \dot{\delta}_{2s}) \cos \omega_1(t - t_s) \right.$$
$$- \Omega_2 (\Omega_1 \dot{\delta}_{1s} - \dot{\delta}_{2s}) \cos \omega_2(t - t_s)$$
$$\left. - \Omega_1 \Omega_2 \delta_{1s}[\omega_1 \cos \omega_1(t - t_s) - \omega_2 \cos \omega_2(t - t_s)] \right\}.$$

The preceding equations apply so long as $\delta_2 < 0$.

Initial contact at C_2 ceases at time t_r when the state of the system is $\delta_1(t_r) \equiv \delta_{1r}, \delta_2(t_r) \equiv 0, \dot{\delta}_1(t_r) \equiv \dot{\delta}_{1r}, \dot{\delta}_2(t_r) \equiv \dot{\delta}_{2r}$. The subsequent motion with $F_3 = 0$ must be checked to determine if there is a second impact. With $F_3 = 0$ the system reduces to a single DOF with a characteristic frequency $\bar{\omega} = \omega_0 \sqrt{b_{12}}$. Together with the initial conditions this gives[7]

$$\delta_1 = \delta_{1r} \cos \bar{\omega}(t - t_r) + \bar{\omega}^{-1} \dot{\delta}_{1r} \sin \bar{\omega}(t - t_r)$$

$$\delta_2 = (t - t_r) b_{11}^{-1}[b_{12} \dot{\delta}_{1r} + b_{11} \dot{\delta}_{2r}]$$
$$+ b_{12} b_{11}^{-1} \left\{ \delta_{1r}[1 - \cos \bar{\omega}(t - t_r)] - \bar{\omega}^{-1} \dot{\delta}_{1r} \sin \bar{\omega}(t - t_r) \right\}$$

$$\dot{\delta}_1 = -\bar{\omega} \delta_{1r} \sin \bar{\omega}(t - t_r) + \dot{\delta}_{1r} \cos \bar{\omega}(t - t_r)$$

$$\dot{\delta}_2 = b_{11}^{-1}[b_{12} \dot{\delta}_{1r} + b_{11} \dot{\delta}_{2r}] + b_{12} b_{11}^{-1} \left\{ \bar{\omega} \delta_{1r} \sin \bar{\omega}(t - t_r) - \dot{\delta}_{1r} \cos \bar{\omega}(t - t_r) \right\}.$$

[6] This general form of initial conditions is required when there are a sequence of impacts at C_2; for the first impact, $t_s = 0$, $\delta_1(0) = 0$, $\dot{\delta}_1(0) = 0$.

[7] The equations of motion presented here have assumed any changes in configuration are infinitesimal. These solutions consider velocity changes during impact only and not during the subsequent phase of motion involving rotation of the pendulum.

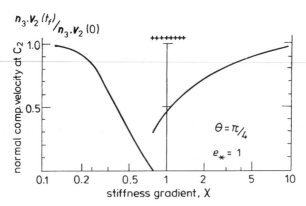

Figure 8.17. Ratio of rebound to incident normal relative velocity $-V_3(t_f)/V_3(0)$ at impact point C_2 as a function of stiffness ratio χ for a compound pendulum with $\theta = \pi/4$.

For a specific example of a compound pendulum composed of a slender uniform bar and assuming that the bar is at an angle θ from $\mathbf{n}_3$ when it collides with a massive elastic body at C_2, one obtains the inertia coefficients $b_{11} = 1 + (3\cos^2\theta)/(1 + 3\sin^2\theta)$, $b_{12} = (6\sin\theta\cos\theta)/(1 + 3\sin^2\theta)$, and $b_{22} = (12\sin^2\theta)/(1 + 3\sin^2\theta)$.

The normal relative velocity at C_2 as a function of stiffness ratio χ is illustrated in Fig. 8.17 for $\theta = \pi/4$. For stiffness ratios that are either very large or small compared with unity, the magnitudes of the incident and the rebound speed are almost the same. For ratios near unity, however, the impact forces have transformed a substantial part of the energy into a transverse translational mode; e.g., for $\chi = 0.78$ at separation we have $\delta(t_f) = 0$, since the rotational velocity $\dot{\theta}(t_f) = 0$. In this elastic system no energy is dissipated at either contact point, but an eccentric impact configuration together with a compliant support can transform the mode of motion and thereby transfer energy into adjacent contacts.

The previous example demonstrates that if contact compliance is considered and the impact configuration is noncollinear, the energetic and kinematic coefficients of restitution are not equivalent, irrespective of friction.

8.2.4   Applicability of Simultaneous Impact Assumption

In a system where rigid bodies are linked by several compliant contacts, energy propagates away from any point of external impact as a dispersive wave. This wave travels at a speed c_g that depends on inertia, the contact arrangement and the local compliance at each contact. Since the points of contact are ordinarily much more compliant than the body between them, it is contact compliance that determines the speed of propagation for hard bodies that are compact (stocky) in shape. Calculations based on assuming that external impact at one contact point results in a sequence of independent collisions at successive contacts can give a good approximation only if the system has a series arrangement and the speed of propagation c_g decreases with increasing distance from the impact point. Calculations based on simultaneity of collisions give a good approximation only if the speed of propagation c_g has a substantial increasing gradient with distance from the contact point.

PROBLEMS

8.1 For a uniform slender rigid bar of length L and mass M with velocity $\mathbf{V}_i$ at the ith end, $i = 1, 2$, derive the following expression for the kinetic energy: $T = (M/6) \times (\mathbf{V}_1 \cdot \mathbf{V}_1 + \mathbf{V}_1 \cdot \mathbf{V}_2 + \mathbf{V}_2 \cdot \mathbf{V}_2)$ (Bahar, 1994).

8.2 A set of three uniform bars of equal length L and mass M are joined end to end by two frictionless joints. The bars are collinear and at rest with the ends numbered sequentially 1, 2, 3, 4 beginning from one end. The bars are stationary until time $t = 0$ when an impulse p acts in the transverse direction at point 1.

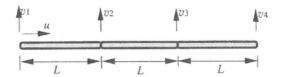

(a) Identify the number of degrees of freedom, and show that the equations of motion can be expressed as

$$
\begin{bmatrix} 2 & 1 & 0 & 0 \\ 1 & 4 & 1 & 0 \\ 0 & 1 & 4 & 1 \\ 0 & 0 & 1 & 2 \end{bmatrix} \begin{Bmatrix} dv_1 \\ dv_2 \\ dv_3 \\ dv_4 \end{Bmatrix} = \frac{6}{M} \begin{Bmatrix} dp \\ 0 \\ 0 \\ 0 \end{Bmatrix}.
$$

(Note that axial velocity u_1 is an ignorable coordinate – it is a constant.)

(b) Solve for the transverse velocities v_i^+ immediately following the impulse, and show that your solution gives a final translational momentum for the system equal to p.

(c) Suppose a sphere with mass $M' = \alpha M$ is traveling in a direction transverse to the bar and moving with speed V_0'. If the sphere strikes the side of the linkage at end 1 and the coefficient of restitution is e_*, find the terminal impulse p_f at separation.

8.3 Two uniform slender rigid bars, each with mass M and length L, are connected by a frictionless joint at one end of each bar. The bars are mutually perpendicular and lying on a smooth level table when an axial impulse p is applied to the end of one bar. Show that immediately thereafter, the other bar has an angular velocity $\omega_2^+ = 2p/ML$.

8.4 Two uniform slender rigid rods, each with length L and mass M, are connected by a frictionless joint at an end of each bar. Initially the assembly lies at rest on a smooth surface with an angle θ between the axes of the rods. An impulse p acts at the end of one rod in a direction inclined by an angle ϕ to the axis.

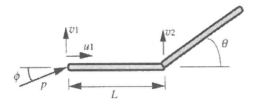

(a) Identify the number of degrees of freedom, and show that the kinetic energy can be expressed as $T = (M/6)[6u_1^2 + v_1^2 + v_1 v_2 + 4v_2^2 + 3(v_2 c\theta - u_1 s\theta)L\dot\theta + L^2\dot\theta^2]$.

(b) Write equations of motion for the response.

8.5 Let AC be a uniform rigid rod with mass M and length $2L$; the rod has a uniform transverse velocity $V_0 \mathbf{n}_3$ when end C strikes against an inelastic stop at a normal angle of obliquity. The stop has coefficient of restitution e_*. For the free end of the rod A find the velocity V_A^+ immediately after impact. Show that the ratio of final to initial kinetic energy equals $T_f/T_0 = (3 + e_*^2)/4$.

8.6 A uniform rigid bar of mass M and length $3L$ lies centered across two elastic rails that are separated by distance L before a sphere with mass M/α strikes the bar transversely at a distance λ from the center. The center of the bar has radius of gyration $\hat{k}_r$. Suppose at contact point C_1 between the bar and a rail the elastic stiffness is κ, while at C_1, where the sphere strikes the rail, the stiffness is $\chi\kappa$. Obtain coefficients b_{11}, b_{12}, b_{22} if the equations for relative displacement at C_1 and C_2 are expressed as

$$\begin{Bmatrix} 0 \\ 0 \end{Bmatrix} = \begin{Bmatrix} \ddot\delta_1 \\ \ddot\delta_2 \end{Bmatrix} + \omega_0^2 \begin{bmatrix} b_{11} & -b_{12}\chi \\ -b_{12} & b_{22}\chi \end{bmatrix} \begin{Bmatrix} \delta_1 \\ \delta_2 \end{Bmatrix}, \qquad \omega_0^2 = \frac{\kappa}{M}.$$

8.7 In example 8.3 of a double compound pendulum striking an inelastic half space (Sect. 8.1.2), for a coefficient of friction $\mu = 0.2$ and coefficient of restitution $e_* = 0.5$, obtain the impulse for compression p_c, the work done by the normal contact force during compression $W_3(p_c)$ and the total work of both friction and normal force during contact $W_1(p_f)$, $W_3(p_f)$. The coefficient of friction $\mu = 0.2$ is relatively small; explain why in this case such a large part of the initial kinetic energy T_0 is dissipated by friction during contact.

8.8 A collinear stack of two elastic balls B_1 and B_2 with masses M and M/α respectively is illustrated in Fig. 8.14(a) an instant before B_1 strikes against an elastic half space at contact point C_1. Before impact the balls have a common velocity $-V_0\mathbf{n}$. Assume sequential impacts, first at C_1 and secondly at C_2.

(a) Show that the final velocity of the balls can be expressed as

$$\frac{V_1^+}{V_0} = \frac{\alpha - 3}{\alpha + 1}, \qquad \frac{V_2^+}{V_0} = \frac{3\alpha - 1}{\alpha + 1}, \qquad 3 < \alpha$$

$$\frac{V_1^+}{V_0} = \frac{-\alpha + 3}{\alpha + 1}, \qquad \frac{V_2^+}{V_0} = \frac{3\alpha - 1}{\alpha + 1}, \qquad 1 < \alpha \le 3.$$

Sketch these relations, and explain why the sign of V_1^+/V_0 changes at $\alpha = 3$.

(b) Consider $\alpha < 1$, and obtain relations for the velocities similar to those above. Also find the range of values of mass ratio α wherein these relations are valid.

8.9 Perform an experiment, dropping a stack of two balls onto a hard surface, and make rough measurements of the maximum rebound height of each ball. Do this experiment with two identical balls (tennis balls or basketballs) and then with two dissimilar balls (tennis ball and basketball), first with one on the top and then the other. Explain the results, and comment on the applicability of the assumption of sequential impacts. Some of these questions have been addressed by Kerwin (1972), Newby (1984) and Spradley (1987).

CHAPTER 9

Collision against Flexible Structures

> We are all agreed that your theory is crazy. The question which di-
> vides us is whether it is crazy enough to have a chance of being
> correct. My own feeling is that it is not crazy enough.
>
> Niels Bohr, alleged comment at close of seminar
> where Pauli had presented a new theory

Problems of impact against a flexible structure or of a collision in a system that contains flexible structures will involve dynamic deformation of the structure in addition to local deformation of the contact region. In comparison with impact between compact bodies, global deformations in the structure generally prolong the contact period, reduce the maximum contact force and transfer significant energy into structural vibrations. Structural deformations also can result in the more complex interactions that are associated with multiple degree of freedom dynamic systems, e.g. repetitive impacts or chattering at contacts of the type introduced in Chapter 8.

9.1   Free Vibration of Slender Elastic Bodies

9.1.1   Free Vibration of a Uniform Beam

Transverse displacements $w(x, t)$ of a beam with mass per unit length $\rho A(x)$ and cross-sectional bending stiffness $EI(x)$ are represented by an equation of motion[1]

$$-\frac{\partial^2}{\partial x^2}\left(EI\frac{\partial^2 w}{\partial x^2}\right) = \rho A\frac{\partial^2 w}{\partial t^2}.$$

If the beam is uniform, this linear, fourth order partial differential equation can be expressed as

$$0 = \frac{\partial^2 w}{\partial t^2} + \gamma^2\frac{\partial^4 w}{\partial x^4}, \qquad \gamma^2 = \frac{EI}{\rho A}. \tag{9.1}$$

It has a separable solution

$$w(x, t) = X(x)(A_1\cos\omega_0 t + A_2\sin\omega_0 t)$$

---

[1] This equation for a beam neglects rotary inertia and the effect of transverse shear; thus, it applies only for wavelengths that are long in comparison with cross-sectional dimensions of the beam.

where ω_0 is the angular frequency of free vibration. Substitution into (9.1) gives

$$0 = \frac{d^4 X}{dx^4} - \frac{\omega_0^2}{\gamma^2} X.$$

This fourth order equation has a solution with repeated roots. The solution can be expressed in terms of two pairs of trigonometric functions that are respectively symmetric and antisymmetric,

$$\begin{aligned} X(x) = {} & C_1(\cos kx + \cosh kx) + C_2(\cos kx - \cosh kx) \\ & + C_3(\sin kx + \sinh kx) + C_4(\sin kx - \sinh kx) \end{aligned} \tag{9.2}$$

where $k \equiv \sqrt{\omega_0/\gamma}$ is the wave number.

9.1.2   Eigenfunctions of a Uniform Beam with Clamped Ends

Solutions of the form (9.2) with coefficients that satisfy the boundary conditions for the beam are known as *eigenfunctions*; these depend on the end constraints (boundary conditions) appropriate to a particular case. For a beam of length L with clamped (i.e. encastré) ends the boundary conditions are

$$X(0) = X(L) = 0, \qquad dX(0)/dx = dX(L)/dx = 0. \tag{9.3}$$

With clamped boundary conditions (9.3), the constants C_1, C_2, C_3, C_4 are related by

$$C_1 = C_3 = 0, \qquad \frac{C_2}{C_4} = \frac{\cos kL - \cosh kL}{\sin kL + \sinh kL} = -\frac{\sin kL - \sinh kL}{\cos kL - \cosh kL}.$$

The last equality gives an equation for the eigenvalue k,

$$\cos kL \cosh kL = 1. \tag{9.4}$$

The first five eigenvalues are listed in Table 9.1, while the associated mode shapes are sketched in Fig. 9.1b. The eigenvalues are closely approximated by $k_j L \approx (j + 1/2)\pi$, $j = 1, 2, \ldots, \infty$.

The jth eigenfunction (i.e. natural mode of vibration) for a uniform bar with both ends clamped can be expressed as the sum of an antisymmetric and a symmetric function of $k_j x$,

$$X_j = C_j(\sin k_j L - \sinh k_j L) \left\{ \frac{\sin k_j x - \sinh k_j x}{\sin k_j L - \sinh k_j L} - \frac{\cos k_j x - \cosh k_j x}{\cos k_j L - \cosh k_j L} \right\} \tag{9.5}$$

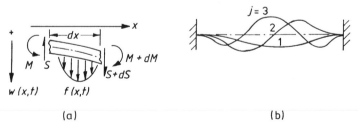

(a)   (b)

Figure 9.1. (a) Differential element used to derive equation of motion, and (b) first, second, and third mode shapes for a beam clamped at both ends.

where the constant C_j can be chosen in order that eigenfunctions are normalized; i.e.

$$1 = \int_0^L X_j X_j \, dx. \tag{9.6a}$$

Alternatively, the constant C_j can be chosen such that the maximum amplitude $X_j|_{\max}$ is unity,

$$1 = X_j|_{\max}. \tag{9.6b}$$

The circular frequency ω_j for the jth mode is obtained from the principle of virtual work, which gives[2]

$$\frac{\omega_j^2}{\gamma^2} = \int_0^L \frac{d^2 X_j}{dx^2} \frac{d^2 X_j}{dx^2} \, dx \div \int_0^L X_j X_j \, dx = k_j^4. \tag{9.7}$$

For the linear equation (9.1), the transverse displacements are obtained by superposition of the response in independent modes,

$$w(x, t) = \sum_{j=0}^{\infty} X_j (A_{1j} \cos \omega_j t + A_{2j} \sin \omega_j t) \tag{9.8}$$

where the coefficients A_{1j}, A_{2j} are obtained from the initial conditions.

9.1.3 Rayleigh–Ritz Mode Approximation

When only a few modes are required to represent the structural response, a set of shape functions $X_j(x)$ that approximate the mode shapes are sufficiently accurate to give a good estimate of modal frequencies.

For a structural element with uniform properties, the functions employed to approximate the deflected shape must be continuous functions that satisfy the natural boundary conditions. Furthermore, it simplifies the analysis if the set of functions which approximate distinct modes are orthogonal so that the frequency equations are decoupled. Let the approximation be expressed as

$$w(x, t) = \sum_{j=0}^{\infty} A_j X_j \sin(\omega_j t - \phi_j) \tag{9.9}$$

with phase angle ϕ_j and amplitude A_j. This gives a partial kinetic energy T_j for the jth approximate mode,

$$T_j = \frac{1}{2} \int_0^L \rho A \dot{w}_j^2(x, t) \, dx = \frac{[A_j \omega_j \cos(\omega_j t - \phi_j)]^2}{2} \int_0^L \rho A X_j^2(x) \, dx.$$

The partial kinetic energy associated with the jth mode is a maximum when the mode deflection is a maximum,

$$T_j|_{\max} = \frac{A_j^2 \omega_j^2}{2} \int_0^L \rho A X_j^2(x) \, dx. \tag{9.10}$$

---

[2] The eigenfunctions are orthogonal; i.e., $0 = \int_0^L X_j X_k \, dx$, $j \neq k$. Hence $0 = \int_0^L (d^4 X_j/dx^4) X_k \, dx = \int_0^L (d^2 X_j/dx^2)(d^2 X_k/dx^2) \, dx$ if $j \neq k$.

Table 9.1. *Modal Eigenvalues and Frequencies for Uniform Bar with Clamped Ends*

Mode no.	Wave number $k_j L$ (rad)	C_2/C_4	Exact	Modal Frequency $L^2\omega_j/\gamma$ Rayleigh Approx.	Single DOF Approx.
1	4.730	−0.98	22.37	22.80	22.80
2	7.853	−1.00	61.67	78.98	–
3	10.996	−1.00	120.91	–	–
4	14.137	−1.00	199.85	–	–
5	17.279	−1.00	298.56	–	–

Similarly, associated with mode X_j there is a potential $U_j(t)$, and for a beam in bending, this part of the potential energy is written as

$$U_j = \frac{1}{2}\int_0^L \mathcal{M}(x,t)\frac{d^2w(x,t)}{dx^2}\,dx = \frac{1}{2}\int_0^L EI\left[\frac{d^2w_j(x,t)}{dx^2}\right]^2 dx$$

where curvature d^2w/dx^2 is related to the bending moment $\mathcal{M}(x,t)$ by $\mathcal{M}=EI(d^2w/dx^2)$. The maximum value of this potential occurs when the modal displacement is maximum,

$$U_j|_{\max} = \frac{A_j^2}{2}\int_0^L EI\left(\frac{d^2X_j}{dx^2}\right)^2 dx. \tag{9.11}$$

In a conservative system the maximum values of kinetic and potential energies are equal; i.e. for any mode, $U_j|_{\max} = T_j|_{\max}$. This gives an equation for the modal frequency ω_j,

$$\omega_j^2 = \int_0^L EI\left(\frac{d^2X_j}{dx^2}\right)^2 dx \div \int_0^L \rho A X_j^2(x)\,dx. \tag{9.12}$$

Example 9.1 A uniform beam of length L, with Young's modulus E and cross-sectional second moment of area I_s has both ends clamped. This beam satisfies the equation of motion (9.1) and boundary conditions (9.3). Estimate the frequencies of the first symmetric and antisymmetric modes.

Solution Mode approximations (not normalized):

$$X_1 = 1 - \cos\frac{2\pi x}{L}, \qquad\qquad \text{symmetric}$$

$$X_2 = \sin\frac{2\pi x}{L} - \frac{1}{2}\sin\frac{4\pi x}{L}, \qquad \text{antisymmetric.}$$

It is worth noting that in addition to satisfying the boundary conditions, these functions are orthogonal. Substitution of the mode approximations into (9.12) and integration gives

$$L^2 \omega_1 / \gamma = 22.80$$
$$L^2 \omega_2 / \gamma = 78.98.$$

In Table 9.1 these approximations are compared with exact values. With these approximating shape functions, Rayleigh's method gives errors of 2% for the frequency of the lowest mode and 28% for the frequency of the second mode. In general the approximate frequency is always in excess of the exact value because the mode approximation represents a variant of the actual deformation mode shape – a variant that is constrained to satisfy the specified displacement function.

9.1.4 Single Degree of Freedom Approximation

If the mass of the colliding body, M', is small in comparison with the mass of the structure, M, a single degree of freedom analysis for the dynamic response can give a useful estimate of the structural response to impact. The single degree of freedom approximation essentially is an approximation of the mode shape $\tilde{X}(x)$. In this approximation, the structure displacement $w(x, t)$ is assumed to be a separable function, $w(x, t) = q_0(t)\tilde{X}(x)$. For the dynamic model shown in Fig. 9.2, the equivalent spring stiffness $\tilde{\kappa}$ is obtained from the strain energy at maximum deflection with the proviso that the shape function $\tilde{X}(x)$ satisfies the normalization condition (9.6b). For example, if the structure is a uniform beam in bending, this gives

$$\tilde{U}|_{\max} = \frac{1}{2}q_0^2 \int_0^L EI\left(\frac{d^2\tilde{X}}{dx^2}\right)^2 dx = \frac{1}{2}\tilde{\kappa}q_0^2\Big|_{\max}. \tag{9.13}$$

By normalizing the mode shape such that the maximum displacement is unity; i.e. $\tilde{X}|_{\max} = 1$, the equivalent stiffness is obtained as

$$\tilde{\kappa} = \int_0^L EI\left(\frac{d^2\tilde{X}}{dx^2}\right)^2 dx. \tag{9.14a}$$

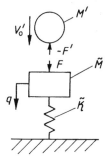

Figure 9.2. Single degree of freedom model of a compact, perfectly plastic body with mass M' striking a flexible system represented by effective mass $\tilde{M}$ and equivalent stiffness $\tilde{\kappa}$.

Similarly, an equivalent mass $\tilde{M}$ is obtained from the maximum kinetic energy in the mode:

$$\tilde{M} = \int_0^L \rho A \tilde{X}^2 \, dx. \tag{9.14b}$$

These parameters give an approximate circular frequency $\tilde{\omega}^2 = \tilde{\kappa}/\tilde{M}$.

Example 9.2   A uniform beam of length L and mass $M = \rho A L$ is clamped at both ends. For the mode approximation $\tilde{X} = 0.5[1 - \cos(2\pi x/L)]$ find the equivalent mass $\tilde{M}$ and the modal frequency $\tilde{\omega}$ for free vibrations of the beam.

Solution   Let the transverse displacement $w(x, t) = q_0(t)X(x) = 0.5q_0(t)[1 - \cos(2\pi x/L)]$. The kinetic energy T of the mode approximation is obtained as

$$T = \frac{1}{2} \int_0^L \rho A [\dot{q}_0 X(x)]^2 \, dx = \frac{3}{16} M \dot{q}_0^2.$$

Since $T = 0.5 \tilde{M} \dot{q}_0^2$ and $M = \rho A L$, this gives an equivalent mass $\tilde{M} = 3M/8$ for a uniform clamped–clamped beam. The maximum potential energy U for this mode approximation is

$$U = \frac{1}{2} \int_0^L EI \left[q_0 \frac{d^2 X}{dx^2} \right]^2 \, dx = \frac{\pi^4 EI}{L^3} q_0^2.$$

The equivalent single degree of freedom system (Fig. 9.2) has an equivalent spring stiffness $\tilde{\kappa}$ obtained from $U = 0.5 \tilde{\kappa} q_0^2$; hence, $\tilde{\kappa} = 2\pi^4 EI/L^3$. This gives a frequency $\tilde{\omega}$ for the mode approximation,

$$\tilde{\omega} = \frac{4\pi^2}{L^2} \sqrt{\frac{EI}{3\rho A}}.$$

Since for this example the mode approximation is geometrically similar to the deformation profile used in the Rayleigh mode approximation (Ex. 9.1), the frequencies obtained by the single degree of freedom and the Rayleigh method are identical.

9.2   Transverse Impact on an Elastic Beam

For transverse impact on a flexible structure, the analytical method to be employed depends on the mass of the colliding body in comparison with the mass of the structure, the location of the impact in relation to boundaries or other points where displacements of the structure must satisfy boundary conditions, and the deformability of the contact region in comparison with the structural compliance. These factors determine the number of structural modes which significantly affect the structural response and consequently the duration of the period of contact. If the structure is very stiff in comparison with the contact region, the results of Chapter 5 are little affected by structural deformations. On the other hand, if the structure is compliant in comparison with the contact region, structural vibrations play a major part in the response to impact.

9.2.1   Forced Vibration of a Uniform Beam

For a uniform beam in bending that is acted on by a distributed transverse force $f(x, t)$ the partial differential equation for transverse displacement $w(x, t)$ is written as

$$\gamma^2 \frac{\partial^4 w}{\partial x^4} + \frac{\partial^2 w}{\partial t^2} = \frac{f(x, t)}{\rho A}. \tag{9.15}$$

A separable solution $w(x, t) = q(t)X(x)$ can be obtained in terms of the eigenfunctions $X_j(x)$ obtained for free vibration where $f(x, t) = 0$. For forced vibration, the principle of virtual work and a set of orthogonal eigenfunctions X_j gives

$$\sum_{j=1}^{n} \left\{ \gamma^2 q_j(t) \int_0^L \left(\frac{d^2 X_j}{dx^2} \right)^2 dx + \frac{d^2 q_j}{dt^2} \int_0^L X_j^2 \, dx \right\} = \int_0^L X_j \frac{f(x, t)}{\rho A} \, dx \tag{9.16}$$

Recalling from (9.7) that $\int_0^L \left(d^2 X_j / dx^2 \right)^2 dx = k_j^4 \int_0^L X_j^2 \, dx$, the previous equation yields an ordinary differential equation for the time-dependent function $q_j(t)$ associated with each eigenfunction. If in addition the force acts at point x_0 only – i.e., $f(x, t) = F(t)\delta(x_0)$, where $\delta(x)$ is the Dirac delta function – then the differential equation becomes

$$\omega_j^2 q_j + \frac{d^2 q_j}{dt^2} = \frac{F(t)}{\rho A} \frac{X_j(x_0)}{\int_0^L X_j^2(x) \, dx}, \qquad \omega_j^2 = \gamma^2 k_j^4. \tag{9.17}$$

This differential equation is solved by convolution of the impulse response function. For a transient force applied to an initially quiescent structure, it has a solution

$$q_j = \frac{X_j(x_0)}{\rho A \omega_j} \frac{\int_0^t F(t') \sin \omega_j (t - t') \, dt'}{\int_0^L X_j^2 \, dx}. \tag{9.18}$$

Finally, the partial displacements from the eigenfunctions are summed to obtain the distribution of bar displacement in response to the applied force $F(t)$,

$$w(x, t) = \sum_{j=1}^{\infty} \frac{X_j X_j(x_0)}{\rho A \omega_j} \frac{\int_0^t F(t') \sin \omega_j (t - t') \, dt'}{\int_0^L X_j^2(x) \, dx}. \tag{9.19}$$

9.2.2   Impact of a Perfectly Plastic Missile

If a 1D structure (e.g. a beam) is struck transversely by a perfectly plastic missile at a point located at x_0 and local deformation (indentation) in the contact region is negligible, then during the contact period the missile and the point of contact on the beam will undergo the same acceleration. Let the missile with mass M' initially be traveling in a direction normal to the surface of the structure with speed V_0'. After initiation of contact, the contact force $F(t)$ that acts on the beam is related to the rate of change of momentum of the missile, $F = -M' \, dV'/dt$. If at the contact point the local indentation of both missile and beam are negligible, then the transverse acceleration of the missile equals the transverse acceleration of the contact point on the beam, $F = -M' X(x_0) d^2 q / dt^2$. Consequently (9.17) can be written as

$$\omega_j^2 q_j + \frac{d^2 q_j}{dt^2} = \frac{-M'}{\rho A} \frac{d^2 q_j}{dt^2} \frac{X_j(x_0)}{\int_0^L X_j^2 \, dx} \tag{9.20}$$

which gives $q_j = A_j \sin \bar\omega_j t$ where $\bar\omega_j$ is a modified modal frequency that takes into account the effect of additional mass from the missile resting on the beam during contact,

$$\bar\omega_j^2 = \frac{\omega_j^2}{1 + \dfrac{M' X_j(x_0)}{\rho A \int_0^L X_j^2 \, dx}}. \tag{9.21}$$

For the beam with an attached mass M' the equation of motion for modal response has a solution

$$q_j = A_j \sin(\bar\omega_j t + \bar\phi_j).$$

The beam displacement and velocity of this modified system can be expressed as

$$w(x, t) = \sum_{j=1}^{\infty} A_j \sin(\bar\omega_j t + \bar\phi_j) \, X_j(x) \tag{9.22a}$$

$$\frac{dw(x, t)}{dt} = \sum_{j=1}^{\infty} A_j \bar\omega_j \cos(\bar\omega_j t + \bar\phi_j) \, X_j(x). \tag{9.22b}$$

Initial conditions are used to obtain the coefficients for amplitude A_j and phase angle $\bar\phi_j$. Suppose an initially quiescent body is struck by a rigid missile of mass M' and before it strikes the missile is traveling with speed V_0'; this gives initial conditions that can be represented by $w(x, 0) = 0$, $\dot w(x, 0) = M' V_0' \delta(x_0)$. Multiplying each side of the expressions above by the mode shape X_j and then integrating over the volume of the structure gives

$$0 = A_j \sin \bar\phi_j, \qquad \int_0^L \dot w(x, 0) X_j(x) \, dx = A_j \bar\omega_j \cos \bar\phi_j \int_0^L X_j^2 \, dx.$$

The coefficients that satisfy conservation of linear momentum at impact are as follows:

$$\bar\phi_j = 0 \qquad A_j = \frac{M' V_0' X_j(x_0)}{\bar\omega_j \rho A \int_0^L X_j^2 \, dx}. \tag{9.23}$$

Example 9.3 A uniform elastic beam of length L, cross-sectional area A, elastic modulus E, and second moment of area I is clamped at both ends. The beam is initially at rest before it is struck by a rigid missile of mass M' that is traveling with speed V_0' in a direction perpendicular to the axis of the beam. Using the two mode Rayleigh–Ritz approximation obtained in Ex. 9.1, find the response of the beam to impact for impact points $x_0/L = 0.25$ and 0.5; in each case obtain an estimate of the time of separation.

Solution Mode approximations obtained in Ex. 9.1:

$$X_1 = 1 - \cos \frac{2\pi x}{L}, \qquad \text{symmetric}$$

$$X_2 = \sin \frac{2\pi x}{L} - \frac{1}{2} \sin \frac{4\pi x}{L}, \qquad \text{antisymmetric.}$$

Recall from Ex. 9.1 that $\omega_1 = 22.8\gamma/L^2$ and $\omega_2 = 78.98\gamma/L^2$, so that the modified modal

Table 9.2. *Modal Frequencies and Amplitudes as Functions of Mass Ratio M'/M for a Uniform Clamped–Clamped Beam Hit at x_0/L by a Rigid Missile*

	$x_0/L = 0.25$				$x_0/L = 0.5$			
M'/M	$\dfrac{L^2\bar{\omega}_1}{\gamma}$	$\dfrac{A_1 L^2}{\gamma V_0'}$	$\dfrac{L^2\bar{\omega}_2}{\gamma}$	$\dfrac{A_2 L^2}{\gamma V_0'}$	$\dfrac{L^2\bar{\omega}_1}{\gamma}$	$\dfrac{A_1 L^2}{\gamma V_0'}$	$\dfrac{L^2\bar{\omega}_2}{\gamma}$	$\dfrac{A_2 L^2}{\gamma V_0'}$
0.1	22.1	0.003	75.3	0.001	21.4	0.006	79.0	0
1.0	17.7	0.038	55.8	0.018	14.9	0.089	79.0	0
10.0	8.2	0.808	23.8	0.420	6.0	2.211	79.0	0

frequencies obtained from (9.21) are

$$\bar{\omega}_1 = \frac{22.8\gamma/L^2}{\sqrt{1 + \dfrac{2M'X_1(x_0)}{3M}}}, \qquad \bar{\omega}_2 = \frac{78.98\gamma/L^2}{\sqrt{1 + \dfrac{M'X_2(x_0)}{M}}}.$$

The amplitude coefficients A_j are obtained as

$$\frac{A_1 L^2}{\gamma V_0'} = \frac{2M'X_1(x_0)}{3M}\sqrt{1 + \frac{2M'X_1(x_0)}{3M}} \times \frac{1}{22.8}$$

$$\frac{A_2 L^2}{\gamma V_0'} = \frac{M'X_2(x_0)}{M}\sqrt{1 + \frac{M'X_2(x_0)}{M}} \times \frac{1}{78.98}.$$

Amplitudes calculated for various mass ratios M'/M are listed in Table 9.2.

Unless the colliding mass is very large in comparison with the mass of the structure, the analysis in this section gives too long a period of contact and thus underestimates the maximum contact force. Usually the structural compliance will be larger than the compliance at the contact point, so that the period of contact is controlled by local contact rather than structural compliance.

9.2.3 Effect of Local Compliance in Structural Response to Impact

In general the compliance of the structure prolongs the contact duration and reduces the maximum force in any impact. The magnitude of these effects depends on the difference between the contact period for local deformations and the period of the fundamental mode of structural vibration. To illustrate these interactions, consider a uniform elastic beam with length L and cross-sectional second moment of area I that is composed of material with Young's modulus E. The beam is struck by a colliding missile of mass M' that is initially traveling in a direction transverse to the axis of the beam. The missile is collinear. At the impact point x_0 there is a contact force $F(t)$, and this results in flexural deflection of the beam $w(x_0, t)$, for which Eq. (9.19) gives

$$w(x_0, t) = \sum_{j=1}^{\infty} \frac{X_j^2(x_0)}{\rho A \omega_j} \frac{\int_0^t F(t')\sin\omega_j(t - t')\,dt'}{\int_0^L X_j^2(x)\,dx}.$$

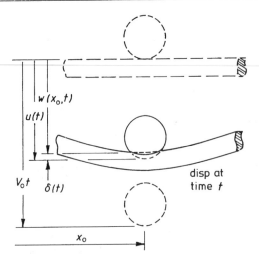

Figure 9.3. Local indentation $\delta(t) = u(t) - w(x_0, t)$ during collision between a sphere and an initially stationary beam or plate. The displacements of contact points on the colliding mass $u(t)$ and beam $w(x_0, t)$ do not include displacement due to local deformation.

Simultaneously, an equal but opposite contact force is decelerating the colliding body. Here we consider that the contact force acts in a direction normal to the initial contact surface and has a line of action passing through the center of mass of the colliding body; consequently, the colliding body is decelerated from an impact speed V_0, but it suffers no rotation. During contact the center of mass of the colliding body undergoes a displacement $u(t)$ where

$$u(t) = V_0 t - \frac{1}{M'} \int_0^t (t - t')F(t') \, dt'.$$

The combined local indentation of missile and beam is represented by $\delta(t) = u(t) - w(t)$ as shown in Fig. 9.3. This local indentation is related to the contact force through a force–indentation relation that depends on the geometry of the contacting surfaces and the magnitude of the indentation:

$$\delta = \kappa^{-1} F, \qquad\qquad \text{linear approximation} \qquad\qquad (9.24a)$$

$$\delta = \kappa_c^{-1} F, \quad \delta < \delta_Y, \qquad \text{cylindrical elastic contact surfaces} \qquad (9.24b)$$

$$\delta = \kappa_s^{-2/3} F^{2/3}, \quad \delta < \delta_Y, \qquad \text{spherical elastic contact surfaces.} \qquad (9.24c)$$

With any particular force–indentation relation, compatibility of displacements at x_0 gives an equation for the contact force $F(t)$ acting on the beam,

$$\delta(t) = V_0 t - \frac{1}{M'} \int_0^t (t - t')F(t') \, dt'$$

$$- \sum_{j=1}^{\infty} \frac{X_j^2(x_0)}{\rho A \omega_j} \frac{\int_0^t F(t') \sin \omega_j (t - t') \, dt'}{\int_0^L X_j^2 \, dx}. \qquad (9.25)$$

This integral equation for the contact force $F(t)$ was obtained by Timoshenko (1913) who used quadrature to obtain a numerical solution to the equation for particular cases of spherical contact; i.e., he discretized the integrals and solved for F at each successive time step.

Lennertz (1937) developed an approximate solution to the equation based on the structural response of the lowest mode of vibration only. This method was improved and extended by E.H. Lee (1940). The method gives reasonable results if the contact period (Eq. 6.31) is short in comparison with the modal period of vibration $2\pi/\omega_1$.

Lee's method assumes that during contact *the reaction force is a pulse that varies sinusoidally with time*. The pulse has period Ω and amplitude F_c that are obtained from structural response of the compliant system, i.e. $F(t) \approx F_c \sin \Omega t$, $0 < t < \pi/\Omega$. After substitution into (9.26) and integration, the indentation $\delta(t)$ is obtained as

$$\delta(t) = V_0 t - \frac{F_c}{\Omega^2 M'}(\Omega t - \sin \Omega t) - \frac{F_c}{\omega_1 \tilde{M}_1(x_0)} \left\{ \frac{\Omega \sin \omega_1 t - \omega_1 \sin \Omega t}{\Omega^2 - \omega_1^2} \right\}$$

$$\tilde{M}_1(x_0) \equiv \rho A \int_0^L X_1^2 \, dx \div X_1^2(x_0)$$

where $\tilde{M}_1(x_0)$ is an equivalent mass for a beam that is hit at impact point x_0.[3] Because the contact period π/Ω is short in comparison with the period of the fundamental mode of vibration, the function $\sin \omega_1 t$ can be approximated by $\sin \omega_1 t \approx \omega_1 t$ while the colliding missile is in contact. After substituting this approximation into the equation above, we have

$$\delta = V_0 t - \frac{F_c}{M'\Omega^2} \left\{ \frac{(\tilde{M}_1 + M')\Omega^2 - \tilde{M}_1 \omega_1^2}{\tilde{M}_1(\Omega^2 - \omega_1^2)} \right\} (\Omega t - \sin \Omega t). \tag{9.26}$$

At this point the force–indentation relation for a particular contact geometry and indentation depth δ is substituted into the equation and the functions of time t and $\sin \Omega t$ are equated separately.

Linear compliance, $\delta = \kappa^{-1}F$.
We have

$$\frac{M'\Omega^2}{\kappa} = \frac{(1+\alpha)\Omega^2 - \omega_1^2}{\Omega^2 - \omega_1^2} \equiv B(\Omega)$$

$$\frac{M'V_0}{F_c\Omega^{-1}} = B(\Omega)$$

where mass ratio $\alpha = M'/\tilde{M}_1$. These expressions give

$$\Omega^2 = \frac{1}{2}\left[\omega_1^2 + \frac{(1+\alpha)\kappa}{M'} \right]$$

$$+ \frac{1}{2} \left\{ \left[\omega_1^2 + \frac{(1+\alpha)\kappa}{M'} \right]^2 - \frac{4\kappa\omega_1^2}{M'} \right\}^{1/2} \tag{9.27a}$$

$$F_c = \kappa V_0/\Omega. \tag{9.27b}$$

(The expression for Ω^2 contains a choice of sign as $+$; this choice can be explained by considering the limit as $\omega_1 \to 0$.)

Hertz compliance, $\delta = \kappa_s^{-2/3}F^{2/3}$.
To solve the nonlinear equation resulting from beam dynamics with Hertz contact compliance, Lee (1940) adopted an approximation for the sinusoidal function, $(\sin \Omega t)^{2/3} \approx$

---

[3] Note that if the impact point approaches a fixed boundary where the displacement vanishes, the equivalent mass becomes large without limit.

1.093 sin Ωt. For the period of contact this approximation gives an impulse equal to that of the Hertzian pulse. With this approximation, Eq. (9.26) can be rearranged to give

$$\frac{1.093 M' \Omega^2}{\kappa_s^{2/3} F_c^{1/3}} = \frac{M' V_0 \Omega}{F_c} = B(\Omega)$$

This equality gives frequency Ω and force amplitude F_c,

$$\Omega = \left\{\frac{\kappa_s^2 B^2(\Omega) V_0}{1.306 M'^2}\right\}^{1/5}, \qquad F_c = \kappa_s \left(\frac{V_0}{1.093\Omega}\right)^{3/2} \qquad (9.28)$$

For a spherical contact between elastic bodies, the stiffness $\kappa_s = \frac{4}{3} E_* R_*^{1/2}$ [Eq. (6.8)]. If the period of structural response is much longer than the contact period $\Omega \gg \omega_1$ the expressions above give $B \approx 1 + \alpha$ and

$$\Omega \approx \left[\frac{\kappa_s^2 (1 + \alpha)^2 V_0}{1.306 M'^2}\right]^{1/5}, \qquad F_c = 1.437 \kappa_s \left[\frac{T_0'}{(1 + \alpha)\kappa_s}\right]^{3/5} \qquad (9.29)$$

where T_0' is the incident kinetic energy of relative motion. While the approximation to the trigonometric function $(\sin \Omega t)^{2/3}$ can be improved by a method of successive approximation, in most cases this effort is not justified.

For either linear or Hertz compliance, at the impact point x_0 the displacement of the beam $w(x_0, t)$ and that of the center of mass of the sphere $u(t)$ are given by

$$w(x_0, t) = \frac{F_c}{(\Omega^2 - \omega_1^2)\tilde{M}_1(x_0)} \left\{\frac{\Omega}{\omega_1} \sin \omega_1 t - \sin \Omega t\right\} \qquad (9.30)$$

$$u(t) = V_0 t - \frac{F_c}{\Omega^2 M'}(\Omega t - \sin \Omega t)$$

$$= \frac{V_0 t}{1 + \alpha}\left\{\alpha + \frac{\sin \Omega t}{\Omega t}\right\}, \qquad t < t_f. \qquad (9.31)$$

Example 9.4 A simply supported steel beam of length $L = 153.5$ mm, depth $h = 10$ mm and width $b = 10$ mm is struck at midspan by a steel sphere of radius $R' = 10$ mm with an incident velocity of 0.01 m s^{-1}. Compare the maximum force F_c at the interface with the force obtained from a similar impact against an elastic half space, and use a single degree of freedom analysis to estimate the dynamic deflection $w(x_0, t)$ of the impact point on the beam.

Solution

Material properties: $E = 220$ GPa, $\rho = 7900$ kg m^{-3}.
Section properties: $EI = 183.3$ N m, $\rho A = 0.78$ N s^2 m^{-2}, $\gamma = 15.23$ m^2 s^{-1}.
Beam properties: mass $M = 0.120$ kg, mode equivalent mass $\tilde{M}_1 = M/2 = 0.060$ kg, frequency of fundamental mode $\omega_1 = 6381$ rad s^{-1}.
Sphere properties: mass $M' = 0.033$ kg, mass ratio $\alpha = M'/\tilde{M}_1 = 0.55$.
Contact conditions: stiffness $\kappa_s = 29.3 \times 10^9$ N m$^{-3/2}$, frequency $\Omega = 27.07 \times 10^3$ rad s^{-1}, contact force $F_c = 5.76$ N, contact period $t_f = 0.116 \times 10^{-3}$ s.

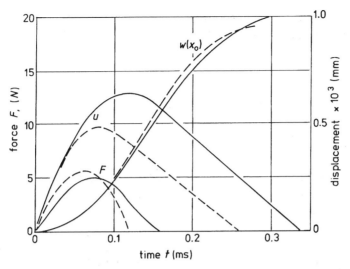

Figure 9.4. Transverse impact of steel sphere, with radius $R' = 0.01$ m, at midspan of simply supported steel beam, $b = h = 0.01$ m, $L = 0.1535$ m, which gives mass ratio $\alpha = 0.55$. Incident speed of sphere, 0.01 m s^{-1}, yields solely elastic deformation. Contact force $F(t)$, beam displacement $w(x_0, t)$ and sphere displacement $u(t)$ for single degree of freedom approximation (dashed curves) are compared with a numerical solution that includes higher order modes of deformation (solid curves: numerical solution from Timoshenko, 1913).

The maximum indentation δ_c of the half space is obtained from equating the work during compression to the change in kinetic energy, $\int_0^{\delta_c} \kappa_s \delta^{3/2}\, d\delta = M'V_0^2/2$. Hence

$$\delta_c = \left(\frac{5M'V_0^2}{4\kappa_s}\right)^{2/5} = 0.456 \times 10^{-6}\,\text{m}$$

so that the maximum contact force is

$$F_c = \kappa_s \delta_c^{3/2} = 9.03\,\text{N}$$

and the contact period is

$$t_f(\pi) = 2.94\frac{\delta_c}{V_0} = 0.135 \times 10^{-3}\,\text{s} \qquad [\text{Eq. (6.31)}].$$

Figure 9.4 compares the contact force during impact obtained from this single degree of freedom modal approximation with the displacements obtained from a numerical solution of Eq. (9.26) (see Timoshenko, 1913).[4] The numerical solution includes the effect of higher order modes. This comparison shows that the fundamental mode approximation gives an increased magnitude and decreased period for the pulse of contact force; these are consequences of excessive beam stiffness. The higher order modal deformations (which are neglected by the single mode solution) become more important as slenderness ratio L/h increases and as the impact point becomes more off-center.

---

[4] Lee, Hamilton and Sullivan (1983) developed a higher order lumped parameter method and used comparison with this example problem to demonstrate that their calculated result converges to Timoshenko's numerical solution.

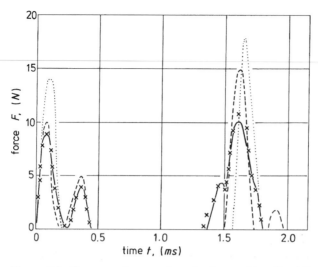

Figure 9.5. Transverse impact of steel sphere, with radius $R' = 0.02$ m, at midspan of simply supported steel beam, $b = h = 0.01$ m, $L = 0.307$ m, which gives mass ratio $\alpha = 2.2$. Incident speed of sphere, 0.01 m s^{-1}. The contact force $F(t)$ converges to the numerical solution as the number of degrees of freedom increases. Modal approximations used in these calculations are as follows: mode 1, dotted curves; modes 1 and 3, dashed curves; modes 1, 3, 5 and 7, solid curves (Lee, Hamilton and Sullivan, 1983).

During collision of an elastic sphere against a slender elastic beam, the maximum contact force is substantially reduced in comparison with the maximum force during collision of the same sphere against an elastic half space. The force reduction is due to the larger compliance of the beam; this effect increases as the beam compliance becomes larger in comparison with compliance of the contact region.

For mass ratio $\alpha > 1$, the response is complicated by multiple strikes occurring at the impact point before separation at time t_f. Figure 9.5 illustrates the variation of contact force for a case of multiple impact in a system that is but a slight modification of that in Ex. 9.4, i.e. direct impact in a transverse direction by an elastic sphere on a slender simply supported beam. In this case the mass ratio has been increased to $\alpha = 2.2$. The dotted and dashed curves were calculated using only low order modes; this set of curves indicate convergence to the continuum solution (crosses in Fig. 9.5) as the number of modes in the approximate solution is increased.

For a light mass striking a slender simply supported beam ($\alpha = 0.205$) at midspan, Fig. 9.6 compares the maximum force calculated from Eq. (9.29) with experimental measurements by Schwieger (1970). Both the ball and beam are steel, and the impact speed ranges from 0 to 1.5 m s^{-1}. For a light mass, in order to obtain an accurate approximation for the contact force it is necessary to consider the effect of local indentation, e.g. the approximation by Lee.

9.2.4   Impact on Flexible Structures – Local or Global Response?

A light mass striking a stiff and heavy structure results in little motion of the structure during impact. The response to impact by a light mass is accurately represented

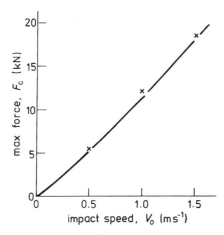

Figure 9.6. Maximum force F from central impact of mass $M' = 0.885$ kg striking steel beam of length $L = 0.86$ m, depth $h = 0.0051$ m and width $b = 0.00254$ m. For this impact the mass ratio $\alpha = 0.205$, while the stiffness $\kappa_s = 24.2 \times 10^9$ N m$^{-3/2}$. The curve is the analytical approximation, while crosses indicate experimental data by Schwieger (1970).

by the Hertzian local indentation model of Chapter 5, since during the contact period there is little motion of the structure; consequently the contact period is brief, the contact force is large, and there is little energy lost to structural vibrations.

On the other hand, a heavy missile striking a slender beam or plate results in substantial structural deformations that limit the contact force; consequently, the indentation is small and the inertia of the missile is the dominant factor, so that the response is quasistatic as described in Sect. 9.2.2. Quasistatic structural response gives a relatively long contact period (roughly half the period of the lowest mode of structural vibration), a reduced contact force, energy loss to structural vibrations and a likelihood of multiple impact.

For inertia and stiffness parameters that are between these limits, an analysis is required that incorporates both local indentation and structural deformation, i.e. an analysis such as that in Sect. 9.2.3. What, however, is sufficiently light and stiff or heavy and compliant in order for one approximate method or another to be acceptable?

Christoforou and Yigit (1998) used elastic–plastic contact relations to analyze impact of spherical missiles against beams and rectangular plates having a range of stiffness. The characterizing feature of their linearized analysis was a nondimensional contact force

$$\bar{F}(\tau) \equiv F(t)/M'V_0'\Omega_0$$

where Ω_0 is the frequency for local indentation of an elastic half space (linearized stiffness)

$$\Omega_0^2 = \kappa_s \delta_Y^{1/2}/M'.$$

For a light mass striking a stiff and heavy structure, the response is local, so the Hertzian contact relation gives the correct contact force, viz. the maximum nondimensional force $\bar{F}_c = 1$. At the other end of the spectrum, for a heavy mass striking a light and compliant structure the quasistatic approximation essentially can be represented as a body of mass M' striking a pair of springs arranged in series. This quasistatic limit gives a maximum nondimensional force which depends on the ratio $K = \tilde{\kappa}/\kappa$ of the structural stiffness $\tilde{\kappa}$ to the linearized stiffness of the local contact region, $\kappa = \frac{4}{5}\kappa_s\delta_Y^{1/2} = \frac{4}{5}\pi\vartheta_Y Y R_*$ (for a spherical indentor); viz.

$$\bar{F}_c = \left(\frac{K}{1+K}\right)^{1/2}$$

For impact on thin plates and cylinders, Swanson (1992) has shown that the error of this approximation is less than 5% if the colliding body is sufficiently heavy so that the mass ratio $\alpha > 10$.

To determine the effect of impact on systems with intermediate mass or stiffness ratios, Christoforou and Yigit (1998) investigated the impact response of infinitely long beams and plates of thickness h. These solutions for unbounded systems are valid only so long as the contact terminates before shear waves emanating from the impact can be reflected from boundaries and return to the impact site. The Christoforou–Yigit solution for infinitely large structures depends on an additional nondimensional parameter – the *vibration energy loss factor* ζ. This factor represents the energy transformed to elastic vibrations of the beam or plate during the contact period; it depends on the mass of the colliding body, M'; the mass per unit area of the plate, ρh (or for a beam $\rho b h$); the local contact stiffness κ; and the beam or plate bending stiffness $D_{11} = (h^3/12)[E/(1 - \nu^2)]$. The vibration energy loss factor acts like a damping ratio; for a beam or plate it has the following representations:

Uniform beam:

$$\zeta = \frac{1}{2\sqrt{2\pi}} \left(\frac{M'}{\rho b^2 h}\right)^{1/2} \left(\frac{M'\kappa}{\rho h D_{11}}\right)^{1/4}.$$

Uniform plate:

$$\zeta = \frac{1}{16} \left(\frac{M'\kappa}{\rho h D_{11}}\right)^{1/2}.$$

Figure 9.7 distinguishes domains for local and quasistatic behavior based on the maximum value of the nondimensional contact force $\bar{F}_c$. In this figure the curves for impact on bodies of infinite extent are obtained from the infinite elastic beam or plate solutions, while if the vibration energy loss factor is large ($\zeta > 1$), one has the branches shown for impact on beams or plates with finite length and width. The solutions for finite size structural elements asymptotically approach quasistatic solutions as the loss factor becomes large, $\zeta \gg 1$.

For impact near a boundary where displacements are constrained, the quasistatic solution for a heavy mass on a flexible structure results in an increasing frequency and contact force as the boundary is approached. Figure 9.8 illustrates that near a boundary the apparent coefficient of restitution (based on assuming that at separation the plate velocity is negligible) approaches unity as the energy transformed into elastic vibrations diminishes. The width of the region wherein the coefficient of restitution (or energy absorbed by elastic vibrations) is affected by boundary conditions is roughly equal to half the distance traversed by shear waves during the contact period (Sondergaard, Chaney and Brennen, 1990).

In general the effect of a resilient structure is to prolong the contact period and reduce the maximum contact force in comparison with the Hertz solution for an elastic half space. For elastic–plastic solids, the compliance of the structure effectively increases the normal impact speed for yield. Thus the energetic coefficient of restitution depends on the

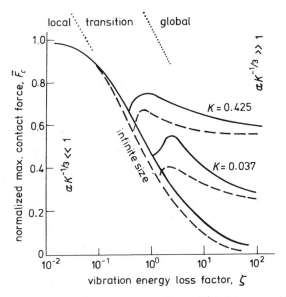

Figure 9.7. Maximum contact force divided by contact force for half space, $\bar{F}_c$, as a function of both the vibration energy loss factor ζ and the stiffness ratio K: solid curves, uniform beam; dashed curves, uniform plate (Christoforou and Yigit, 1998).

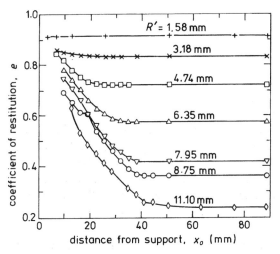

Figure 9.8. Coefficient of restitution e for impact of different size solid spheres near a clamped support on a 12.7 mm thick Lucite plate. Impact speed $V'_0 = 3.5$ m s^{-1} of a steel sphere with radius R' as specified (Sondergaard, Chaney and Brennen, 1990).

compliance of the structure at the point of impact in addition to the material properties, impact speed and relative mass of the colliding bodies.

PROBLEMS

9.1 A uniform beam of length L, cross-sectional area A, and second moment of area I is composed of material with density ρ and Young's modulus E. The beam is simply supported. Find the mode shapes and modal frequencies.

9.2 For a uniform beam with simple supports, use the Rayleigh–Ritz method with sinusoidal shape functions to obtain estimates of the lowest symmetric and antisymmetric modes. Compare the modal frequency approximations with exact values obtained in Problem 9.1.

9.3 The simply supported beam in Problem 9.1 is struck at midspan by a rigid missile of mass M'. Initially the beam is at rest while the missile is moving transverse to the axis at speed V_0'. For eigenfunctions $X_j(x) = \sin k_j L$, $k_j = j\pi/L$, find the equivalent mass $\tilde{M}_j$, the equivalent stiffness $\tilde{\kappa}$ and the amplitude A_j of response of the beam as a function of the mass ratio M/M'.

9.4 Suppose the beam in Problem 9.3 has a uniform cross-section $10 \text{ mm} \times 10 \text{ mm}$ and length $L = 153.5 \text{ mm}$ and is composed of steel ($\rho = 7.9 \times 10^3 \text{ kg m}^{-3}$, $E = 210 \text{ GPa}$). Show that the modal frequencies of free vibration are $\omega_j/2\pi \approx j^2 \times 10^3 \text{ Hz}$. Find the amplitude of vibration if the stationary beam is struck transversely at midspan by a 20 mm diameter steel sphere traveling with an initial speed 10 mm s^{-1} and the impact is perfectly plastic.

9.5 A slender elastic beam is struck transversely at low speed by a spherical elastic missile.
    (a) Assuming the contact period is short in comparison with the fundamental period of vibration, use the terminal missile velocity $\dot{u}(t_f)$ to show that for *elastic* impact on a flexible body, the single mode approximation gives a coefficient of restitution $e_* = (1-\alpha)/(1+\alpha)$. Explain the physical significance of a coefficient $e_* < 0$ if the mass ratio $\alpha > 1$.
    (b) Obtain an expression for the displacement of the sphere (and the contact point on the beam) at termination of contact ($\alpha < 1$).

CHAPTER 10

Propagating Transformations of State in Self-Organizing Systems

Molecules far from equilibrium have far reaching sensitivity whereas
those near equilibrium are sensitive to local effects only,

Ilya Prigogine, Cambridge Lecture, 1995

A ball that falls in a gravitational field before colliding against a flat surface will rebound from the surface with a loss of energy that depends on the coefficient of restitution. If the ball is free, it will continue bouncing on the surface in a series of collisions; these arise because in each collision the ball is partly elastic and during the period between collisions the ball is attracted towards the surface by gravity. In Chapter 2 it was shown that an inelastic ball $(0 < e_* < 1)$ which is bouncing on a level surface in a gravitational field has both a period of time between collisions and a bounce height that asymptotically approach zero as the number of collisions increases. In other words, with increasing time this dissipative system asymptotically approaches a stable *attractor* – the equilibrium configuration where the ball is resting on the level surface.

Some other systems can experience energy input during each cycle of impact and flight; consequently these systems exhibit more complex behavior. For example, a pencil has a regular hexagonal cross-section with six vertices. If the pencil rolls down a plane, the mean translational speed of the axis asymptotically approaches a steady mean speed of rolling where the kinetic energy dissipated by the collision of a vertex against the plane equals the loss in gravitational potential energy as the pencil rolls from one flat side to the next. Sequential toppling of dominoes is another system where a gravitational potential drives a series of dissipative collisions. Here again there is a natural speed of propagation (toppling) where the energy dissipated by each collision equals the change in gravitational potential as the wavefront moves forward one domino in a uniform set. The equations representing sequential toppling will be shown to be directly analogous to those for a rolling pencil; i.e., there is an intrinsic speed of toppling that depends on the domino spacing and size but is independent of the initial conditions. A third system involving a sequence of collisions is a ball bouncing on a vibrating table. Here, however, there are excitation frequencies where steady bouncing develops and other frequencies where the bounce period is chaotic. The key to classifying these alternative behaviors is to identify the steady state solutions, i.e. the solution attractors.

In each of these systems, the release of energy from some source is triggered by activity at a wavefront. Typically the potential energy that can be released is uniformly distributed, so that the rate of energy added to the system increases linearly with speed of propagation.

These systems also have some source of energy loss or dissipation. In any system which exhibits an intrinsic wave speed for propagation of activity, the dissipation is always a non-linear function of speed of propagation; e.g., for domino toppling the energy dissipation rate from collisions between dominoes depends on the cube of the speed of propagation. Hence these systems always satisfy a kinetic or evolution equation of the form

> rate of change of active energy
> $= -$ (rate of change of potential energy)
> $-$ (rate of change of dissipation).

In the case of a mechanical system such as domino toppling or progressive collapse of warehouse racking, the active energy is kinetic energy. For nerve signals in a neuron system, the active energy is an electrical potential (voltage), while for propagation of an infectious disease the active energy is infection.

10.1   Systems with Single Attractor

10.1.1   Ball Bouncing down a Flight of Regularly Spaced Steps

A ball of mass M is dropped from height h_0 onto the top of a flight of regularly spaced steps. Gravity produces a steady downward force $-Mg$ acting on the ball. The ball falls onto the top step with a vertical relative velocity $-v_0$, then rebounds and continues to bounce just once on each successive step in the set. The path of the ball is illustrated in Fig. 10.1. At each impact the coefficient of restitution is e_*.

In this representation of a bouncing ball, each level step is an equilibrium configuration for the ball. At each bounce the ball loses a part $1 - e_*^2$ of the kinetic energy that it possessed just before impact; after each collision it falls to the next step down, thereby gaining an amount of kinetic energy equal to the negative of the loss in potential $-Mgb$, where b is the height of each step. If the energy gain equals the energy dissipated, then every cycle is identical and the ball bounces steadily down the flight of stairs.

Height of Fall for Steady Bouncing, h_*

If the ball bounces steadily down the stairs, then at each bounce the ball attains the same height above the next step as the initial drop height h_0; i.e., $h_* = h_0$. After each collision

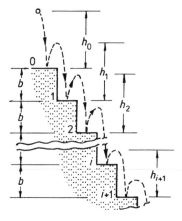

Figure 10.1. Inelastic ball bouncing down a flight of steps with one bounce per step.

the ball that was dropped from an initial height h_* attains a maximum height $h_* e_*^2$ above the step that it bounced from; hence if the height of fall for each successive step is identical,

$$h_* = h_* e_*^2 + b.$$

This gives a drop height for steady bouncing,

$$\frac{h_*}{b} = \frac{1}{1 - e_*^2}. \tag{10.1}$$

Evolution in Bounce Height for General Initial Conditions

Ordinarily the initial drop height is not equal to the height for steady bouncing (h_*), so that the height of each bounce differs as the ball bounces down the stairs. To obtain an expression for the evolution of drop height with number of bounces, consider the cycle of fall, impact and rebound for any step. Let h_i denote the height of fall onto the ith step. Then the height of fall for any two successive steps is

$$h_{i+1} = h_i e_*^2 + b = h_i e_*^2 + h_*(1 - e_*^2).$$

We note that

$$dh_i/di \approx h_{i+1} - h_i.$$

The difference equation has a solution

$$h_i = h_* + B e_*^{2i}.$$

The constant of integration B is evaluated from the initial condition $h_i = h_0$ for $i = 0$; thus $B = h_0 - h_*$ and the height of fall h_i onto the ith step is given by

$$h_i = h_* + (h_0 - h_*)e_*^{2i}. \tag{10.2}$$

Equation (10.2) represents bouncing where the bounce height asymptotically approaches the height for steady bouncing independently of whether the initial drop height h_0 is larger or smaller than the height h_* for steady bouncing. This asymptotic approach is illustrated in Fig. 10.2. The system is stable, and with increasing time (or number of

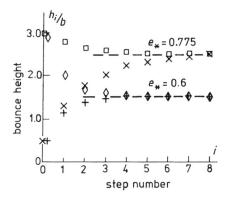

Figure 10.2. Change in bounce height as ball progresses down steps for coefficients of restitution $e_* = 0.6$ and 0.775 and initial drop heights $h_0/b = 0.5$ and 3.0.

bounces) the solution asymptotically approaches the single attractor where the bounce height above each step equals h_*.

Example 10.1   A sphere is dropped onto a level anvil from an initial height h_0.

(a) Find the rebound height h_i after i bounces if each impact is represented by a coefficient of restitution e_*,

(b) Find the time t_f when bouncing ceases.

Neglect the effect of air drag on the sphere.

Solution   After falling a distance h_0 in a gravitational field with intensity g, the relative speed at incidence $v_0 \equiv v(0)$ can be obtained from conservation of energy,

$$v_0^2 = 2gh_0.$$

The coefficient of restitution then gives the separation speed $v_f \equiv v(t_f)$ for this impact and subsequently the rebound height h_1 for the first bounce:

$$v_f^2 = e_*^2 v_0^2 = 2e_*^2 gh_0$$

$$h_1 = e_*^2 h_0.$$

After the ith bounce,

$$h_i = e_*^2 h_{i-1} = e_*^{2i} h_0. \tag{a}$$

Thus the rebound height is reduced to $h_i = \alpha h_0$ after $i = 0.5 \ln \alpha$ bounces, i.e., after i bounces the maximum height is reduced to $h_i/h_0 = e_*^{2i}$.

The time required to complete i bounces is

$$t_i = \sqrt{\frac{2h_0}{g}} \left\{ -1 - e_*^i + 2 \sum_{j=0}^{i} e_*^j \right\}, \qquad 0 < e_* < 1.$$

For a coefficient of restitution in the range $0 < e_* < 1$, the time when bouncing terminates can be obtained by taking the limit of this expression as $i \to \infty$. Letting $x = e_*$ and noting that $(1-x)^{-1} = 1 + x + x^2 + \cdots$ gives a terminal time

$$t_f = \lim_{i \to \infty} t_i = \frac{1 + e_*}{1 - e_*} \sqrt{\frac{2h_0}{g}}. \tag{b}$$

10.2   Systems with Two Attractors

10.2.1   Prismatic Cylinder Rolling down a Rough Inclined Plane

A regular prismatic cylinder with an angle 2Ψ between the sides can be in equilibrium with any side resting against a rough inclined plane if the angle of inclination θ is such that $\theta < \Psi$ and the coefficient of friction $\mu > \tan \theta$. Suppose a pencil (i.e. a regular prismatic cylinder with hexagonal cross-section) rolls down a rough plane without bouncing from the surface. The hexagonal cylinder passes through a series of possible equilibrium configurations, each with a smaller potential energy than the preceding equilibrium configuration – in this respect the rolling prismatic cylinder is similar to the

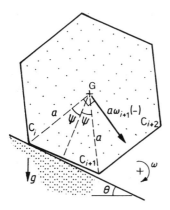

Figure 10.3. Hexagonal prism rolling down a rough inclined plane. The velocity of the center of mass is shown an instant before impact at vertex C_{i+1}.

sphere bouncing down regularly spaced steps. Both the bouncing ball and the rolling prismatic cylinder exhibit steady motion if the kinetic energy gained during each cycle of motion equals the energy dissipated by each collision and other dissipative processes.

Consider a prismatic cylinder with a polygonal cross-section having N equal sides, where $N \geq 4$, as shown in Fig. 10.3. Between adjacent vertices a regular polygon has a central angle 2Ψ, where $\Psi = \pi/N$. Let the sides of the cylinder be slightly convex, so that when each side collides with the plane, the reaction impulse acts at the corner C_{i+1}. Let the prismatic cylinder of mass M have a radius a from the center G to each vertex C_i. Hence the cylinder has a polar moment of inertia $\hat{I}$ for the center of mass G, where

$$\hat{I} = \frac{Ma^2}{6}(2 + \cos 2\Psi).\tag{10.3}$$

From the parallel axis shift theorem, the polar moment of inertia for any vertex can be obtained as

$$I_C = \hat{I} + Ma^2.$$

Equations of Motion

Assume that friction is sufficiently large that there is no sliding during each collision. As the cylinder rotates about a vertex from one side to the next in a gravitational field with intensity g, the decrease in the potential energy equals $Mga \sin \Psi \sin \theta$. This decrease in potential energy increases the kinetic energy of the cylinder during the interval $t_{i+1} - t_i$ (i.e. during the period of time between the collision at vertex C_i and the collision at vertex C_{i+1}). During this interval the cylinder rolls about the vertex C_i, so the only active force is the conservative force of gravity. Hence for this period, conservation of energy gives the change in angular velocity $\omega(t_i)$ as

$$I_C \omega_{i+1}^2(-) - I_C \omega_i^2(+) = 4Mga \sin \Psi \sin \theta$$

where the angular speed of the prism just after the collision at vertex C_i is denoted by $\omega_i(+) \equiv \omega(t_i+)$, and the angular speed just before the same collision by $\omega_i(-) \equiv \omega(t_i-)$.

At this point it is useful to introduce a parameter representing inertia and the active forces during the rolling phase of motion. Suppose the prismatic cylinder were pendu- luming about vertex C_i. For small angular deflections ϕ the equation for free oscillations in the gravitational field gives

$$0 = \ddot{\phi} + \omega_g^2 \phi, \qquad \omega_g^2 \equiv Mga/I_C.$$

With this definition of a natural frequency ω_g the previous equation for conservation of energy during rotation about the ith corner can be expressed as

$$\omega_{i+1}^2(-) - \omega_i^2(+) = 4\omega_g^2 \sin \Psi \sin \theta. \tag{10.4}$$

The other part of each cycle of rotation is the collision of the next vertex C_{i+1} with the inclined plane at time t_{i+1}. During the instant of collision the active force is simply the impulsive reaction at C_{i+1} and there is no impulsive couple, so the moment of momentum about C_{i+1} is conserved[1]:

$$(\hat{I} + Ma^2)\omega_{i+1}(+) = (\hat{I} + Ma^2 \cos 2\Psi)\omega_{i+1}(-).$$

Hence the ratio φ of the angular speed before each impact $\omega_{i+1}(-)$ to that immediately afterward $\omega_{i+1}(+)$ depends on the geometry and inertia properties:

$$\varphi \equiv \frac{\omega_{i+1}(-)}{\omega_{i+1}(+)} = \frac{\hat{I} + Ma^2}{\hat{I} + Ma^2 \cos 2\Psi} = \frac{8 + \cos 2\Psi}{2 + 7\cos 2\Psi} \tag{10.5}$$

where $\varphi > 1$. Combining (10.4) and (10.5), the ratio of speeds just after two successive impacts gives an iteration equation for evolution of the rolling speed,

$$\frac{\omega_{i+1}^2(+)}{\omega_i^2(+)} = \frac{1}{\varphi^2} \left\{ 1 + \frac{4\omega_g^2 \sin \Psi \sin \theta}{\omega_i^2} \right\}. \tag{10.6}$$

Steady State and Transient Solutions

A *steady rolling speed*[2] $\omega_*(+)$ implies that $\omega_{i+1}(+) = \omega_i(+) \equiv \omega_*(+)$. Substituting this condition into (10.6), we obtain

$$\frac{\omega_*^2(+)}{\omega_g^2} = \frac{4\sin \Psi \sin \theta}{\varphi^2 - 1} \tag{10.7}$$

Apparently rolling persists at the steady speed if the initial conditions are consistent with this motion. In general however the rolling cylinder will have an initial speed $\omega_0(+)$ which is not the same as the steady state solution. The evolution of the rolling speed can be obtained by replacing the parameter ω_g^2 in the iteration equation (10.6) with expression (10.7):

$$\omega_{i+1}^2(+) = \varphi^{-2} \left\{ \omega_i^2(+) + (\varphi^2 - 1)\omega_*^2(+) \right\}.$$

---

[1] Before the $(i + 1)$th impact the velocity of the center of mass G is perpendicular to line GC_i as shown in Fig. 10.3; therefore at the instant $t_{i+1}(-)$ the velocity of the center of mass is not perpendicular to line GC_{i+1}.

[2] Of course the speed of rolling varies during each cycle, but in steady rolling the variation in speed is periodic with N cycles per revolution. Thus for steady rolling at the instant just after each collision, the angular speed is $\omega_*(+)$.

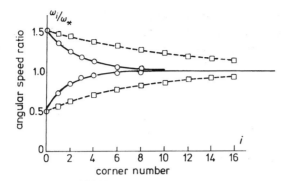

Figure 10.4. Evolution of rolling speed from initial states $\omega_0(+)$ that are either larger or smaller than the speed of steady rolling, $\omega_*(+)$. Solid lines are for $N = 6$ ($\varphi = 1.545$), while dashed lines are for $N = 12$ ($\varphi = 1.10$).

For the initial rolling speed $\omega_0(+)$ this iteration equation has a solution

$$\omega_i^2(+) = \omega_*^2(+) + \varphi^{-2i}\left\{\omega_0^2(+) - \omega_*^2(+)\right\}, \qquad i = 0, 1, 2, \ldots . \tag{10.8}$$

Here it is apparent that rolling is a stable process where the speed asymptotically approaches the steady rolling speed from either above or below, as shown in Fig. 10.4; i.e., the steady rolling speed is an attractor.

Minimum Initial Speed for Rolling, $\omega_{\min}(+)$

If the inclination of the plane is sufficiently large ($\theta > \Psi$), the process of rolling is self-starting from an initial condition $\omega_0(+) = 0$. For somewhat smaller angles of inclination, rolling still asymptotically approaches the steady state if the initial speed is larger than a minimum value that is necessary to initiate rolling, $\omega_{\min}(+)$. The minimum initial speed for continuous rolling is obtained from the smallest kinetic energy which is sufficient to bring the center of mass to a position vertically above the contact point,

$$I_c\omega_{\min}^2(+) = 2Mga[1 - \cos(\Psi - \theta)], \qquad \Psi > \theta.$$

This gives

$$\frac{\omega_{\min}^2(+)}{\omega_*^2(+)} = \frac{[1 - \cos(\Psi - \theta)](\varphi^2 - 1)}{2\sin\Psi\sin\theta}. \tag{10.9}$$

Minimum Angle of Inclination for Rolling, θ_{cr}

If the angle of inclination of the plane is too small $\theta < \theta_{cr} < \Psi$ the steady state rolling speed is inaccessible. For small angles of inclination where $\omega_0(+) > \omega_*(+)$ an initial rolling speed $\omega_0(+) > \omega_{\min}(+)$ will slow at each successive impact until $\omega_*(+) < \omega_i(+) < \omega_{\min}(+)$. At this point the cylinder will no longer have sufficient energy to continue rolling but instead will rock back and forth, with decreasing amplitude of rotation, between two vertices on the same face as shown in Fig. 10.5. The inclination θ_{cr} where the steady state rolling becomes inaccessible is given by the condition $\omega_{\min}(+) = \omega_*(+)$; thus it is a root of

$$0 = 2\sin\Psi\sin\theta_{cr} - (\varphi^2 - 1)[1 - \cos(\Psi - \theta_{cr})] \tag{10.10}$$

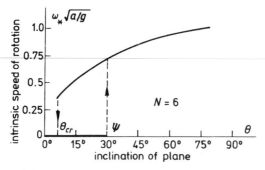

Figure 10.5. Attractors $\omega_i(+) = 0$ and $\omega_*(+)$ for rolling of a hexagonal prism ($N = 6$) as functions of the angle of inclination of the rough plane. There are two attractors for angles of inclination in the range $\theta_{cr} < \theta < \alpha$; in this range the attractor which is approached by $\omega_i(+)$ depends on the initial rolling speed (Abeyaratne, 1989).

Hence if the angle of inclination θ is in the range $0 < \theta < \Psi$, the stationary state $\omega_i(+) = 0$ is an attractor. The state $\omega_i(+) = \omega_*(+)$ is an attractor for $\theta > \theta_{cr}$. These two regions overlap and there are two attractors for $\theta_{cr} < \theta < \Psi$. In Fig. 10.5 this behavior is illustrated for a hexagonal prism. Suppose the prism (e.g. a pencil) is resting on a rough plane and the angle of inclination of the plane is slowly increased. The hexagonal prism does not begin to roll until the angle of inclination equals Ψ, but then the rolling speed rapidly accelerates to the steady rolling speed $\omega_*(+)$. Now assume that the angle of the plane decreases slowly. Steady rolling continues until the angle of inclination equals θ_{cr}, where rolling ceases and the hexagonal prism oscillates to rest.

Table 10.1 lists the minimum inclination for rolling and the steady state rolling speeds for prismatic cylinders with an increasing number of sides. The minimum number of sides for rolling is four, as was noted by Abeyaratne (1989). For a triangular prism ($N = 3$), the moment of momentum equation (10.5) shows that when any vertex strikes the plane, the direction of rotation reverses rather than the cylinder rolling onto the next side.

Energy Dissipated by Collisions

In this system the loss of energy is solely due to the perfectly plastic collisions which occur as each vertex strikes the plane. From the equation preceding (10.8) the change of

Table 10.1. *Minimum Inclination for Rolling (θ_{cr})
and Steady Rolling Speed $\omega_*(+)$ for Prismatic
Cylinders with n Sides*

N	Ψ (deg)	φ	θ_{cr} (deg)	$\dfrac{\omega_*}{\omega_g\sqrt{\sin\theta}}$
4	45	4.000	27.9	0.434
5	36	1.996	12.3	0.887
6	30	1.545	6.6	1.201
9	20	1.191	1.8	1.808
90	2	1.002	<0.1	6.588
180	1	1.0014	<0.1	24.91

kinetic energy T per cycle is obtained as

$$dT_i/di = d\omega_i^2(+)/di = \omega_{i+1}^2(+) - \omega_i^2(+) = (1 - \varphi^{-2})\left[\omega_*^2(+) - \omega_i^2(+)\right]$$

where T has been multiplied by $2/I_C$. Likewise the change of a similarly modified potential energy per cycle, $dU_i(+)/di$, equals

$$dU_i/di = -4\omega^2 \sin \Psi \sin \theta = -(\varphi^2 - 1)\omega_*^2(+)$$

where ω^2 is a parameter and the second equality comes from (10.7). Since the rate of collision is proportional to the angular speed $\omega_i(+)$, the rate of change of kinetic energy and rate of change of potential energy with time can be expressed as

$$\frac{dT_i}{dt} = \frac{dT_i}{di}\frac{di}{dt} = (1 - \varphi^{-2})\left[\omega_*^2(+)\omega_i(+) - \omega_i^3(+)\right]$$

$$\frac{dU_i}{dt} = -(\varphi^2 - 1)\omega_*^2(+)\omega_i(+). \tag{10.11}$$

With these expressions the rate of change of dissipation D_i can be expressed as

$$dD_i/dt = -dT_i/dt - dU_i/dt$$
$$= (1 - \varphi^{-2})\left[(\varphi^2 - 1)\omega_*^2(+)\omega_i(+) + \omega_i^3(+)\right]. \tag{10.12}$$

Figure 10.6 compares the rate of energy dissipation dD_i/dt with the rate that the loss of potential energy adds kinetic energy to the system, $-dU_i/dt$. If the rolling speed is below the intrinsic speed for rolling $[\omega_i(+) < \omega_*(+)]$, the rate of decrease of potential energy is larger than the rate of dissipation, so the speed of rolling increases. On the other hand, if $\omega_i(+) > \omega_*(+)$, the rate of dissipation exceeds the rate of decrease of potential energy, so the kinetic energy and the rolling speed decrease. Whether the rolling speed is above or below the intrinsic speed, with increasing time the system asymptotically approaches the intrinsic speed of rolling $\omega_*(+)$ in a stable manner. This indicates that

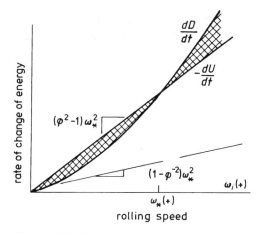

Figure 10.6. Rate of energy dissipation by collisions (dD_i/dt) and rate of decrease of potential energy $(-dU_i/dt)$ are equal when the rolling speed $\omega_i(+)$ equals the intrinsic speed for rolling, $\omega_*(+)$.

the speed $\omega_*(+)$ is an attractor. Since the rate of dissipation and the rate of change of potential energy also are equal for $\omega_i(+) = 0$, this too is an attractor.

10.2.2   The Domino Effect – Independent Interaction Theory

The domino effect describes a wave of impacts and toppling that propagates through a periodic array of regularly spaced elements where each element is marginally metastable in the initial state. The effect is readily observed in an array of slender blocks (dominoes) that initially stand on end with a regular spacing between faces of adjacent elements; the blocks stand in a gravitational field with the faces vertical and parallel. In this array, toppling one element can initiate a sequence of collisions where each toppling block knocks over its neighbor; an impact that imparts sufficient kinetic energy to knock over the first block is sufficient to knock over the entire array in a wave of destabilizing collisions.

The term "domino effect" has been used to describe disparate phenomena such as sequential collapse of periodic frame structures, blowdown of trees in ice-laden forests and propagation of neuron firing in a synapse-coupled neural network. The effect is a discrete counterpart of propagation phenomena in exothermic chemical reactions that are represented by reaction–diffusion equations (Nicholas and Prigogine, 1977; Bimping-Bota et al., 1977). In the case of dominoes however, the energy loss at the wavefront is due to dissipation by successive collisions rather than diffusion.

Toppling and Collision of Dominoes
Consider a regularly spaced array of uniform slender blocks that initially stand on end on a rough level surface as shown in Fig. 10.7. Each block has mass M, length L and thickness h, and they are spaced at distance $\lambda + h$ along the surface. Each block is initially vertical, so the center of mass is displaced by a small angle $\Psi_0 = \arctan(h/L)$ from the vertical line passing through any edge in contact with the supporting surface. If the block rotates about the edge, the potential energy increases until the rotation angle equals Ψ_0. At this angle of rotation the block becomes unstable and for even larger angles of rotation the potential energy begins to decrease. Since there is no sliding on the supporting surface, an unstable domino topples into a collision with its neighbor after rotating through an angle Ψ_1; thus at collision there is a total rotation $\Psi_0 + \Psi_1 = \arcsin(\lambda/L)$. With the present assumptions, the collision is the only coupling between neighboring elements in a domino array.

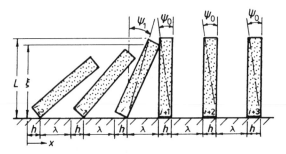

Figure 10.7. Geometry of toppling in a regularly spaced array of dominoes.

At the head of the group of toppled dominoes, the domino at the wavefront is rotating with angular speed $\omega(t)$. Let the angular speed of this leading domino be denoted by $\omega(t_i+) \equiv \omega_i(+)$ at the instant just after impact against the ith domino and $\omega(t_{i+1}-) \equiv \omega_{i+1}(-)$ at the instant just before impact against the $(i+1)$th domino. Following the impact of the $(i-1)$th domino against the ith domino, the latter has an initial kinetic energy $T_i(-) = (ML^2/6)\omega_i^2(-) \sec^2 \Psi_0$. During the subsequent motion of the ith domino in the gravitational field there is a change in potential energy $\Delta U_i = U_{i+1}(-) - U_i(+)$ so that for the rotation phase of each cycle the principle of conservation of total mechanical energy gives an energy ratio

$$\frac{\Delta U_i}{T_i(+)} = \frac{2\omega_g^2}{\omega_i^2(+)}(\cos \Psi_1 - \cos \Psi_0) \tag{10.13}$$

where $\omega_g \equiv [(3g \cos \Psi_0)/(2L)]^{1/2}$ is the natural frequency of small oscillations for the block rotating as a pendulum about the supporting edge.[3] This oscillation frequency for penduluming is the same idea used in the example of a rolling pencil to obtain a parameter that represents the ratio of active force to inertia for the prescribed motion.

Conservation of energy and this change in potential energy during toppling give for the ratio of angular speed at impact to initial angular speed

$$\omega_{i+1}(-)/\omega_i(+) = [1 - \Delta U_i/T_i(+)]^{1/2}. \tag{10.14}$$

The collision against each block occurs at distance ξ above the supporting plane, where $\xi/L = \cos(\Psi_0 + \Psi_1)$. The impact against the $(i+1)$th domino generates an impulse p_{i+1} normal to the surface and if the dominoes are rough, there is also a component of impulse tangent to the surface μp_{i+1}, where μ is the coefficient of Coulomb friction. The frictional impulse acts in a direction opposed to sliding. The following analysis assumes that sliding continues throughout the contact region; this is a valid assumption for colliding bodies if the angle of incidence is substantially larger than the angle of friction (Maw, Barber and Fawcett, 1981). The analysis also assumes that during each collision the sliding is continuously in the same direction; the case of slip reversal during collision occurs only if both the coefficient of friction and the coefficient of restitution are large (Stronge, 1987).

The normal impulse p_i that initiates toppling in the ith domino can be expressed as $p_i = 2T_i(+)/(\xi - \mu h)\omega_i(+)$. For block $i-1$ the component of impulse normal to the longitudinal axis is $p_i(\xi + \mu\lambda)/L$. Then for each block the change in angular velocity is related to the moment of the impulse about the supporting edge. The changes in angular velocity are also related to the work done in changing the kinetic energy of relative motion of the bodies by the coefficient of restitution e_*. Thus we obtain the ratio of angular speed for block $i-1$ at the instant before impact, $\omega_i(-)$, to the initial angular speed of the neighbor, $\omega_i(+)$, as

$$\frac{\omega_i(+)}{\omega_i(-)} = \varphi^{-1}(1 + e_*), \qquad \varphi \equiv 1 + \frac{\xi + \mu\lambda}{\xi - \mu h} \tag{10.15}$$

where φ is a geometric parameter and $\varphi \geq 2$. Multiplying this expression with (10.14) and

---

[3] Like the natural frequency defined following (10.3), this characteristic frequency depends on gravity and inertia properties of the system for rotation about a particular instantaneous center – the supporting edge of the domino.

noting that $\omega_{i+1}(+)/\omega_i(+) = \omega_{i+1}(-)/\omega_i(-)$ gives an iteration equation for evolution of the angular speed of the domino at the wavefront,

$$\frac{\omega_{i+1}(+)}{\omega_i(+)} = \frac{1+e_*}{\varphi}\left[1 - \frac{\Delta U_i}{T_i(+)}\right]^{1/2}. \tag{10.16}$$

This expression is analogous to (10.6) in the example of a rolling prism. (Recall that the analysis of the rolling prism in Sect. 10.2.1 assumed perfectly plastic collisions, i.e. $e_* = 0$.)

As toppling propagates from one domino to the next, each collision dissipates energy. The loss of energy per collision can be evaluated from theorem (3.21):

$$\frac{\Delta T_{i+1}}{T_i(+)} = \left[\frac{1-e_*}{1+e_*}(\varphi - 1)^2 + \frac{2}{1+e_*}(\varphi - 1) - 1\right]\frac{\omega_{i+1}^2(+)}{\omega_i^2(+)}.$$

This result can be obtained from the difference in kinetic energies at the inception and termination of each collision.

Steady Speed of Toppling

Equation (10.16) has a steady state solution where each successive domino has identical motion when it is at the wavefront; i.e., the interval between impacts is constant. At the *intrinsic speed of propagation* the initial angular speed of every domino is the same; hence setting (10.16) equal to unity after substituting from (10.13) gives the initial angular speed at the intrinsic or steady speed for propagation ω_*, where

$$\frac{\omega_*}{\omega_g} = (1+e_*)\left[\frac{2(\cos\Psi_0 - \cos\Psi_1)}{\varphi^2 - (1+e_*)^2}\right]^{1/2}. \tag{10.17}$$

Since $\varphi \geq 2$ and $e_* \leq 1$, there is a real intrinsic speed if $\Psi_1 > \Psi_0$; (i.e. $\lambda/h > \sqrt{2}\cos\Psi_0$). Thus a real intrinsic speed of propagation can exist only if the domino spacing is larger than a minimum. For slender dominoes this minimum spacing is slightly smaller than the domino thickness.

Transient Solution for Approach to Steady State

Toppling of an array of dominoes is initiated by some external impulse which is sufficient to topple one element. Ordinarily the impulse does not initiate toppling at the intrinsic speed. If $\lambda/h > \sqrt{2}\cos\Psi_0$ and toppling is initiated at a speed that is smaller than the intrinsic speed [$\omega_i(+) < \omega_*$], then the speed of propagation will steadily increase towards the intrinsic speed; likewise, if the initial speed is larger than the intrinsic speed the speed of propagation will steadily decrease towards the intrinsic speed. An equation describing the approach to the intrinsic speed ω_* can be obtained directly from a linear recurrence equation resulting from Eqs. (10.13), (10.16) and (10.17):

$$\omega_{i+1}^2(+) = \omega_*^2 + \varphi^{-2}(1+e_*)^2\left[\omega_i^2(+) - \omega_*^2\right].$$

For an initial angular speed $\omega_1(+)$ of the first block, this equation has a solution

$$\omega_i^2(+) = \omega_*^2 + \left[\omega_1^2(+) - \omega_*^2\right][\varphi^{-1}(1+e_*)]^{2i-2}. \tag{10.18}$$

Because $\varphi > 2$, the factor $(1+e_*)/\varphi < 1$, so that for a sufficiently large number i of toppled dominoes the initial angular speed of each successive domino asymptotically approaches the intrinsic angular speed for steady toppling.

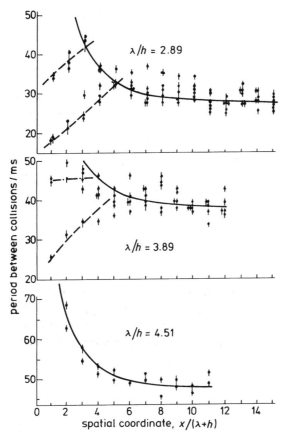

Figure 10.8. Intercollision period during toppling of dominoes with array spacings λ/h of (a) 2.89, (b) 3.89 and (c) 4.51. Each test had an initial speed ω_0/ω_* that was lower than, about the same as or higher than the natural speed of propagation.

Figure 10.8 shows measurements of the period between successive collisions during propagation of toppling in domino arrays with three different spacing. In these experiments, toppling was initiated by imparting a specified impulse to the end domino. The data points for the interval of time between successive collisions show that at each spacing the intercollision period rapidly settles to a roughly constant value representative of steady propagation. That steady propagation speed increases with domino spacing. If toppling is not initiated at the steady speed, it closely approaches the steady speed by the time that 5–10 dominoes have been toppled; thus, at the spacing described in Fig. 10.8 steady propagation is quite stable.

Closely Spaced Dominoes, $\Psi_1 < \Psi_0$

For closely spaced dominoes ($\lambda/h < \sqrt{2}\cos\Psi_0$) it is clear from (10.17) that real steady wave speeds do not exist. To determine the evolution of angular speeds as the wavefront passes through the array of dominoes, first consider the minimum initial speed for toppling of an individual block, $\omega_{\min}$, where

$$\omega_{\min}^2/\omega_g^2 = 2(1 - \cos\Psi_0).$$

If the initial speed of the ith block is $\omega_i(+) = \beta\omega_{\min}$ where $\beta > 1$ then a ratio of initial angular speeds between neighboring blocks will be

$$\frac{\omega_{i+1}(+)}{\omega_i(+)} = \frac{1 + e_*}{\varphi}\left\{1 + \beta^{-2}\left[\frac{\cos\Psi_0 - \cos\Psi_1}{1 - \cos\Psi_0}\right]\right\}^{1/2}. \tag{10.19}$$

For closely spaced dominoes where $\cos\Psi_0 < \cos\Psi_1$ this expression shows that $\omega_{i+1}(+) < \omega_i(+)$; i.e., the wave of collisions continually slows until there is insufficient momentum to topple the next domino.

Experiments

A series of high speed photographs of a wave of collisions in a widely spaced array of dominoes ($\lambda/h = 2.89$) are shown in Fig. 10.9. These photographs cover the period in which the toppling wavefront progresses a distance of about 1.5 domino spacings along the array. The size of the dominoes and measured values of their characteristic parameters are given in Table 10.2. For each spacing the theoretical value for speed of the wavefront has been compared with the measured value; the theoretical value was obtained by integrating the angular speed [obtained from (10.14)] with respect to time to determine the period between collisions if toppling is initiated at the intrinsic steady wave speed $\omega_i(+)$. The table indicates satisfactory agreement only if the spacing is large.

For dominoes with a spacing $\lambda/h < 4$, the high speed photographs show that this single collision model is not satisfactory because there are multiple collisions. For more closely spaced dominoes multiple collisions are even more common, as shown in Fig. 10.10 for a spacing $\lambda/h = 1.7$. Multiple collisions on the domino leading the toppling group drive the system at a larger speed of propagation than would occur if there were solely single collisions.

Table 10.2. *Domino Toppling Experiment*

Spacing Ratio λ/h	Initial Angle Ψ_0 (deg)	Collision Angle Ψ_1 (deg)	Collision Height ξ/L	Exp. Steady Collision Period Δt (ms)	Exp. Steady Collision Speed V_* (m s^{-1})	Theor. Steady Wave Speed V_* (m s^{-1})
2.89	10.3	21.3	0.851	28.4	1.04	0.65
3.89	10.3	34.6	0.708	38.3	0.97	0.80
4.51	10.3	44.6	0.573	47.8	0.87	0.86

Domino dimensions: $L = 41.78$ mm, $h = 7.58$ mm, $w = 21.90$ mm.
Natural frequency: $\omega = 18.95$ rad s^{-1}.
Coefficient of restitution: $e_* = 0.846 \pm 0.03$ (edge impact on domino surface). In this experiment the coefficient of restitution was measured for an edge of one plastic domino colliding with the face of a similar domino at a speed in the range 0.5–1.0 m s^{-1}.
Coefficient of friction: $\mu = 0.176^{+0.0}_{-0.04}$ (edge sliding on domino surface).

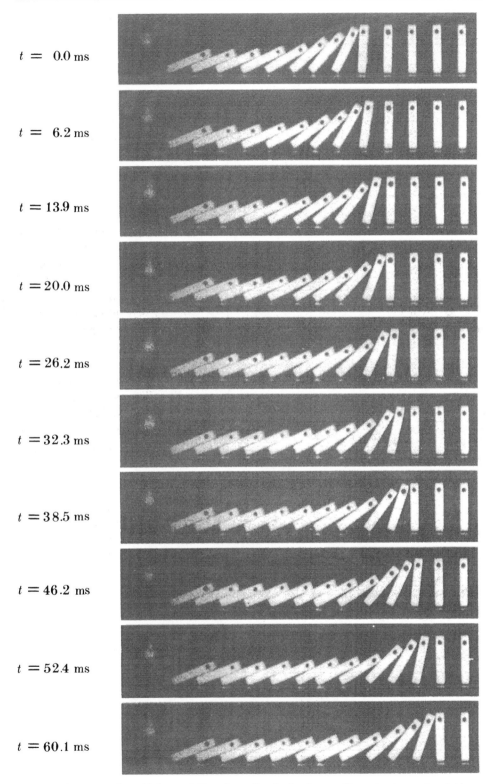

Figure 10.9. Frames from high speed film of toppling in dominoes with moderate spacing, $\lambda/h = 2.89$.

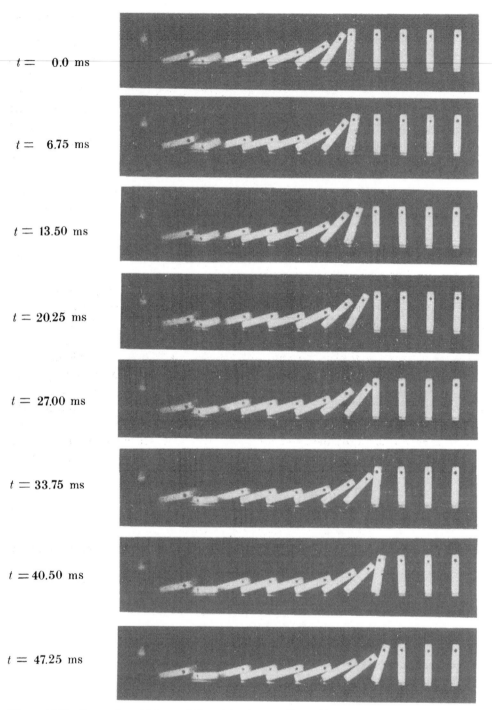

$t = \quad 0.0 \text{ ms}$

$t = \quad 6.75 \text{ ms}$

$t = 13.50 \text{ ms}$

$t = 20.25 \text{ ms}$

$t = 27.00 \text{ ms}$

$t = 33.75 \text{ ms}$

$t = 40.50 \text{ ms}$

$t = 47.25 \text{ ms}$

Figure 10.10. Frames from high speed film of domino toppling, $\lambda/h = 1.7$.

10.2.3 Domino Toppling – Successive Destabilization by Cooperative Neighbors

The following analysis considers domino toppling caused by a group of neighboring elements that interact as they topple. The toppled dominoes behind the wavefront are assumed to each lean against their neighbor, and they maintain contact during toppling rather than just striking once and then separating. The dominoes all lean in the same direction towards a leading element at the wavefront. Ahead of the wavefront, all dominoes are stationary and vertical, while behind the wavefront, the dominoes all rotate in the same direction. Toppling of the next undisturbed element by the *cooperative group* has an intrinsic speed of propagation whenever the weight of the leaning dominoes is sufficient to destabilize an undisturbed element at the wavefront. The natural speed for toppling by a cooperative group is larger than the natural speed for toppling by single collisions unless the loss of energy at each collision is small and the spacing between elements is large.

Collision and Toppling by Cooperative Neighbors
Consider a uniformly spaced row of identical slender blocks (dominoes) that initially stand on end on a rough level surface in a gravitational field. Each block has length L, thickness h, width w and mass M and is spaced a distance λ from its nearest neighbors as shown in Fig. 10.11. In this analysis the toppling blocks are numbered backwards from the block at the wavefront $i = 1$, so that the first block is the leading block in the toppling group. Before toppling, the long edges of each block are vertical; thus the center of mass is displaced from vertical by a small angle $\Psi_0 = \arctan(h/L)$. The friction between each block and the rough level surface is assumed to be large enough to prevent any sliding as the block pivots about the edge; hence a typical block simply rotates about the supporting edge until it collides with the next block. At any instant the ith block behind the wavefront has rotated through an angle θ_i. When the leading domino has rotated through an angle $\theta_1 = \arcsin(\lambda/L)$, the top edge collides against the face of the next domino. Let the angular speed of this leading domino an instant before the collision against the 0th block be $\dot{\theta}_1(-) \equiv \omega_1(-)$.

Several features of toppling by *united action of a cooperative group* are illuminated by an analysis which considers the following hypothetical interaction conditions. This analysis assumes:

(a) there are an indefinitely large number of toppling dominoes behind the collision wavefront;
(b) each domino behind the wavefront leans forward against its neighbor;

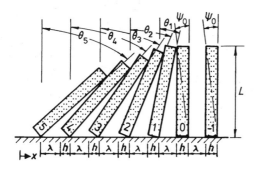

Figure 10.11. Geometry for toppling by cooperative neighbors.

(c) at the sliding contact between each pair of neighboring dominoes, friction is negligible;

(d) the coefficient of restitution is negligible, so that after a collision the colliding blocks remain in sliding contact.

With these hypotheses, the rotations of two neighboring dominoes behind the wavefront are related by the kinematic condition that sliding contact is maintained. Thus,

$$L \sin(\theta_{i+1} - \theta_i) = (\lambda + h) \cos \theta_i - h.$$

In order to obtain the ratio of angular speeds at any instant, the above equation can be differentiated with respect to time to give

$$\frac{\dot{\theta}_{i+1}}{\dot{\theta}_i} = \frac{\omega_{i+1}}{\omega_i} = 1 - \left(\frac{\lambda + h}{L}\right) \frac{\sin \theta_i}{\cos(\theta_{i+1} - \theta_i)}. \tag{10.20}$$

Hence the angular speed $\omega_i(\theta_i)$ of each block can be determined from a function that depends solely on the angular speed of the first block. Let $\omega_1(0) \equiv \omega_1(+)$ be the angular speed of the first block at the instant when it begins to move. At this instant the kinetic energy T of the group of toppling dominoes can be expressed as $T = \vartheta T_1$, where an energy ratio ϑ is calculated from (10.20) as

$$\vartheta \equiv \sum_{i=1}^{\infty} \frac{T_i}{T_1} = 1 + \sum_{j=1}^{\infty} \left\{ \prod_{i=1}^{j} \left[\frac{\omega_{i+1}(+)}{\omega_i(+)} \right]^2 \right\}$$

and $T_1 = (ML^2/6)\omega_1^2(+) \sec^2 \Psi_0$ is the kinetic energy of the first domino an instant after the impact that sets it in motion.

If a block is far behind the wavefront, its rotation θ_i approaches an angle $\hat{\theta} = \arccos(1 + \lambda/h)$ where each block lies flat against its neighbor.[4] As the wavefront moves forward one element in the array, the change in potential energy of the group of toppling dominoes, $-\Delta U$, equals the change in potential energy of a single block rotating from $\theta = 0$ to $\theta = \hat{\theta}$. A ratio of the change of potential energy to the kinetic energy of the leading domino gives

$$\frac{-\Delta U}{T_1} = \frac{2\omega_g^2}{\omega_1^2(+)} [\cos \Psi_0 - \cos(\hat{\theta} - \Psi_0)] \tag{10.21}$$

where $\omega_g = [(3g \cos \Psi_0)/(2L)]^{1/2}$ is the natural frequency of a block penduluming freely about a supporting edge.

If there are no energy losses due to friction, the sum of the kinetic and potential energies is constant during the toppling phase of motion that precedes the next collision. Consequently the leading element in the group of toppling dominoes has an angular speed at collision, $\omega_1(-)$, that is related to the initial angular speed of this element $\omega_1(+)$:

$$\frac{\omega_1(-)}{\omega_1(+)} = \left\{ \left(\frac{\vartheta}{\vartheta - 1}\right)\left(1 - \frac{\Delta U}{k T_1}\right) \right\}^{1/2}. \tag{10.22}$$

---

[4] The rotation angle approaches this bound in the limit as the number of toppling dominoes becomes indefinitely large.

At impact against the next stationary domino, an impulse is imparted to the face of the 0th domino and an equal but opposite impulse acts on a top edge of the first domino. This impulse initiates motion of the 0th domino at an angular speed $\dot{\theta}_0(+) \equiv \omega_0(+)$. With no friction between sliding blocks, the entire energy dissipation per cycle, D_0, occurs at the collision,

$$\frac{D_0}{T_1} = \frac{\omega_0(+)\omega_1(-)}{\omega_1^2(+)}.$$

Hence as the collision wavefront travels through the array, the change of total mechanical energy per cycle gives a ratio of initial angular speeds of adjacent elements equal to

$$\frac{\omega_0(+)}{\omega_1(+)} = \left\{ \left(\frac{\vartheta - 1}{\vartheta} \right) \left(1 - \frac{\Delta U}{\vartheta T_1} \right) \right\}^{1/2}. \tag{10.23}$$

Intrinsic Speed of Propagation for Toppling by a Cooperative Group
From Eq. (10.23) it is evident that the wave of destabilizing collisions moves from one element to the next at a speed that depends on the initial angular speed of the leading block and the change in this speed during the toppling that precedes the next collision. A natural or intrinsic speed of propagation for successive toppling implies that there is a speed where every element has an identical motion but this is displaced in time by a common period between collisions; i.e., $\omega_0(+) = \omega_1(+)$. The initial angular speed for steady propagation, $\omega_*(+)$, is obtained from (10.21) and (10.23) with this condition of identical motion for every element:

$$\omega_*(+)/\omega_g = \{2(1 - \vartheta^{-1})[\cos \Psi_0 - \cos(\hat{\theta} - \Psi_0)]\}^{1/2}. \tag{10.24}$$

There is a real natural speed of propagation whenever the limiting angle of rotation $\hat{\theta} > 2\Psi_0$. This condition is satisfied if the spacing $\lambda/h > 2[(L/h)^2 - 1]^{-1}$. For typical dominoes with $L/h \approx 5$ the required spacing is very small. This minimum spacing for steady propagation is required for potential energy to be supplied to the system by the toppling of each domino. If the spacing is very small, so that $\hat{\theta} < 2\Psi_0$, an initial disturbance imparted at one end propagates into the array with a speed that decreases steadily until it vanishes after a finite number of dominoes have been displaced.

Logistic Map for Transient Phase of Propagation
Generally toppling is initiated at a speed other than the steady speed for toppling. Suppose that after toppling has been initiated, the leading domino in the toppling group has an initial angular speed $\omega_1(+) = \beta\omega_*(+)$. Then the angular speed ratio (10.23) can be used to obtain the initial speed for the neighbor that is next at the wavefront,

$$\omega_0(+)/\omega_1(+) = [1 - \vartheta^{-1}(1 - \beta^{-2})]^{1/2}. \tag{10.25}$$

This ratio is illustrated in Fig. 10.12 in the form of a logistic map (Collet and Eckmann, 1980). Such maps are sometimes used by ecologists for analyzing the dynamic effect of interactions on the population of successive generations of animals (May, 1986). A typical analysis for $\lambda/h = 2$ and a ratio of initial to intrinsic angular speeds $\beta = 0.7$ is indicated by the broken line in the figure. Successive dominoes have angular speed ratios $\beta = 0.7, 0.82, 0.89$, etc.; the initial angular speed of successive dominoes monotonously approaches the intrinsic speed $\omega_*(+)$. This map shows stable behavior for dominoes since

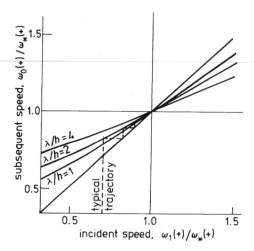

Figure 10.12. Logistic map for change in initial angular speed between successive collisions ($h/L = 0.12$). The broken line indicates a typical trajectory for an initial speed less than the intrinsic speed ω_*. This map shows a stable approach to the intrinsic speed from initial speeds either above or below.

as sequential toppling progresses, the wavefront travels at a speed that asymptotically approaches the intrinsic speed for initial conditions either above or below this natural speed of propagation.

Stable behavior which results in an asymptotic approach to the steady speed is also evident from the finite difference equation for initial angular speed of successive blocks. Equations (10.21), (10.23) and (10.24) give a change of kinetic energy per cycle, $T_0 - T_1$, equal to the loss in potential energy minus the dissipation per cycle,

$$\vartheta\left[\omega_0^2(+) - \omega_1^2(+)\right] = \omega_*^2(+) - \omega_1^2(+).$$

Average Translational Speed of Wavefront

The translational wavefront speed $V(t)$ varies as each successive domino is toppled; nevertheless the average speed V_* can be calculated by integrating the motion to obtain the period between collisions and dividing this period into the distance traversed. Thus for toppling that initiates at the intrinsic angular speed $\omega_*(+)$,

$$V_* = \frac{\lambda + h}{\int_0^{\arcsin(\lambda/h)} \omega_1^{-1}\, d\theta}$$

where

$$\omega_1(\theta) = \omega_*(+)\left\{\left[\frac{\vartheta(\theta)}{\vartheta(\theta) - 1}\right]\left[1 - \frac{U(\theta)}{\vartheta(\theta)T_1(\theta)}\right]\right\}^{1/2}.$$

Figure 10.13 compares the intrinsic speed for steady toppling by a cooperative group with that calculated for single collisions between neighbors. For almost every spacing the cooperative group gives a larger change in potential per cycle so that at any spacing it has a larger natural speed.

Table 10.3. *Steady Speed of Toppling for Two Sets of Dominoes*

Spacing λ/h	(Exp. Speed)/$\sqrt{gL}$ Tufnol	Perspex	(Theor. Speed)/$\sqrt{gL}$, $V_*/\sqrt{gL}$
0.78	1.55	1.49	1.99
1.0	–	1.50	1.91
2.0	1.57	1.51	1.68
3.0	1.47	1.45	1.59
4.0	1.42	1.40	1.52
5.0	1.36	–	1.46
6.0	1.23	1.20	1.42

Material	L (mm)	h (mm)	w (mm)	Ψ_0 (rad)	e_*	μ_{dyn}
Tufnol (T)	80.0	9.6	50.0	0.12	0.62	0.15
Perspex (P)	80.0	9.9	50.0	0.12	0.55	0.25

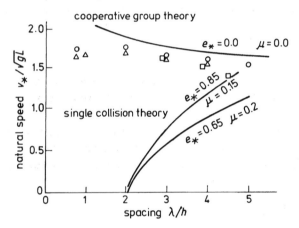

Figure 10.13. Experimental measurements of nondimensional intrinsic speed of propagation compared with theory over a range of domino spacings. Experiments were made on three sets of dominoes made from different materials: Opalene, Tufnol, Perspex.

Toppling Experiments

In addition to the experiments using Opalene dominoes that were described in Table 10.2, toppling experiments were performed with sets of thin rectangular blocks manufactured from Perspex and Tufnol. These blocks had better uniformity and much smaller edge radii than commercially available dominoes. Dimensions of these blocks, a coefficient of restitution e_* and a coefficient of sliding friction μ are listed in Table 10.3. The coefficients of restitution and friction were measured in separate experiments that closely simulated contact conditions during a collision at the intrinsic speed.

Toppling of each set was initiated by a low speed collision at the center of percussion of the domino at the end of the array. Usually there was an initial phase of propagation where the speed approached the steady speed for toppling – typically this required the wavefront to pass through 6–15 blocks. Thereafter the speed of the wavefront settled down to a small but regular variation about a constant value. Figure 10.13 compares the theoretical nondimensional speed of propagation $V_*/\sqrt{gL}$ with measurements obtained from the three sets of dominoes. These nondimensional speeds are almost the same as that obtained for Opalene dominoes (Stronge, 1987) and slightly larger than the speed measured by McLachlan et al. (1983).

Frames from a segment of high speed film taken of the toppling process (Fig. 10.10) indicate that successive toppling never settles down to a single type of interaction between every pair of dominoes; when the dominoes are closely spaced, there are some pairs of blocks near the wavefront with sliding contact and others with multiple collisions. For more widely spaced dominoes there are some single collisions and other multiple collisions. The agreement with theory is best for an intermediate range of spacing where the cooperative neighbor hypotheses are most representative. If the spacing is small ($\lambda/h < 2$), there is a relatively large impulsive reaction at the pivot point on the colliding block when it is at the wavefront. Friction does not prevent closely spaced blocks from sliding on the supporting surface, and this sliding dissipates energy. There is friction also where the corner of one block slides down the face of its neighbor. These sources of friction dissipate energy that is not accounted for, so that for small spacing the theory underestimates the intrinsic speed of propagation.

Domino toppling is a nondispersive wave that has an intrinsic speed of propagation. In this wave, particle velocity and pulse shape are properties of the system. Other systems having similar characteristics also can propagate changes of state where destabilization reduces the potential energy. Figure 10.14 shows photographs of sequential collapse in heavily loaded warehouse racking. In heavily loaded periodic frame structures and icy forests the source of distributed potential energy that drives progressive collapse is gravity, whereas neuron firing is driven by an electrical potential. The release of part of the potential energy in a neuron is chemically triggered by incoming action potential pulses. In all of these marginally stable systems, a wave of destabilization has a natural speed of propagation where the rate of change of energy per cycle vanishes (Scott, 1975).

10.2.4 Wavefront Stability for Multidimensional Domino Effects

The domino effect is a solitary wave of reaction in a periodic dissipative system with two stable states of potential energy U. The reaction transforms potential energy into *active energy T* that propagates from sequentially triggered initiation sites into unreacted medium; this propagated energy triggers subsequent reactions at neighboring sites. The traveling reaction is self-sustaining if active energy supplied by each reaction is sufficient to trigger reactions at nearest neighbors.

In domino toppling the active energy is kinetic energy, and this is transported by convection; in this case the propagating phase transformation can be represented by either differential or difference forms of the Bloch equation (Harris and Stodolsky, 1981; Leggett et al., 1987). Similar traveling reactions occur in other periodic, dissipative systems; e.g. involuntary spasms in striated muscle tissue propagate from one sacromere to the next when contraction triggers the release of Ca^{2+} ions in the adjoining sacromere (Regirir, 1989).

Domino toppling is powered by loss in gravitational potential U and transported by convection; the transportable or active energy is the kinetic energy T. The energy dissipation D is predominantly related to collisions at the reaction wavefront; dissipation increases as the cube of wavefront speed. Since the distribution of energy in the leading

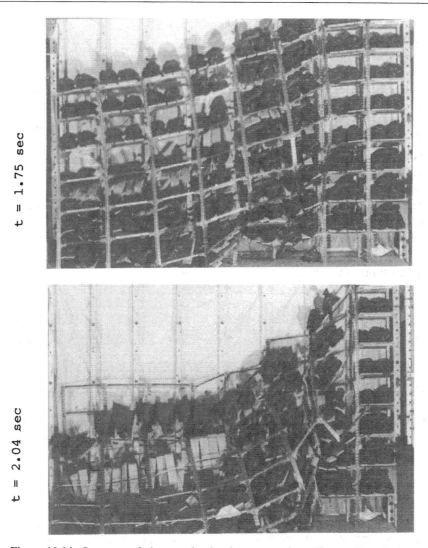

t = 1.75 sec

t = 2.04 sec

Figure 10.14. Sequence of photographs showing progressive collapse of heavily loaded warehouse racking. Unbraced, 3 m high cold-rolled steel racks are collapsing at a load $F = 0.6F_c$, where the buckling load $F_c = 5.6$ kN per leg (Bajoria, 1986).

group of toppling dominoes is independent of the wavefront speed,

$$dT/dt = -dU/dt - dD/dt. \tag{10.26}$$

For planar (1D) reaction waves in a homogeneous medium, domino toppling has a intrinsic or natural speed of propagation where reaction rate equals dissipation rate. For cylindrical and spherical wavefronts (2D and 3D) or radius r, however, reaction and dissipation rates increase as r^j, where the dimensionality of the wavefront equals $j + 1$ (i.e., $j = 0, 1, 2$ for 1D, 2D, 3D). Consequently, radiating 2D and 3D reaction wavefronts naturally propagate more slowly than the 1D intrinsic wave speed. As the wavefront radius becomes indefinitely large, the speed of propagation asymptotically approaches the intrinsic speed for 1D propagation. In contrast, converging cylindrical or spherical wavefronts propagate

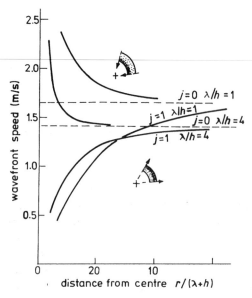

Figure 10.15. Speed of radiating and converging cylindrical wavefronts as function of radius r for domino spacings $\lambda / h = 1$ and 4.

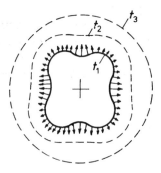

Figure 10.16. Cylindrical wavefront of 2D domino toppling shown at successive times. An initially irregular wavefront becomes more circular (smoother) as it propagates.

at a speed that increases as the radius decreases. For 2D domino toppling, these trends are illustrated in Fig. 10.15.

For a radiating wavefront with uneven radius, the perturbations with a smaller radius of curvature progress more slowly than adjacent parts which have a larger radius of curvature. This stabilizes or smoothes the reaction wavefront as the reaction progresses. This smoothing is illustrated in Fig. 10.16.

10.3   Approach to Chaos – an Unbounded Increase in Number of Attractors

10.3.1   Periodic Vibro-impact of Single Degree of Freedom Systems

Vibrating systems that strike a barrier during a cycle of motion are inherently nonlinear. If such a nonlinear system is subjected to harmonic excitation, it can exhibit

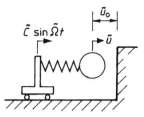

Figure 10.17. Forced vibration of single degree of freedom system with unilateral displacement constraint $\tilde{u} \leq \tilde{u}_0$.

a rich variety of periodic responses depending of the excitation force and frequency, the clearance of the barrier and the coefficient of restitution. Such problems have been investigated both analytically and with the aid of numerical simulations by Holmes (1982), Shaw and Holmes (1983), Shaw (1985), Thompson and Stewart (1986) and Gontier and Toulemonde (1997). In this system the period of the response can be the same as that of the excitation (period 1 orbits), twice as long (period 2 orbits), three times as long (period 3 orbits), etc. In each case the number of impacts in one period is indicated by the numeral, but these impacts are not necessarily regularly spaced in time during the period. The approach to chaos is marked by a very rapid doubling of the period of the stable response with small increases in excitation frequency.

The single degree of freedom oscillator shown in Fig. 10.17 is subjected to a sinusoidal base displacement $\tilde{C} \sin \tilde{\Omega}\tilde{t}$. The displacement $\tilde{u}(\tilde{t})$ of the system is limited by a stop that provides a unilateral constraint, $\tilde{u} < \tilde{u}_0$. When the mass collides against the stop, the change in velocity is given by a coefficient of restitution e_*. This *vibro-impact system* has an equation of motion for the two phases of motion,

$$d^2\tilde{u}/d\tilde{t}^2 + \omega^2\tilde{u} = \tilde{C} \sin \tilde{\Omega}\tilde{t}, \qquad \tilde{u} < \tilde{u}_0$$

$$d\tilde{u}(\tilde{t}_i+)/dt = -e_* \, d\tilde{u}(\tilde{t}_i-)/dt, \qquad \tilde{u}(\tilde{t}_i) = \tilde{u}_0$$

where ω is the natural frequency of this undamped oscillator and e_* is a coefficient of restitution which applies to each impact against the stop. For this system one can define nondimensional variables

$$u \equiv \tilde{u}/\tilde{u}_0, \qquad \Omega \equiv \tilde{\Omega}/\omega, \qquad C = \tilde{C}/u_0\omega^2, \qquad t \equiv \omega\tilde{t}.$$

In terms of these nondimensional variables the equations of motion can be expressed as two first order ordinary differential equations in the dependent variables for displacement u and velocity $\dot{u}$,

$$du/dt = \dot{u} \tag{10.27a}$$

$$d\dot{u}/dt + u = C \sin \Omega t, \qquad u < 1 \tag{10.27b}$$

$$\dot{u}(t_i+) = -e_*\dot{u}(t_i-), \qquad u(t_i) = 1. \tag{10.27c}$$

For $0 < e_* < 1$ this is a nonconservative oscillatory system with energy dissipated at every impact.

10.3.2 Period 1 Orbits

On the ith cycle let the impact velocity be denoted by $v_i \equiv \{\dot{u}(t_i-), 1, t_i-\}$. The subsequent motion can be expressed as

$$u(v_i, t) = \left(1 - \frac{C}{D}\sin\Omega t_i\right)\cos(t - t_i)$$

$$- \left(e_*\dot{u}_i + \frac{C\Omega}{D}\cos\Omega t_i\right)\sin(t - t_i) + \frac{C}{D}\sin\Omega t \qquad (10.28a)$$

$$\dot{u}(v_i, t) = -\left(1 - \frac{C}{D}\sin\Omega t_i\right)\sin(t - t_i)$$

$$- \left(e_*\dot{u}_i + \frac{C\Omega}{D}\cos\Omega t_i\right)\cos(t - t_i) + \frac{C\Omega}{D}\cos\Omega t \qquad (10.28b)$$

where $D = 1 - \Omega^2$. From (10.28a) the time t_{i+1} of the $(i + 1)$th impact can be obtained, and this is subsequently used in (10.28b) to determine the impact velocity v_{i+1}. Thus $v_{i+1} = \phi(v_i)$.

10.3.3 Poincaré Section and Return Map

A *Poincaré section* Σ is defined as a plane at a constant value of u in the solution space. This plane can be chosen as the impact surface $u = 1$, and on it the solutions of (10.28a) and (10.28b) appear as points where the solution pierces the surface (see Fig. 10.18). The section Σ is designated as

$$\Sigma = \{(u, \dot{u}, t) : u = 1, \ \dot{u} > 0\}.$$

A point $(\dot{u}_i, t_i)$ maps to $(-e_*\dot{u}_i, t_i)$ due to the impact rule and then follows a free flight motion to $\dot{u}_{i+1}, t_{i+1}$. A *Poincaré map* P is a rule that transforms points on a section Σ into other points on the section; i.e.

$$P : \Sigma \to \Sigma, \qquad \text{or} \quad P : (\dot{u}_i, t_i) \to (\dot{u}_{i+1}, t_{i+1}). \qquad (10.29)$$

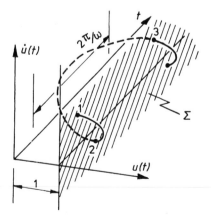

Figure 10.18. Solution trajectory in phase space $(u, \dot{u}, t)$ and points $i = 1, 2, 3$ where the solution pierces the Poincaré section Σ.

If before the initial conditions are repeated there are j impacts in an orbit, the impact velocity at the end of the jth impact is denoted by v_j:

$$v_j = P^j(v_0), \qquad P^j \equiv P \cdot P \cdot \cdots \cdot P \quad (j \text{ times }) \tag{10.30}$$

where in a set the initial impact velocity is denoted as $v_0 = (\dot{u}_0, t_0)$. Hence a j-impact *periodic motion* is described by

$$P^j(v_0) = v_0.$$

This orbit has period $t_1 - t_0 = 2j\pi/\Omega$.

10.3.4 Stability of Orbit and Bifurcation

When a periodic orbit exists, it can be either stable or unstable; i.e., neighboring orbits may be either attracted or repelled from the periodic orbit. The stability is determined by considering a small perturbation $v = v_0 + dv$ from a periodic orbit v_0. At the end of a period this gives, to first order,

$$P(v) = P(v_0 + dv) \approx v_0 + (DP)\,dv.$$

The derivative DP is the Jacobian of the Poincaré map,

$$DP = \begin{bmatrix} \partial t_1/\partial t_0 & \partial t_1/\partial \dot{u}_0 \\ \partial \dot{u}_1/\partial t_0 & \partial \dot{u}_1/\partial \dot{u}_0 \end{bmatrix}$$

and its elements are calculated using implicit differentiation:

$$DP = \begin{bmatrix} \dfrac{-\partial u_1/\partial t_0}{\partial u_1/\partial t_1} & \dfrac{-\partial u_1/\partial \dot{u}_0}{\partial u_1/\partial t_1} \\[2ex] \dfrac{\partial \dot{u}_1}{\partial t_0} + \dfrac{\partial \dot{u}_1}{\partial t_1}\dfrac{\partial t_1}{\partial t_0} & \dfrac{\partial \dot{u}_1}{\partial \dot{u}_0} + \dfrac{\partial \dot{u}_1}{\partial t_1}\dfrac{\partial t_1}{\partial \dot{u}_0} \end{bmatrix}. \tag{10.31}$$

For the impact oscillator this Jacobian has a determinate and trace:

$$\det(DP) = e_*^2 \dot{u}_0/\dot{u}_1$$

$$\mathrm{tr}(DP) = -\frac{2e_*\dot{u}_0 \cos(2j\pi/\Omega) + (1 + e_*)[1 - C\sin(\Omega t_0)]\sin(2j\pi/\Omega)}{\dot{u}_0}.$$

The eigenvalues λ_1 and λ_2 of DP are roots of the characteristic equation,

$$\lambda^2 - \mathrm{tr}(DP)\lambda + \det(DP) = 0. \tag{10.32}$$

Since DP is real, the eigenvalues are either real or a complex conjugate pair.

If both eigenvalues are inside the unit circle, the periodic orbit is stable. On the other hand, if one eigenvalue is outside the unit circle, the orbit is unstable. Bifurcation occurs when an eigenvalue crosses the unit circle. For periodic orbits $\dot{u}_0 = \dot{u}_1$ of the vibro-impact system, the determinant $\det(DP) < 1$, so that the single period orbit becomes unstable only if $\lambda_i = \pm 1$ (Budd and Lee, 1996; Shaw and Holmes, 1983). If $\lambda = 1$, the orbit becomes unstable because of a saddle point bifurcation. Gontier and Toulemonde (1997) show that this occurs when $C < C_c$, which implies that the exciting force is insufficient to

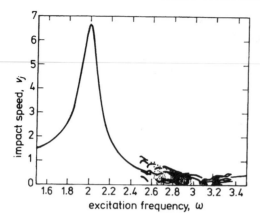

Figure 10.19. Frequency response of impact speed v_j for a sphere vibrating against a stop. For ($\tilde{u}_0 = 0$, $e_* = 0.8$) chaotic motion occurs at frequencies a little larger and smaller than $\omega = 3$.

give an amplitude large enough for impact. If $\lambda = -1$, the orbit undergoes a subharmonic, or *flip*, bifurcation.

Figure 10.19 shows the frequency response of the single degree of freedom oscillator with a unilateral displacement constraint. For the specified parameters ($\tilde{u}_0 = 0$, $e_* = 0.8$) the velocity of impact v_j has been plotted for 100 successive impacts. The frequency response exhibits resonance at $\omega \approx 2, 4, \ldots$, and there is a period-doubling bifurcation at $\omega \approx 2.65$. In the ranges $2.7 < \omega < 2.95$ and $3.05 < \omega < 3.35$ the response seems chaotic and of small amplitude. A small neighborhood around $\omega = 3.0$ exhibits stable behavior with two impacts per period. Additional detals are available in Budd and Lee (1996).

PROBLEMS

10.1 A particle dropped onto a hard level surface has a coefficient of restitution e_*. The surface is then inclined at angle θ from horizontal, and the particle is dropped from a height h_0 above the first impact point. Find the period of the ith bounce, and show that this can be expressed as $t_i = 2e_*^i (2h_0/g)^{1/2}$. Hence find expressions for the time t_f when bouncing ceases and the distance along the plane, x_f, where bouncing ceases. Explain the motion of the particle for $t > t_f$.

10.2 Let the staircase in Sect. 10.1.1 have a thin vertical wall of height h_b at the edge of each step of height b, $h_b < b$. Find the initial drop height h_* to initiate bouncing which continues down the flight of stairs.

10.3 For a pencil (i.e. a hexagonal prism with $N = 6$) rolling at the steady speed $\omega_*(+)$, find an expression for the time period between successive collisions. Hence obtain an expression for the mean translational speed of steady rolling as a function of the radius a.

10.4 Consider a prism with N sides that is rolling down a plane inclined at angle θ from horizontal. Find the asymptotic steady state rolling speed as the number of sides becomes indefinitely large ($N \to \infty$), and compare this solution with that for a circular cylinder. (Rolling of a circular cylinder involves no energy loss and hence no source of energy dissipation.)

10.5 For a regular prism with N sides that is rolling down a rough plane, find that to prevent sliding the coefficient of friction μ must satisfy

$$\mu \geq \tan^{-1}\left[\frac{(1 - \varphi^{-1})\cot\Psi}{1 + \varphi^{-1}}\right].$$

Hence find the limiting coefficient of friction for $N = 6$ and $N = 9$. Note that this limit is independent of the angle of inclination θ.

Role of Impact in the Development of Mechanics During the Seventeenth and Eighteenth Centuries

He that will not apply new remedies must expect new evils, for time
is the greatest innovator.

Francis Bacon

Before publication of Newton's *Principia* (1687), dynamics was an empirical science; i.e., it consisted of propositions that described observed behavior without any explanation for the forces that caused motion. For example, Kepler's laws are kinematic relations that describe orbital motion. Kepler (1571–1630) discovered these relations by laboriously fitting various possibilities to the voluminous measurements of planetary motion that had been recorded by Tycho Brahe. Likewise Galileo stated propositions describing the motion of freely falling bodies. The propositions are based on relating transit times for different drop heights to clear ideas of distance, time and translational velocity.[1] These savants, however, possessed only the vaguest notion of force.

While Galileo recognized that there must be some extended cause for acceleration or retardation, he did not realize that a uniform acceleration was a consequence of a steady force; he recognized that there was a cause for acceleration but was not able to relate cause and effect. Indeed, it is difficult to imagine how there could be progress in this direction before the creation of calculus.

At the time of Galileo, the topics at the forefront of dynamics were percussion, projectile ballistics and celestial mechanics. Each of these topics had technological importance for warfare, industrial development or navigation. Percussion in particular was concerned with the terminal ballistics of musket balls as well as the effect of a forging hammer on a workpiece. There were many examples of the powerful effects of percussion, and it was comparatively simple to measure the apparently instantaneous changes in velocity that resulted from impact. At that time demonstrations of impact phenomena focused on collisions of shot fired from cannon rather than the work done by a blacksmith's hammer.

Galileo Galilei (1564–1642)

La Dynamique est la science des forces accelratrices or retardatrices,
et des mouvements varis qu'elles doivent produire. Cette science est
due entirement aux modernes, et Galile est celui qui en a jet les pre-
miers fondemens.

Lagrange, Mec. Anal. I. 221.

---

[1] The concept of velocity at an instant of time is a cornerstone of kinematics that was identified by scholars of Merton College, Oxford, during the fourteenth century (Truesdell, 1968, p. 30).

On the basis of a treatise on the center of gravity of solid bodies, Galileo secured a position as lecturer of mathematics at the University of Pisa; he was then 24 years of age. At Pisa he conducted the famous experiments on uniformly accelerated motion in which spheres of different weights were simultaneously dropped from the Tower onto a sounding board. Similar experiments had been performed since the time of Philoponus (sixth century A.D.),[2] but Galileo improved on these by simultaneously measuring the time of fall from different heights by means of a water clock (clepsydra); these experiments provided data that he used to draw correct deductions regarding the *rate of change of distance traversed* by a falling (uniformly accelerated) body. In year 1591 Galileo accepted the professorship at Padua. He soon became known for inventions such as the thermometer and the construction of telescopes. With his telescopes he discovered sunspots and four satellites of Jupiter – this was the first discovery of bodies orbiting another planet. He also was renowned for his experiments and teaching in mechanics and Ptolemaic astronomy. His teaching in celestial mechanics, however, was considered heretical by the Church, and in 1615 the Holy Office in Rome condemned him for claiming that the sun was the center of the world.[3] In that trial Galileo was forbidden to "hold, teach or defend" his theories, and he retained his freedom only by promising to obey. His next book, *Dialogo sopra i due massimi sistemi dei monda Tolemaico e Copernicano (Dialogue Concerning the Two Chief World Systems – Ptolemaic and Copernican)* was published at Florence in 1632. The book took the form of a dialogue between three characters – Salviati, an expert who usually expresses Galileo's view; Sagredo, an intelligent layman; and Simplicio, a slow-witted student. In order to obtain the inquisitor's imprimatur for publication, Galileo was required to include an argument against the Copernican system. He rashly put this into the mouth of Simplicio. Subsequently he was condemned to prison for having flouted the Inquisitor's order; the Vatican's Inquisitors declared Galileo's writings "an absurd and false proposition, that the Sun is at the center of the world and does not move from east to west, and the earth moves and is not the center of the world."[4] In order to gain release from prison he was compelled to publicly recant and accept that the Earth was stationary.[5]

To circumvent the Church ban, Galileo's subsequent work had to be published outside the reach of Rome. His next book was a more general exposition on mechanics. *Discorsi e Dimostrazioni Matematiche interno due Nuove Scienze, Mecanica & Movimenti Locali (Dialogue Concerning Two New Sciences)* was published in Leyden in 1638. This remarkable book on statics and dynamics presented a systematic development of mechanics organized into four chapters, or Days. It seems there was a draft for a fifth Day dealing with the *forza della percossa*, but the printer was rushed, so the fifth Day never appeared.[6]

In the *Two New Sciences*, Galileo often gropes for terminology to describe the phenomena that he observed. Lacking any idea of mass, he frequently uses velocity and momentum

---

[2] M.R. Cohen and I.E. Drabkin, *A Source Book in Greek Science*, Harvard Univ. Press (1948).

[3] It was not until 15 March 1616 that the Congress of Cardinal Inquisitors officially condemned the writings of Copernicus.

[4] Finocchiaro (1989).

[5] In 1823 the Vatican quietly authorized publication of a book by G. Settele, a priest, which acknowledged that Galileo was correct, but it was not until 1992 that the Pope admitted that the church had done Galileo an injustice.

[6] Galileo's dialogue on percussion appears in Drake's translation (Galileo, 1638b, pp. 281–303). This material was not printed in the original edition.

synonymously (*velocitatem, impetum seu momentum*).[7] Nevertheless the book contains a clear statement of the principle of inertia (Newton's first law of motion)[8]: "... any velocity once imparted to a moving body will be rigidly maintained as long as external causes of acceleration or retardation are removed, ..." [243].[9]

To examine uniformly accelerated motion, Galileo simultaneously dropped different-sized spheres from the Tower of Pisa. In 1604 he wrongly believed that velocity increased in proportion to distance. By 1638, however, he correctly described uniformly accelerated motion as having velocity that increases in proportion to time, and distance traversed that is proportional to time squared [209].

In regard to impulsive forces in collisions, Galileo states that "the impetus of collision" depends on the relative velocity (*velocit del percuziente*); i.e., it depends on the difference between the velocities of the colliding bodies [291]. This idea of relative velocity would later be successfully picked up by Wallis and Wren and most thoroughly by Huyghens in papers they published in 1668–1669. Galileo's illustration indicates that he recognizes that a significant effect of impact – the impetus of collision – is directly proportional to the *normal component of relative velocity*:

> To what has hitherto been said concerning momenta, blows or shocks of projectiles, we must add another very important consideration; to determine the force or energy of the shock (*forzsa ed energia della percossa*) it is not sufficient to consider only the speed of the projectiles, but we must also take into account the nature and condition of the target which, in no small degree, determines the efficiency of the blow. First of all it is well known that the target suffers violence from the speed (*velocit*) of the projectile in proportion as it partly or entirely stops the motion [291].
>
> Moreover it is to be observed that the account of yielding in the target depends not only upon the quality of the material, as regards hardness, whether it be iron, lead, wool, etc., but also upon the angle of incidence. If the angle of incidence is such that the shot strikes at a right angle, the momentum imparted by the blow (*impeto del colp*) will be a maximum; but if the motion be oblique, that is to say slanting, the blow will be weaker; and more and more so in proportion to the obliquity; for, no matter how hard the material of the target thus situation, the entire momentum (*impeto e moto*) of the shot will not be spent and stopped; the projectile will slide by and will, to some extent, continue its motion along the surface of the opposing body [292].[6]

Despite these observations, Galileo did not decompose the incident velocity into normal and tangential components, nor did he employ the principle of inertia in any analytical manner.

A contemporary of Galileo, Professor Marcus Marci, wrote the treatise *De Proportione Motus* (published in Prague, 1639) which includes some observations on impact. He wrote

---

[7] According to the *Oxford English Dictionary, impeto seu momento* meant both moving power and movement (Jouguet, 1871, Vol. I, p. 106; ref. Dugas, 1957, p. 140). Jouguet observed that Galileo sometimes uses *impeto* to mean "velocity acquired by a body in a given time" and sometimes "the distance travelled in a certain time as a body accelerates from rest"; i.e., he confuses momentum and kinetic energy acquired during uniform acceleration.

[8] The idea of *impetus*, or momentum, that maintains an existing state of motion was conceived by John Buridan, rector of the University of Paris, about 1327. Buridan proposed that a moving body has an *impetus* that is proportional to the velocity and heaviness of the body.

[9] Numbers in square brackets are Galileo's section numbers. This quotation is from the rather free translation by Crew and de Salvio in 1914 (Galilei, 1638a); a later translation by Drake in 1976 (Galilei, 1638b) is more literal.

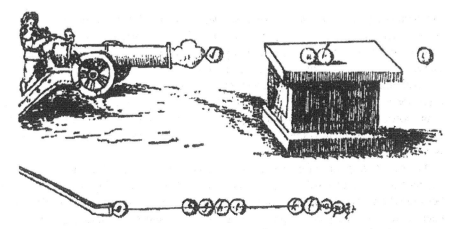

Figure A.1. Illustration from *De Proportione Motus* by Marcus Marci. The illustration of impact of a cannon ball against an identical sphere depicts their positions at equal intervals of time before and after collision. The second illustration shows a cue stick striking a billiard ball *e*, which in its turn will strike a set of tightly packed balls; this is reminiscent of the present day "Newton cradle." (Illustration from Mach, 1883.)

that in an elastic collision between two identical bodies, if before the collision one body moves with speed v while the other is stationary, then after the collision the reverse is true – the body that was moving is now stationary while the second body moves away with speed v. Marci's illustration of this observation is illustrated as Fig. A.1. It shows the positions of the balls at equal intervals of time before and after a collinear collision on the table top.

René Descartes (1596–1650)

Descartes is known as a philosopher, mathematician and physicist; the rectilinear or Cartesian coordinate system was named after him. He is remembered most for teaching an aesthetically satisfying system of world order governed by a minimum number of universal laws of nature. He looked for scientific explanation based on a few fundamental principles. Descartes's greatest failing was to rely solely on reasoning and neglect experience.

Conservation of Quantity of Motion

Independently of Galileo, Descartes also addressed the motion of bodies in the absence of active forces. In 1829 he wrote to Father Mersenne, "I suppose that the motion that is once impressed on a body remains there forever if it is not destroyed by some means. In other words, that something which has started to move in a vacuum will move indefinitely and with the same velocity." Later, in *Principia Philosophiae* (1644), he made the following laws a centerpiece of his mechanics:

> The law of nature: that everything whatever, – so far as depends on it, – always perseveres
>
> in the same state; and thus whatever once moves always continues to move.
>
> The second law of nature: that all motion by itself alone is rectilinear . . . .

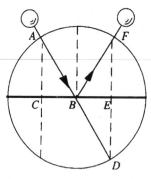

Figure A.2. Descartes's analysis of a sphere A reflecting from a "hard" plane surface CBE during elastic impact. The illustration uses a vector diagram for velocities which conserve momentum in the tangential direction and have a normal component of velocity at separation which is the negative of that at incidence.

> Of these the first [law] is that everything whatever, insofar as it is simple and indivisible, remains, so far as depends on it, always in the same state, nor ever changes [in its state] save by external causes. (Transl. in Cohen and Drabkin, 1948, p. 184.)

Here motion or quantity of motion equals the product of the weight of a body and its velocity; i.e., this is a precursor to or early form of the principle of inertia. Descartes does not say what can cause velocity to change or how it changes when the body is acted on by an "external cause". At that time the word *force* was used quite loosely. By it Descartes usually meant the work required to lift a weight to some height (Dugas, 1957, p. 156).

Impact of Bodies

In his book *Dioptrics*, Descartes relied on an analogy between reflections of light rays and the rebound of a ball that collides with a surface at an angle of incidence. He used a spatial diagram (Fig. A.2) to illustrate the laws of reflection. He considered a ball that collides with a plane surface CBE; the ball moves steadily from point A until it collides with the surface at B and then rebounds. Descartes explicitly neglects "the heaviness, the size and the shape" of the ball, and supposed the earth to be "perfectly hard and flat". He asserts that the ball is reflected on meeting the earth and its "determination to tend to B" is modified "without there being any other alteration of the force of its motion than this".

He states that "the determination to move towards some direction, like the movement, to be divided into all the parts of which it can be imagined that it is composed". The ball is thus driven by two "determinations"; one makes it descend, and the other makes it travel parallel to the surface. Impact with the surface disturbs the first but has no effect on the second, in accord with the conservation of the quantity of motion in the second direction.

Since collisions obviously modified the quantity of motion, the impact event had substantial importance in *Principia Philosophiae*. Descartes proposed seven rules for direct impact of elastic bodies:[10]

> (1) If two equal bodies impinge on one another with equal velocity, they recoil, each with its own velocity.

[10] Dugas (1957, p. 162).

(2) If one of the two is greater than the other, and the velocities equal, the lesser alone will recoil, and both will move in the same direction with the velocity they possessed before impact.

(3) If two equal bodies impinge on one another with unequal velocities, the slower will be carried along in such a way that their common velocity will be equal to half the sum of the velocities they possessed before impact.

(4) If one of the two bodies is at rest and another impinges on it, this latter will recoil without communicating any motion to it.

(5) If a body at rest is impinged on by a greater body, it will be carried along and both will move in the same direction with a velocity which will be to that of the impinging body as the mass of the latter is to the sum of the masses of each body.

(6) If a body C is at rest and is hit by an equal body B, the latter will push C along and, at the same time, C will reflect B. If B has a velocity 4 it gives a velocity 1 to C and itself moves backwards with velocity 3.

(7) (A seventh rule related velocities of two unequal bodies initially traveling in the same direction.)

The greatest importance of these rules for subsequent development of this subject was that many were easily shown to be wrong; it was apparent that they did not represent the results of simple experiments. This stimulated further examination of impact by Huyghens, Newton, Wallis and Wren. Today only one of these rules, (3), can be accepted as having any validity.

Descartes himself apparently suspected that some of these rules were not in accord with experimental evidence. He dismissed his critics, however, with a disdainful comment that his rules applied to ideal collisions and hence they were not entirely representative of anything that could be measured.

John Wallis (1616–1703)

Descartes's speculations led to 10 laws of nature; most are laws of impact for elastic particles. Many of his impact laws are vague, and most are wrong. Having in mind the mistaken concepts proposed by Descartes, The Royal Society, in 1668, initiated an investigation into the laws relating to the collision of bodies. Wallis, Wren and Huyghens replied with papers published in 1668 and 1669. Wallis dealt with perfectly plastic collisions, while Huyghens and Wren dealt with elastic collisions.

In 1668 Wallis was the Savilian Professor of Geometry at Oxford and a founding member of The Royal Society of London. He had been a scholar at Emmanuel College, Cambridge, and then was ordained in the Church of England and later appointed to the chair in Oxford by Oliver Cromwell in recognition of his discovery of a method of deciphering codes during the civil war. He wrote an important book on algebra (*Arithmetica Infinitorum*) and later a treatise on mechanics (*Mechanica sive Tractatus de Motu*, 1670); the historian Duhem calls the latter "the most complete and the most systematic [book on mechanics] which had been written since the time of Stevin". In this treatise Wallis extended the idea of force to include forces other than gravity; moreover his propositions expressed for the first time a relation between force and momentum in bodies being accelerated from rest (although he calculated momentum as the product of velocity and weight rather than mass).

Wallis's Royal Society paper dealt with impact of what he called perfectly *hard bodies*; he explained that this meant that during the compression phase of impact they did not store energy in elastic deformations that subsequently would drive the bodies apart during restitution. He distinguished these collisions from those of soft or elastic bodies where "... part of the contact force is expended in deforming the bodies during compression". Thus Wallis considers inelastic impact where the bodies have a common velocity when impact terminates; he does not however seem to recognize that the reaction that changes the momentum of each of the colliding bodies during compression also changes the kinetic energy of the system – the kinetic energy lost during compression goes into work done in deforming the bodies, whether or not that work is reversible.

For impact of *hard* or perfectly inelastic bodies Wallis wrote, "If a body in motion collides with a body at rest, and the latter is such that it is not moving nor prevented from moving by any external cause, after the impact the two bodies will go together with a velocity which is given by the following calculation. Divide the *momentum* furnished by the product of the weight and the velocity of the body which is moving by weight of the two bodies taken together. You will have the velocity after impact." For two bodies moving collinearly in the same direction, he let body B with weight W' and velocity v' be struck from behind by body A with weight W and velocity v; then if the bodies are hard, he said they had a common final velocity,

$$\frac{Wv + W'v'}{W + W'}.$$

In *Mechanica sive Tractatus de Motu*, Wallis dealt also with elastic impact of *soft* bodies, where he commented on how the reaction force deforms the body during compression. He noted that this elastic deformation causes the same change in momentum during restitution as had occurred previously during compression. Further, he suggests the case of partly elastic collisions where the velocity changes during restitution fall between the elastic and inelastic limits.

Christopher Wren (1632–1723)

Wren had a very distinguished career in science before turning increasingly towards architecture after the age of 30. Today he is most widely known for planning the rebuilding of London after the Great Fire of 1666, and as the architect of St. Paul's Cathedral (London), Trinity Library (Cambridge), the Sheldonian Theatre (Oxford) and many other churches and public buildings. He started out, however, as Fellow in mathematics and astronomy at Gresham College, London, from 1657 to 1661, before becoming Savilian Professor of Astronomy, 1661–1673. Wren and Hooke (an Oxford don), were among the group of mathematicians and natural philosophers who met regularly in rooms at Gresham College, London. A merger of these meetings and those of the Oxford experimental club resulted in the founding of The Royal Society of London in 1660. Wren's scientific work brought him the presidency of The Royal Society from 1680 to 1682. It was as a leading mechanician and founding member of the society that he contributed a paper on elastic collisions that was read before The Royal Society and published (together with Wallis's paper) in the *Proceedings* of November–December 1668.

Like the subsequent paper by Huyghens, Wren's considered a collision between two elastic bodies (spheres) with different weights. He began from a concept of "proper velocity" which, for any body, is inversely proportional to weight. He regarded impact of two bodies, each traveling at its proper velocity, as equivalent to a balance oscillating about its center of gravity; i.e., he saw collisions from the viewpoint of the *law of the lever*. Wren expressed this analogy in diagrams illustrating the correct relations he developed for the speeds at separation, given any incident speeds. The greatest virtue of Wren's paper, however, was that it described experiments that he used to develop his ideas and validate the results. As Newton wrote in the scholium following the laws of motion in his *Principia*, "Wren proved the truth of these rules before the Royal Society by means of an experiment with pendulums." Wren's experiment involved two pendula of equal length but different weights that are released from different heights – the heights were chosen such that each pendulum bob rebounded to the same proportion of its initial release height. He demonstrated that two elastic bodies approaching each other with velocities that are inversely proportional to the weight of the other body result in simple reversal of velocity of both bodies when they collide. While Wren came up with essentially the same rule of impact as Huyghens, he (unlike Huyghens) did not present the general principles that were used to derive these results. As was his wont, this $1\frac{1}{2}$ page paper was the only thing that Wren ever wrote on this subject.

Christian Huyghens, 1629–1695

Huyghens was born in Holland but spent much of his life in Paris. He was an eminent mathematician and physicist who is remembered for his concepts in physical optics and the wave theory of light. Using his knowledge of optics, he constructed telescopes with improved lenses that allowed him to discern that the planet Saturn was surrounded by an annular disk rather than a pair of moons. In 1655 he discovered Titan – the first moon of Saturn to be identified.

In 1652–1657 Huyghens was working on an improved description of motion and treatment of impact.[11] From 1652 he distinguished between momentum ("quantity of movement") and kinetic energy, but did not understand the relationship between these variables. While Galileo is credited with suggesting that if no external forces act on a system, then translational momentum is conserved, Huyghens used this concept and conservation of total mechanical energy (*conservatio vis ascendentis*), without pointing out that these concepts had any special significance (Mach, 1919).[12, 13] (Subsequently Liebnitz named the latter principle as *vis viva* or living force – it is presently known as the conservation of energy.) About 1659 Huyghens distinguished between mass and weight.[14] Most of these investigations did not have as wide an influence as they might have had if Huyghens had not been so reticent to publish. His posthumous book on impact, *Tractus de Motu Corporum ex Percussione*, was written almost entirely before 1659 but not published until 1703. By that time Newton's *Principia* had already been in print for 15 years.

---

[11] Correspondence with Schooten, Gutschoven and Slusius (see Bell, 1953, p. 110).

[12] In Huyghens's application the term *conservatio vis ascendentis* meant that in a conservative system (e.g., a pendulum swinging in a gravitational field), any displacement of a body in a direction parallel to gravity resulted in work that changed the kinetic energy. This work was just sufficient to raise the swinging pendulum back to the height from which it was first released.

[13] Regarding previous statements of a idea of conservation of translational momentum, see footnote 8.

[14] Crew (1935) remarks that this is probably the earliest suggestion of such a distinction. (See "De Vi Centrifuga", *Oeuvres complètes de Huyghens*, Dutch Scientific Society, Vol. XVI, p. 266.)

Figure A.3. Drawing from *De Motu Corporum ex Corporum ex Percussione* illustrating impact of bobs of two pendula supported in a moving boat as viewed from a stationary reference frame (the man on the bank).

Huyghens was widely recognized for his mathematical skills and physical intuition. He was an early member of The Royal Society of London and a founding member of the French Academy. In 1668 he was a late addition to the group of three savants commissioned by The Royal Society to clarify the phenomena of impact – because of their methods, Newton referred to Huyghens, Wallis and Wren as the three geometers. Huyghens's paper on the subject was published in March 1669, and it treated impact of elastic particles. While each of the three authors produced a correct solution, Huyghens's paper had the major distinction of recognizing that all changes in velocity during collision depend on the relative velocity at the instant when the bodies first come into contact. He freely uses translating coordinate systems in order to express velocity changes relative to a coordinate system moving in the normal direction with the center of gravity of the system, as shown in Fig. A.3. Regarding conservation of momentum in a collision between two elastic particles, he says "... the common center of gravity of the bodies advances always equally towards the same side in a straight line before and after impact". In arriving at his propositions, Huyghens uses two concepts:

(i) He explicitly gives a translational velocity to the entire system such that the center of gravity of the system is stationary both before and after the collision.
(ii) He considers that for equal but opposite efforts applied to the two bodies, relative to the coordinate system moving with the center of gravity, each body has a speed of approach that is inversely proportional to its weight; i.e., he gives the bodies equal but opposite initial *quantities of motion* relative to the center of gravity.

The latter proposition is similar to but more general than the idea behind Wren's demonstration experiment. Here, for any initial velocity of each of the colliding bodies, Huyghens calculates the speed of the steadily moving reference frame such that the difference in velocities satisfies proposition (ii).

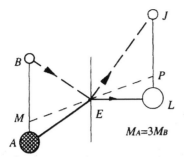

Figure A.4. Marriotte's illustration of direct collision of balls A and B with unequal mass that approach each other along a line. The horizontal axis represents time. The positions of the balls are indicated at equal intervals of time before and after impact at E. The speed of the center of mass (shown as a dashed line MEP) is constant.

Although this book is about impact, it is well to acknowledge that in mechanics, Huyghens is best known as the father (if not the inventor)[15] of the pendulum clock. The clocks which Huyghens invented were intended for use at sea – mainly for determination of longitude. In this connection he went beyond the dynamics of particles and provided an expression to calculate the *moment of inertia* of solid bodies (by summation).[16] In his *magnum opus* on the pendulum clock, *Horologium Oscillatorium* (1673), he named and provided the first correct explanation for centrifugal force. Subsequently Newton acknowledged with chagrin, "What Mr. Huyghens has published since about centrifugal force I suppose he had before me."

Edmé Mariotte (1620–1684)

Mariotte was a Roman Catholic abbé who was best known for experimentation. While not in the first rank of contributors who developed an understanding of principles describing impact, he solved a number of problems of elastic collisions in a manner similar to Huyghens and wrote a useful book, *Traité de la Percussion ou chocq des Corps* (Paris, 1673), that gives a clear presentation of the geometric method of analysis in vogue at that time.[17] Mariotte's analysis was based on a principle of invariability of the quantity of motion (conservation of translational momentum of the system) which clearly resolved the velocities of colliding spheres into components normal and tangential to the plane of the common tangent at the contact point.

Figure A.4 is an example of a collision at a normal angle of incidence between two elastic balls, A and B, where the masses are unequal: $M_A = 3M_B$. The vertical axes on the left and right give the positions of the balls at instants of time τ before and after

---

[15] Galileo had designed a clock escapement based on the pendulum, but it is unlikely that this was ever constructed.

[16] Wallis introduced him to this idea in a letter written 1 January 1659, where he described his method of finding a moment of weight about a certain axis as imagining the body separated into segments and forming a series in which each term was the weight of a segment multiplied by a distance. To find the distance from the axis to the center of gravity Wallis divided the sum of the moments of weight of the individual segments about the axis by the total weight of the body (Bell, 1953, p. 125).

[17] Mariotte redefined Wallis's confusing terminology for elastic and inelastic impact. He used the more accurate descriptions *flexible and resilient* for elastic collisions and *flexible and not resilient* for inelastic collisions.

the impact at E, while the horizontal axis represents time. The two balls approach each other at a speed proportional to length AB, and after impact they separate at a speed proportional to length JL. A dashed line MEP represents the motion of the center of mass during this period; i.e., lengths AM and MB are inversely proportional to the masses of the adjacent balls, A and B, respectively. Hence Mariotte determines the separation velocities by considering that any changes in velocity depend only on velocity differences relative to a reference frame moving with the center of mass; for elastic collisions (the only case he considers) he simply reflects the normal component of incident velocity about the vertical line through E, where changes in velocity are relative to the steadily moving center of mass of the system. It is likely that Mariotte followed Huyghens in recognizing that changes in velocity during collision are independent of the reference frame of the observer.

Marriotte's diagram (Fig. A.5) represents two examples of oblique impact of a moving ball A against a stationary ball B. Here again the lines represent velocity vectors for each ball before and after impact. Following collision, ball B moves in the direction of the common normal with speed proportional to BE, while ball A has velocity proportional to bF. The center of mass of the system has a postcollision velocity bG given by conservation of quantity of motion, so that the speeds are related by $bG/Ab = M_A/(M_A + M_B)$. Mariotte illustrates cases of (i) balls of equal mass ($M_A = M_B$) and (ii) balls of unequal mass ($M_A > M_B$). In these examples he follows a Cartesian principle by decomposing velocities into components normal and tangential to the common tangent plane at the point of contact. In each case Ab is the incident velocity of ball A with components AD and Db in the tangential and normal directions respectively. Moreover, in both cases the vertical (tangential) component of velocity for A is the same before and after collision; i.e., bF = AD, or LF = AD. Since the collision is elastic, the relative normal velocity after impact equals that before impact. Hence Mariotte obtains the postcollision normal component for each ball by separating the incident relative velocity Db into two parts that are inversely proportional to the masses; he then says that following collision this same proportionality applies to the speeds relative to that of the center of mass; i.e.,

$$\frac{IL}{IE} = \frac{DH}{Hb} = \frac{M_B}{M_A}.$$

Thus he says that for case (ii) triangles FEL = AbD and GEI = CbH. Once again this proportionality expresses the conservation of *quantity of motion*.

Isaac Newton (1642–1727)

Newton came from the farming village of Woolsthorpe near Grantham, England. His family were moderately well off, but uneducated, landowners. His father died before his birth, and his mother remarried an elderly vicar who wanted nothing to do with the boy. Young Isaac was raised by his grandmother, who taught him to read and write. In 1661 he entered Trinity College, Cambridge as a sizar (i.e., he earned most of the money for his expenses by performing menial tasks for Fellows, Fellow commoners and affluent student pensioners). After three years he became a scholar, and in 1665 he took his B.A.; that same year, because of the Plague, he returned home for two years to reflect, study and develop his ideas on mathematics, optics and mechanics. This enforced period of solitary study was most fruitful – he worked on many of those concepts that would later become

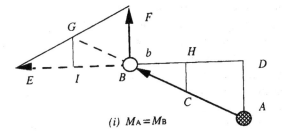

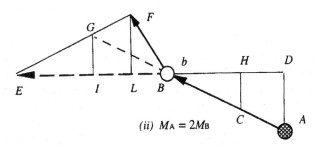

Figure A.5. Oblique elastic impact of moving ball A with stationary ball B that has (i) equal mass or (ii) unequal mass, $M_A = 2M_B$. In this figure Marriotte uses arrows to represent the incident and separation velocities of both A and B. At incidence the normal component of relative velocity is Db. At separation, ball A has velocity bF, ball B has velocity bE, and the center of mass of the system has velocity bG.

the more important contributions in the *Principia*. Following his return to Cambridge in 1667, he was elected a Fellow of his college and subsequently succeeded his mentor, Barrow, as Lucasian Professor of Mathematics. He published his *Theory of Light* in 1671, shortly before being elected a Fellow of The Royal Society. The *Principia* was published in 1687. In later years his extraordinary contribution to science was recognized by his country when he became Master of the Mint and President of The Royal Society. A most thorough and entertaining biography of Newton was written by Westfall (1980).

Leaving aside the individual developments that Newton expressed so clearly in the *Principia*, his major contribution was a new approach to problems – he sought to explain causes of phenomena rather than merely describing them. He was a master of synthesis; e.g., the law of universal gravitation was synthesized as a consequence of the third law of motion. Moreover, he continually tested hypotheses against experimental evidence. Nowhere is use of experiments to validate theory more clearly demonstrated than in his treatment of impact.

Newton's third law of motion states: *To every action there is always an equal and opposed reaction*. In a scholium, or example, illustrating this law, Newton analyzed an experiment verifying that the law applies to collisions also. He provided a drawing (Fig. A.6) describing his experiment which involved two spherical bodies A and B hung at the end of 10 foot long cords, CA and DB. His description of his experimental technique demonstrates concern for eliminating experimental errors; he says "we are to have due regard as well to the resistance of the air as to the elastic force of the concurring bodies."[18]

---

[18] Excerpts from the *Principia* are from Cajori's 1934 revision of an original translation by Motte (1729) (see Newton, I.).

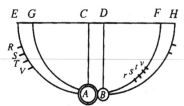

Figure A.6. Illustration from the *Principia* of Newton's experiment where pendulum bobs with different weights collide. Incident and separation speeds were calculated from measured heights of descent and reascent by Galileo's expression of proportionality between square of speed and height of descent. Letters R, S, T, V, etc. were used to correct the measurements in order to allow for air resistance.

In order to correct for air resistance, the free-swinging path EAF of body A is subdivided as follows: if body A, when released from R, returns to V after one free swing, then the translational speed of A at the bottom of its swing (impact point) will be the same as if it had been released *in vacuo* from S, where $RS:RV = 3:8$. Similarly suppose that, after reflection from body B, A comes to place s. Then if a free-swinging pendulum A returns to r when released from v, the corrected height after the collision is t rather than the measurement s. After this demonstration of his care, Newton goes on to say; "Thus trying the thing with pendulums of 10 feet, in unequal as well as equal bodies, and making the bodies to concur after a descent through large spaces, as of 8, 12, or 16 feet, I found always, without an error of 3 inches, that when the bodies concurred together directly, equal changes towards the contrary parts were produced in their motions, and, of consequence, that the action and reaction were always equal."

At this point Newton goes on to describe his results from impact experiments with bodies composed of various materials. This may have been the first time the material used in experiments was actually named; neither Wren nor Mariotte had provided this information. Of course they considered perfectly elastic impacts only. Newton is careful to distinguish his results from those for elastic collisions. He writes,

> ... I must add, that the experiments we have been describing, by no means depending upon that quality of hardness, do succeed as well in soft as in hard bodies. *For if the rule is to be tried in bodies not perfectly hard we are only to diminish the reflection in such a certain proportion as the quantity of the elastic force requires.* By the theory of *Wren* and *Huyghens*, bodies absolutely hard return one from another with the same velocity with which they meet. But this may be affirmed with more certainty of bodies perfectly elastic. In bodies imperfectly elastic the velocity of the return is to be diminished together with the elastic force; because that force (except when the parts of bodies are bruised by their impact, or suffer some such extension as happens under the strokes of a hammer) is (as far as I can perceive) certain and determined, and makes the bodies to return one from the other with a relative velocity, which is in a given ratio to that relative velocity with which they met. This I tried in balls of wool, made up tightly, and measuring their reflection, I determined the quantity of their elastic force; and then, according to this force, estimated the reflections that ought to happen in other cases of impact. And with this computation other experiments made afterwards did accordingly agree; the balls always receding one from the other with a relative velocity, which was to the relative velocity with which they met as about 5 to 9. Balls of steel returned with almost the same velocity; those of cork with a velocity something less; but in balls of glass the proportion was as about 15 to 16.

And thus the third Law, so far as it regards percussions and reflections, is proved by a theory exactly agreeing with experience (Motte).[19]

Thus Newton uses the principle of conservation of translational momentum, but in addition the third law, to find that irrespective of hardness of the colliding bodies, the changes in velocity are always in proportion to the relative velocity at incidence. He finds that the proportion of this relative velocity which is recovered depends on the bodies – i.e., his proportionality is essentially a material property. Here it seems that he was too enamored of large pendulum swings, which could more easily be measured, and missed the fact that his conclusion about this proportionality being independent of relative speed at incidence is logically inconsistent in the limit of indefinitely small incident speeds. Nevertheless, we recognize that in comparison with previous writing on our topic, these passages represent a tremendous leap in style, thoroughness and clarity in regard to relating cause and effect.[20]

Leonhard Euler (1707–1783)

Euler was a student at the University of Basel, where he was tutored by John Bernoulli, professor of mathematics. Truesdell (1968) quotes from a 1720 letter from Bernoulli's son, Daniel: "Mr. Euler ... of Basel is a student of my father who will do him much honour." In 1727 Euler was appointed professor of physics of the Academy of Sciences in St. Petersburg by Catherine I, a patron of science. He went there, joining his friend, Daniel Bernoulli who had taken up the chair of mathematics two years previously. In year 1732, D. Bernoulli moved to Berlin to take up chairs, first in anatomy and later in physics. Euler was elected to fill the vacant chair in mathematics in St. Petersburg. He remained there until 1741, when Frederick the Great called him to Berlin. In 1766 he returned to St. Petersburg at the request of Catherine the Great, and he soon went completely blind, but nonetheless continued to produce important papers. He almost certainly is the most prolific writer of mathematical papers of any period. His work is now readily accessible through *Opera Omnia*, edited by Charles Blanc and published by Societatis Scientiarum Naturalium Helveticae.

The first investigation of dynamics of collisions between rigid bodies as distinct from particle collisions is due to Euler in a paper submitted to the Academy of St. Petersburg in 1737.[21] A root of this work goes back to Euler's mentor, John (Johan or Jean) Bernoulli,

---

[19] Newton's reference to steel is not specific, but steel was produced as early as 1000 B.C. by heating iron objects in contact with charcoal. This process of surface carbonization was initially used by the Chalbyes in Northern Anatolia to improve the edge-holding properties of iron tools (Wertime, T.A. and Muhly, J.D., *Coming of Age of Iron*, Yale Univ. Press, 1980).

[20] Newton could achieve clarity because he firmly grasped the concepts. Barbour (1989, p. 678) remarks that Kepler recognized that dynamic behavior of individual bodies must be characterised by a quantity which measures its resistance to applied force; he introduced the term "laziness", or inertia, for this quantity. It was Newton who took the step of replacing "the concept of laziness with respect to *motion* by laziness with respect to *change in motion*" in order to arrive at inertial mass. Previously Huyghens had expressed a similar idea (see footnote 12).

[21] Herivel describes an unpublished manuscript, predating publication of the *Principia*, where Newton analyzes planar impact of two elastic bodies of arbitrary shape. He does not, however, give a method of calculating the radius of the "equator of reflected circulation" (i.e. the radius of gyration). Herivel has shown how Newton's variables can be defined so that the changes in angular and translational velocity of each body are obtained as functions of the radii of the equators of reflected circulation – this gives the correct result, although Newton's method is obscure.

the Professor of Mathematics at Basel and father of Daniel Bernoulli, who is frequently credited with formulating in 1703 the concept of moment of inertia for a body.[22]

Despite the invention of calculus by Leibnitz and Newton, the problem of mechanics of impact or percussion remained of central importance. The Royal Academy of Sciences in Paris biannually awarded a prize for the most outstanding paper; in 1724 and 1726 the prizes were for papers on percussion. Colin Maclaurin, professor of mathematics at the University of Aberdeen, was awarded the prize in 1724 (over John Bernoulli) for his paper entitled *Demonstration des lois du choc des corps*. He expressed that the interaction forces on colliding bodies are equal in magnitude but opposed in direction (Newton's third law) and used the physical construct of an elastic spring between the contact points in order to obtain changes in velocity during compression and restitution phases of collision. The spring represented a small compliant region – implicitly the tangential compliance was assumed to be negligible.[23] Maclaurin says that he won the prize for clarifying that during a collision the changes in velocity depend on relative velocity only – but omits saying that this point had previously been made by Huyghens. The recipient of the next Academy prize was Père Mazière in 1726. That year the runner-up was again John Bernoulli, who also had his paper published. Bernoulli's paper, *Discours sur les loix de la communication du mouvement*, related velocities before and after collision by means of conservation of energy; consequently, his analysis was unintentionally limited to elastic collisions. The fact that Bernoulli did not comment on or otherwise acknowledge this limitation brought forth jeers of derision from Maclaurin[24] and a pointed criticism from Robins.[25] Despite these setbacks, J. Bernoulli was an indefatigable campaigner, and during his lifetime he won 10 French Academy prizes.

Euler submitted his paper on planar collisions between two rigid bodies, *De communicatione motus in collisione corporum sese non directe percutinentium*, in 1737; it was published in the *Proceedings of the Scientific Academy of St. Petersburg* (1744). In this communication he considers planar collisions between two bodies which can rotate as well as translate; the rotational inertia of each body is represented by a moment of inertia about the center of mass – a parameter he terms the "... *sitque horum factorum aggregatum* = S". In this problem there is negligible friction, so the interaction force at the contact point is perpendicular to the common tangent plane. The bodies have centers of mass that are not on the line of action of the contact force; i.e., the impact configuration is noncollinear. Euler described his analytical method with Fig. A.7.

Euler considers limiting cases of perfectly elastic and perfectly inelastic collisions. His analysis supposes that at initial contact (incidence) the two bodies are separated by an infinitesimal elastic element – an artifice originally introduced by Maclaurin to represent the normal compliance of an infinitesimally small deforming region. With this artifice each body has a contact point. Although these points are coincident, there is a relative velocity between them that varies during the collision. The elastic element is

---

[22] Earlier Wallis and Huyghens had expressed the moment of inertia of a body about an axis of rotation as a property that they calculated by summation over all parts of the body.

[23] Whereas Galileo envisioned elastic deformations of colliding bodies as homogeneous, so that when compressed together, spherical bodies became spheroidal, Euler recognized that stresses decreased rapidly with increasing distance from the contact area, so that only a small part of the body around the contact area is significantly deformed.

[24] Maclaurin (1748, p. 192).

[25] Robins, B., *The Present State of the Republic of Letters*, May 1728.

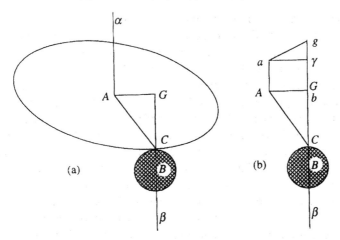

Figure A.7. Euler's illustration of eccentric collision of a sphere B moving at speed βB against stationary elliptical body with center of gravity at A. After an elastic collision the relative velocity between the contact points turns out to be equal but opposite to what it was at incidence; i.e., $C_g = \beta B$. The normal velocity at separation is composed of translational velocities Cb and γg of the centers of mass plus the translational velocity $G\gamma$ at C on body A due to the product of the angular speed of A and the perpendicular distance AG to the line of action. Line segments Cb and γg have lengths that are inversely proportional to the masses the corresponding bodies.

oriented in a direction **n** normal to the common tangent plane at the contact point, so the interaction force **F**, which results from compressing this compliant element, acts solely in the direction of the normal to the common tangent plane, $\mathbf{F} = F_n\mathbf{n}$. Euler was the first to consider interaction forces with a line of action that does not pass through both centers of mass, a consideration that becomes important if the shapes of the colliding bodies are not spherical. The changes in relative velocity are due to changes in both translational velocity of each center of mass and rotational velocity of each body. Euler assumes that the collision period is divided into an initial compression period, where the normal component of relative velocity $v_n = \mathbf{v} \cdot \mathbf{n}$ between the contact points is decreasing as the contact points approach each other, and a subsequent period of restitution, where this component of relative velocity is increasing as the contact points separate.[26] The transition between these periods occurs when the relative velocity at the point of contact has a normal component that vanishes.

Using a time-dependent analysis, Euler obtained the relative velocity at the contact point as a function of work done on the deformable spring by the contact force **F**. His elastic analysis assumes the contact force is reversible, so in an elastic collision $\int_0^{t_f} F_n v_n \, dt = 0$. He relates the work done by the contact force to an expression proportional to changes in the kinetic energy of the system. Euler's determination of the final or separation conditions on the basis of the ratio of kinetic energy restored to the system by contact force during restitution in comparison with the energy absorbed during compression is a major departure from Newton's kinematic law of restitution. Euler simply used this ratio for these limiting cases without commenting on the distinction from the kinematic law of

---

[26] The idea of dividing the impact into separate periods of compression and restitution seems to have originated with Nicolas de Malebranche (1675–1712), a Parisian priest.

restitution. After one takes the step of modeling the collision with an elastic element between the contact points, it seems natural to use the energy stored and subsequently released by this element to define the terminal conditions at separation.

In 1744 Benjamin Robins wrote *New Principles of Gunnery*, which among other things described the use of a ballistic pendulum to measure the momentum of a cannon ball at various distances from the muzzle of a gun. With these measurements he established the significance of air drag in calculating the trajectory of projectiles if the muzzle velocity is larger than 130 ms^{-1}. Also he discussed the aerodynamic lift and consequent curvature of the path of a projectile which is spinning about a transverse axis – what is now termed the *Magnus effect*. Euler annotated and corrected some technical errors in Robins's book; his annotated version (in German) is five times the length of the original. Nevertheless, W. Johnson (1986) suggests that Euler's version contains little that is original. The annotated version was translated into English by Hugh Brown (published as *The True Principles of Gunnery*, London, 1777), and into French by J.L. Lombard (1783).

Historical References

Barbour, J.B. (1989) *Absolute or Relative Motion? Vol. 1: Discovery of Dynamics.* Cambridge Univ. Press.

Bell, E.T. (1953) *Men of Mathematics*, Vol. I, Penguin.

Blanc, C. (1957–58) *Opera Omnia* of E. Euler, Societatis Scientiarum Naturalium Helvetica. Teubner.

Cajori, F. (1934) transl. of *Principia Mathematica* by I. Newton, 1687, Univ. of California Press, Berkeley.

Cohen, M.R. and Drabkin, I.E. (1948) *A Source Book in Greek Science*, Harvard Univ. Press.

Coriolis, G. (1835) *Théorie Mathématique des Effets du Jeu de Billiard*, Carilian-Joeury, Paris.

Crew, H. (1935) *The Rise of Modern Physics*, Bailliere, Tindall and Cox, London.

Dugas, R. (1957) *History of Mechanics* (transl. by J.R. Maddox), Routledge & Kegan Paul, London.

Duhem, P. (1903) *L'Evolution de la Mècanique*, Joanin, Paris

Euler, L. (1737) "De communicatione motus in collisione corporum sese non directe percutientium," *Comment. Acad. Sci. Petropolitanae* **9**, 50–76 (1744); in *Commentationes Mechanicae* (ed. C. Blanc), Societatis Scientiarum Naturalium Helveticae.

Finocchiaro, M.A. (1989) *Galileo Affair: A Documentary History*, Univ. of California Press, Berkeley.

Galilei, Galileo (1632) *Dialogo sopra i due massini Sistemi del Mondo, Tolemaico e Copernicano* (Dialogue Concerning the Two Chief World Systems: Ptolomaic and Copernican) Florence.

Galilei, Galileo (1638a), *Dialogues Concerning Two New Sciences* (transl. by H. Crew and A. de Salvio), Macmillan (1914).

Galilei, Galileo (1638b), *Dialogues Concerning Two New Sciences* (transl. by S. Drake), Univ. of Wisconsin Press (1974).

Herivel, J. (1965) *The Background to Newton's Principia*, Oxford Univ. Press.

Huyghens, C. (1703) *De Moto Corporum ex Corporum ex Percussione* in Ovevres Completes, Vol. 16 (1929) (English transl. by R. J. Blackwell, Isis **68**, 574–97, 1977).

Johnson, W. (1986) "Benjamin Robins: new principles of gunnery", *Int. J. Impact Engng.* **4**, 205–219.

Johnson, W. (1992), "Benjamin Robins (18th century founder of scientific ballistics): European dimensions and past and future perceptions", *Int. J. Impact Engng.* **12**, 293–324.

Jouguet, E. (1871) *Lectures de Mécanique*, Johnson Reprint Corp., New York (1966).

Mach, E. (1883) Die Meckanik in Ihrer Enwicklung Historisch-Kritisch Dargestellt (*The Science of Mechanics*, transl. by T. J. McCormack) Open Court, La Salle, Ill. (1960).

Mach, E. (1919), *The Science of Mechanics* (transl. by T.J. McCormack), 224.

Maclaurin, C. (1748) *An Account of Newton's Philosophical Discoveries*, Patrick Murdoch.

Malebranche, N. (1960) *Oeuvres Complètes de Malebranche: Pièces Jointes, Ècrits Divers. Des Lois du Mouvement* (ed. Pierre Costabel), Vrin, Paris.

Marci, M. (1639) *De Proportione Motus*, Prague.

Newton, I. (1687) *Mathematical Principles of Natural Philosophy and his System of the World* (transl. by A. Motte; revised and annotated by F. Cajori, 1934), Univ. of California Press, Berkeley.

Parsons, W.B. (1939) *Engineers and Engineering in the Renaissance*, MIT Press.

Robins, B. (1742) *New Principles of Gunnery*, J. Nourse, London; Richmond Publ., Richmond, UK (1972).

Truesdell, C. (1968) *Essays in the History of Mechanics*, Springer-Verlag.

Wertime, T.A. and Muhly, J.D. (1980) *Coming of Age of Iron*, Yale Univ. Press.

Westfall, R.S. (1980) *Never at Rest*, Cambridge Univ. Press.

Wilson, J. (1761) *Mathematical Tracts of the Late Benjamin Robins, Esq.*, J. Nourse, London.

Glossary of Terms

This model will be a simplification and an idealization, and consequently a falsification. It is to be hoped that the features retained for discussion are those of greatest importance in the present state of knowledge.

A. M. Turing, 1952

angle of incidence   angle between the direction of the incident relative velocity of the contact points and the common normal direction. Direct or normal collisions have zero angle of incidence, whereas oblique collisions have a nonzero angle of incidence.

angle of rebound   angle between the direction of the relative velocity of the contact points at separation and the common normal direction.

attractor   steady state solution that is approached asymptotically with increasing time if the system has small dissipation.

coefficient of friction   upper limit on ratio of tangential to normal force at contact.

coefficient of stick   geometric parameter specifying ratio of tangential to normal force for stick.

collinear (or central) impact configuration   colliding bodies oriented so that each center of mass is on common normal line passing through the point of initial contact.

common normal direction   normal to common tangent plane that passes through contact point C.

common tangent plane   If at least one of the bodies has a topologically smooth surface at the contact point, this is the plane that is tangent to the surface at the point of initial contact. Usually both bodies have smooth surfaces around their respective points of contact, so they have a common tangent plane.

compression phase of collision   part of the contact period in which the normal component of relative velocity between contact points is negative, i.e., the contact region is being compressed.

configuration   description of position and orientation of each body in a dynamic system.

conforming bodies   two bodies with contact surfaces where the curvature of one is the negative of the curvature of the other so that they initially come into contact over an area rather than at a point.

conforming contact surfaces   surfaces which conform or touch over a finite area.

constitutive relation   equation relating stress at a point to strain or local deformation.

constraint equation kinematic equation that specifies a velocity, relative velocity or range of admissible velocities at point of intersection between two bodies.

contact area area around initial point of contact where surfaces of colliding bodies are coincident.

contact force stress resultant of pressure and tangential surface tractions acting in contact area.

deformation relative displacement between two points on the same body that is due to extension and distortion; obtained by integrating strains along a line at a particular instant of time.

direct collision collision throughout which the relative velocity of contact points is in the normal direction. It is frictionless, since no sliding occurs in the contact region.

dispersion spreading of a pulse with time caused by dependence of speed of propagation on wavelength.

dissipated energy kinetic energy transformed during collision to plastic work, viscous losses, residual vibrations of the bodies, etc.

eccentric (or noncollinear) impact configuration center of mass of at least one body not on the line of common normal that passes through the contact point.

effective mass $m = (M_{\mathbf{B}}^{-1} + M_{\mathbf{B'}}^{-1})^{-1}$, a composite mass term in the equation of relative motion for a pair of colliding bodies.

elastic having a reversible constitutive relation; stress uniquely related to strain.

grazing incidence angle of incidence approaching tangent to contact surface.

incidence time of initial contact of colliding bodies.

incident relative velocity at contact point difference between the velocities of coincident points of contact when contact initiates.

inelastic having an irreversible constitutive relation so that a cycle of loading and unloading exhibits hysteresis.

interference overlap between surfaces of two bodies.

intrinsic speed natural speed that is characteristic of system.

jam self-locking of sliding in eccentric impact with large coefficient of friction.

noncollinear configuration orientation of colliding bodies such that the center of mass is eccentric (not on the common normal passing through the point of initial contact).

normal direction unit vector perpendicular to the common tangent plane.

oblique collision one in which at incidence the relative velocity between the points of contact has a component that is tangential to the common tangent plane, so that the angle of incidence is nonnormal.

reflection coefficient at an interface, the ratio of amplitude of reflected to incident wave.

restitution phase of contact part of the contact period in which the normal component of relative velocity between the contact points is positive; the period in which the bodies are moving apart but have not lost contact.

rough surface one such that friction related to surface roughness opposes sliding at point of contact.

separation instant when contact ceases.

separation relative velocity   final difference at separation between the velocities of the points of contact on the two bodies.

sliding   tangential component of relative velocity between two coincident contact points.

slip   sliding

smooth surface   frictionless surface; i.e., tangential component of contact force is zero.

state of stress   complete set of stress components at a point.

stick   motion in which tangential component of relative velocity remains zero.

topologically smooth surface   region of surface with continuous curvature in every direction. "Smooth" is also used to mean frictionless; hence the cumbersome adjective "topologically" is used to distinguish between these two meanings.

total mechanical energy   sum of kinetic and potential energies (including energy of internal deformation).

traction   vector describing the force per unit area acting at a point on a surface.

transmission coefficient   at an interface, the amplitude ratio of reflected to transmitted wave.

Answers to Some Problems

Genius begins great works, labor alone finishes them.

Joseph Joubert

1.1 $V_f/V_0 = M/(M + M'), (T_0 - T_f)/T_0 = M'/(M + M')$

1.3 $\hat{\rho} = (5, 4, 0)/3; \hat{V} = (5, 2, 0)/6; \hat{I} = \dfrac{2M}{3} \begin{bmatrix} \mathbf{n}_1\mathbf{n}_1 & -2\mathbf{n}_1\mathbf{n}_2 & 0 \\ -2\mathbf{n}_2\mathbf{n}_1 & 4\mathbf{n}_2\mathbf{n}_2 & 0 \\ 0 & 0 & 5\mathbf{n}_3\mathbf{n}_3 \end{bmatrix};$

  $\hat{\mathbf{h}} = (5M/3)(0, 0, 1)$

1.4 (a) $\omega_+/\omega_- = 0.5$; (b) $\omega_+/\omega_- = 0; \hat{V} = V_{B+} = L\omega_-$; (c) $(T_- - T_+)/T_- = 3/4$

1.5 $\omega_+/\omega_- = (2 + 7 \cos 2\alpha)/(8 + \cos 2\alpha)$

2.2 $V_f' = 9.1 \text{ m s}^{-1}, V_f = 44.7 \text{ m s}^{-1}$

2.3 $V_3(p_f)_{max}/V_0 = 4/(1 + \alpha)^2, \lim_{\alpha \to 0}[V_3(p_f)/V_0] = 4$

2.4 $V_n(p_f)/V_0 = 2^{n-1}$

2.5 $M/M' = \cos 2\theta + (\sin 2\theta)/\tan \theta'$

2.6 $\theta = 1.176 \text{ rad.}, V_{1f}/V_0 = 0.934, V_{2f}/V_0 = 0.253$

2.9 $M'/M = 4/3, V_f = (\pi C/M)^{1/2}$

2.10 Largest possible length

3.4 $\hat{V}_1 = 0, \hat{V}_3 = 5V_0/7; x_{3B} = 5V_0^2/49\mu g$

3.5 $\theta_B = 30°, \theta_{B'}(30°) \approx 33.7°;$

  $2\theta_B = \cos^{-1}(5/9) \approx 56° \Rightarrow \theta_{B'max} = \theta_{B'}(28°) \approx 33.7°$

3.8 (b) $\mu = 2\sqrt{3}/21$; (c) $W_3(p_c) = \sqrt{3}MV_0^2/8$

3.11 (b) $-\omega_0/2$

3.12 (a) $\Delta T = 0.5(1 - e_*^2)MR^2\omega_0^2[2h/R - (h/R)^2]$

  (b) $R\omega_0 > \{2gh/[1 - (1 - e_*^2)(2h/R - h^2/R^2)]\}^{1/2}$

6.1 Calculation assumes the period of contact is the same for each pair of like spheres. In a collision between dissimilar spheres the body with a smaller contact period (for impact against an identical sphere) has larger influence than that predicted by this theory.

6.2 (a) $v_Y = 0.64 \times 10^{-3} \text{ m s}^{-1}$; (b) $e_* = 0.33$;

  (c) imperceptible surface indentation.

7.1 $W = Ap_0^2c\tau/6E, U/W = T/W = 1/2$

7.2 Interface $\sigma = 0$ for $t > L/c_0$, but contact lost at $t = 2L/c_0$ when displacement $u(0, 2L/c_0) = \dfrac{L}{c_0} \left(\dfrac{V_2 + V_1}{2}\right) + \dfrac{L}{c_0} V_1 = \dfrac{L}{2c_0}(3V_1 + V_2)$

7.4 $V_a(2L/c_a)/V_0 = -0.493$, $V_s(2L/c_a)/V_0 = 0.17$, momentum conserved;
K.E. + residual strain energy = initial K.E., $V_Y = 24.1$ m s^{-1}

7.5   Deceleration of monkey, $\frac{dV}{dt} = \frac{2c_0 V}{\alpha L} + \left(\frac{c_0 V}{\alpha L} - g\right) \exp\left(\frac{-2}{\alpha}\right)$

7.6   (a) $\dot{u} = V_0 e^{-\alpha(t-x/c_0)}$, $x < c_0 t$; $\dot{u} = 0, x > c_0 t$; (c) $\dot{u} = V_0 e^{-\alpha(t-x/c_0)}$, $x < c_0 t - 2L$;
$\dot{u} = V_0 e^{-\alpha(t-x/c_0)} + B e^{-\alpha(t+x/c_0 - 2L/c_0)}$, $x > c_0 t - 2L$

7.8   (b) $\dot{u}_T = \dot{u}_R + \dot{u}_I$, $-(\alpha L/c)\, d\sigma_T/dt = \sigma_T - (\sigma_R + \sigma_I)$;
(c) $\gamma_R = \exp(-2ct/\alpha L), \gamma_T = 1 - \exp(-2ct/\alpha L)$

7.9   Tension behind locomotive after $t = 2nL/c$

7.10   (a) $u(x,t) = 2u_0 e^{i(\bar{k}x - \bar{\omega}t)} \cos[\Delta k(x - c_g t)]$; (b) $\Delta k = \pi/L$

7.12 $c_g/c_0 = (1 + \bar{k}^2)^{-3/2}$; (a) $\bar{k} > 5$; (b) $\bar{k} < 2 \Rightarrow \lambda < 0.7$

7.13 $\gamma_R = (-1 - i)/2, \gamma_T = (1 - i)/2, \gamma_{ER} = \gamma_{ET} = (1 - i)/2$.

8.2   (b) $(v_1^+, v_2^+, v_3^+, v_4^+)^T = (6p/45M)(+26, -7, +2, -1)^T$
(c) $p_f = \dfrac{1 + e_*}{\alpha^{-1} + 156/45} M V_0'$

8.3 $T = (M/6)\left[4u_1^2 + u_1 u_3 + u_3^2 + v_1^2 + v_1 v_3 + 4v_3^2\right]$

8.4 $\begin{bmatrix} 12 & 0 & 0 & -3s\theta \\ 0 & 2 & 1 & 0 \\ 0 & 1 & 8 & 3c\theta \\ -3s\theta & 0 & 3c\theta & 2 \end{bmatrix} \begin{Bmatrix} du_1 \\ dv_1 \\ dv_2 \\ L\,d\dot{\theta} \end{Bmatrix} = \frac{6\,dp}{M} \begin{Bmatrix} c\phi \\ s\phi \\ 0 \\ 0 \end{Bmatrix}$

8.5 $V_A^+ = (3 + e_*)V_0/2$

8.6 $b_{11} = 1 + L^2/4\hat{k}_r^2, b_{12} = 1 + \lambda L/2\hat{k}_r^2, b_{22} = 1 + \alpha + \lambda^2/\hat{k}_r^2$

8.7 $p_c = 0.1021 ML, W_3(p_c) = -0.0069 ML^2, W_1(p_f) = -0.0038 ML^2, W_3(p_f) = -0.0051 ML^2$;
effect of friction large because initial speed of slip is large in comparison the normal component of incident relative velocity at C.

8.8   (a) If $1 < \alpha \le 3$ then $V_1/V_0 < 0$ after initial impact between balls at B_2, so second impact is at C_1.
(b) $V_1^+/V_0 = (-\alpha^2 + 10\alpha - 5)/(\alpha + 1)^2$,
$V_2^+/V_0 = (-5\alpha^2 + 10\alpha - 1)/(\alpha + 1)^2, 0.528 < \alpha \le 1$;
$V_1^+/V_0 = -(-\alpha^2 + 10\alpha - 5)/(\alpha + 1)^2$,
$V_2^+/V_0 = (-5\alpha^2 + 10\alpha - 1)/(\alpha + 1)^2, 0.333 < \alpha \le 0.528$.
Two impacts between balls at B_2 if $0.333 < \alpha \le 1.0$.

9.1 $\sin(k_j x/L), k_j = j\pi, j = 1, 2, 3, \ldots; L^2\omega_j/\gamma = \pi^2, 4\pi^2, 9\pi^2, \ldots$

9.2   Frequencies the same, since mode shapes are identical

9.3 $\tilde{M}_j = M/2, \bar{k}_j = j^4\pi^4 EI/2L^3, \omega_j = j^2\pi^2\gamma/L^2; A_j\bar{\omega}_j = 2(M'/\tilde{M})V_0'$,
$\bar{\omega}_j = \omega_j/\sqrt{1 + M'/\tilde{M}}, j = 1, 3, 5, \ldots$

9.4   hint: $\tilde{M} = M/2 = 60$ g, $M' = 32.8$ g

9.5   (a) $\alpha > 1$ results in multiple hits (contact chattering).
(b) $u(t_f) = \alpha V_0 t_f/(1 + \alpha)$.

10.1 $t_f = (1 + e_*)(1 - e_*)^{-1}\sqrt{2h_0/g}, x_f = (1 + e_*)^2(1 - e_*)^{-2} h_0 \cos\theta$,
$t > t_f \Rightarrow$ rolling

10.2 $\dfrac{h_*}{b} = \begin{cases} (1 - e_*^2)^{-1}, & e_*^2 > h_b/h_* \\ 1 + h_b/b, & e_*^2 < h_b/h_* \end{cases}$

References

In which a thousand trifles are recounted, as nonsensical as they are
necessary to the understanding of this great history.

Cervantes, *Don Quixote*, 1605

Abeyaratne, R. (1989) "Motion of a prism rolling down an inclined plane", *Int. J. Mech. Engng. Educ.* **17**, 53–61.

Achenbach, J., Hemann, J. H. and Ziegler, F. (1968) "Tensile failure of interface bonds in a composite body subjected to compressive loads", *AIAA J.* **6**, 2040–2043.

Adams, G.G. (1997) "Imperfectly constrained planar impact – a coefficient of restitution model", *Int. J. Impact Engng.* **19**(8), 693–701.

Adams, G.G. and Tran, D.N. (1993) "The coefficient of restitution for a planar two-body eccentric impact", *ASME J. Appl. Mech.* **60**, 1058–1060.

Andrews, J.P. (1931) "Experiments on impact", *Proc. Phys. Soc. Lond.* **43**, 8–17.

Bahar, L.Y. (1994) "On use of quasi-velocities in impulsive motion", *Int. J. Engng. Sci.* **32**, 1669–1686.

Bajoria, K.M. (1986) "Three dimensional progressive collapse of warehouse racking", PhD Dissertation, University Engineering Dept., Cambridge Univ., Cambridge, UK.

Batlle, J.A. (1993) "On Newton's and Poisson's rules of percussive dynamics", *ASME J. Appl. Mech.* **60**, 376–381.

Batlle, J.A. (1996) "The sliding velocity flow of rough collisions in multibody systems", *ASME J. Appl. Mech.* **63**, 168–172.

Batlle, J.A. (1998) "The jam process in 3D rough collisions", *ASME J. Appl. Mech.* **63**(3), 804–810.

Bilbao, A., Campos, J. and Bastero, C. (1989) "On the planar impact of an elastic body with a rough surface", *Int. J. Mech. Engng. Educ.* **17**, 205–210.

Bimping-Bota, E.K., Nitzan, A., Ortoleva, P. and Ross, J. (1997) "Cooperative instability phenomena in arrays of catalytic sites", *J. Chem. Phys.* **66**, 3650–3658.

Boley, B.A. (1955) "Some solutions of the Timoshenko beam equations", *ASME J. Appl. Mech.* **22**, 579–586.

Bouligand, G. (1959) *Mécanique Rationnelle*, 477–483, Vuibert, Paris.

Brach, R.M. (1989) "Tangential restitution in collisions", *Computational Techniques for Contact Impact, Penetration and Perforation of Solids*, ASME AMD 103 (ed. L.E. Schwer, N.J. Salamon and W.K. Liu), 1–7.

Brach, R.M. (1993) "Classical planar impact theory and the tip impact of a slender rod", *Int. J. Impact Engng.* **13**, 21–33.

Brogliato, B. (1996) *Nonsmooth Impact Mechanics: Models, Dynamics and Control*, LNCIS 220, Springer-Verlag, Heidelberg.

Budd, C.J. and Lee, A.G. (1996) "Impact orbits of periodically forced impact oscillators", *Trans. Roy. Soc. Lond.* **452**, 2719–2750.

Calladine, C.R. and Heyman, J. (1962) "Mechanics of the game of croquet", *Engineering* **193**, 861–863.

Chang, L. (1996) "Efficient calculation of the load and coefficient of restitution of impact betwen two elastic bodes with a liquid lubricant", *ASME J. Appl. Mech.* **63**, 347–352.

Chang, W.R. and Ling, F.F. (1992) "Normal impact model of rough surfaces", *ASME J. Appl. Mech.* **114**, 439–447.

Chapman, S. (1960) "Misconception concerning the dynamics of the impact ball apparatus", *Am. J. Phys.* **28**, 705–711.

Chatterjee, A. (1997) "Rigid body collisions: some general considerations, new collision laws and some experimental data", PhD Dissertation, Cornell Univ., Ithaca, NY.

Christoforou, A.P. and Yigit, A.S. (1998) "Effect of flexibility on low velocity impact response", *J. Sound & Vibration* **217**(3), 563–578.

Cochran, A.J. and Farrally, M.R., eds. (1994) *Science and Golf II*, E&FN Spon, London.

Collett, P. and Eckmann, J.P. (1980) "Iterated maps on the interval as dynamical systems", *Progress in Physics*, Vol. I, Birkhauser, Boston.

Cremer, L. and Heckl, M. (1973) *Structure Borne Sound* (transl. E.E. Ungar), Springer-Verlag, New York.

Cundall, P.A. and Strack, O.D.L. (1979) "A discrete numerical model for granular assemblies", *Geotechnique* **29**, 47–65.

Daish, C.B. (1981) *The Physics of Ball Games*, Hodder and Stoughton, London.

Davies, R.M. (1949) "The determination of static and dynamic yield stresses using a steel ball", *Proc. Royal Soc. Lond. A* **197**, 416–421.

Deresiewicz, H. (1968) "A note on Hertz's theory of impact", *Acta Mech.* **6**, 110.

Engel, P.A. (1976) *Impact Wear of Materials*, Elsevier Science, New York.

Evans, A.G., Gulden, M.E. and Rosenblatt, M. (1978) "Impact damage in brittle materials in the elastic–plastic response régime", *Proc. Roy. Soc. Lond. A* **361**, 343–365.

Follansbee, P.S. and Sinclair, G.B. (1984) "Quasi-static normal indentation of an elastic–plastic half-space by a rigid sphere", *Int. J. Solids Struct.* **20**, 81.

Glockner, C. and Pfeiffer, F. (1995) "Multiple impacts with friction in rigid multibody systems", *Advances in Nonlinear Dynamics* (ed. A.K. Bajaj and S.W. Shaw), Kluwer Academic.

Goldsmith, W. (1960) *Impact, the Theory and Physical Behaviour of Colliding Solids*, Edward Arnold, London.

Goldsmith, W. and Lyman, P.T. (1960) "Penetration of hard steel spheres into plane metal surfaces", *ASME J. Appl. Mech.* **27**, 717–725.

Gontier, C. and Toulemonde, C. (1997) "Approach to the periodic and chaotic behaviour of the impact oscillator by a continuation method", *Eur. J. Mech. A/Solids* **16**(1), 141–163.

Greenwood, J.A. (1996) "Contact of rough surfaces", *Solid–Solid Interactions: Proceedings 1st Royal Society–Unilever Indo-UK Forum in Materials Science and Engineering* (ed. M.J. Adams, S.K. Biswas and B.J. Briscoe), Imperial College Press, London, 41–53.

Haake, S.J. (1991) "Impact of golf balls on natural turf: a physical model of impact", *Solid Mechanics IV* (ed. A.R.S. Ponter and A.C.F. Cocks) Elsevier, London, UK, 72–89.

Haake, S.J. (1996) *The Engineering of Sport*, Balkema, Rotterdam.

Han, I. and Gilmore, B.J. (1993) "Multi-body impact motion with friction – analysis, simulation and experimental verification", *ASME J. Mech. Design* **115**, 412–422.

Hardy, C. Baronet, C.N. and Tordion, G.V. (1971) "Elastoplastic indentation of a half-space by a rigid sphere", *J. Num. Methods Engng.* **3**, 451.

Harris, R.A. and Stodolsky, L. (1981) "On time dependence of optical activity", *J. Chem. Phys.* **74**(4), 2145–2155.

Harter, W.G. (1971) "Velocity amplification in collision experiments involving Superballs", *Amer. J. Phys.* **39**, 656–663.

Hertz, H. (1882) "Über die Berührung fester elastischer Körper (On the contact of elastic solids)," *J. Reine Angew. Math.* **92**, 156–171.

Hill, R. (1950) *A Mathematical Theory of Plasticity*, Oxford Univ. Press.

Hogue, C. and Newland, D. (1994) "Efficient computer simulation of moving granular particles", *Powder Tech.* **78**, 51–66.

Holmes, P.J. (1982) "The dynamics of repeated impacts with a sinusoidally vibrating table", *J. Sound & Vibration* **84**(2), 173–189.

Hopkinson, B. (1913) "Method of measuring the pressure produced in the detonation of high explosives or by the impact of bullets", *Proc. Roy. Soc. Lond. A* **89**, 411–413.

Hopkinson, J. (1872) "On rupture of iron wire by a blow", *Proc. Manchester Lit. Phil. Soc.* **XI**, 40–45.

Horak, Z. (1948) "Impact of rough ball spinning round its vertical diameter onto a horizontal plane", Vyskoká škola strojního a elektrotechnického inženýrstvi při Českém vysokém učení technickém v Praze (*Trans. Fac. Mech. Elec. Engng. Tech. Univ. Prague*).

Horak, Z. and Pacakova (1961) "Theory of spinning impact of imperfectly elastic bodies", *Czech. J. Phys. B* **11**, 46–65.

Hunt, K.H. and Crossley, F.R.E. (1975) "Coefficient of restitution interpreted as damping in vibro-impact", *ASME J. Appli. Mech.* **97**, 440–445.

Hunter, S.C. (1956) "Energy absorbed by elastic waves during impact", *J. Mech. Phys. Solids* **5**, 162–171.

Ivanov, A.P. (1995) "On multiple impact", *J. Appl. Math. Mech.* **59**, 930–946 (transl. of *Prik. Mat. Mekh.* **59**, 930–946).

Ivanov, A.P. (1997a) "The problem of constrained impact," *J. Appl. Math. Mech.* **61**, 341–353 (transl. of *Prik. Mat. Mekh.* **61**, 355–368).

Ivanov, A.P. (1997b) "A Kelvin theorem and partial work of impulsive forces", *ASME J. Appl. Mech.* **64**, 438–440.

Johnson, K.L. (1968) "Experimental determination of contact stresses between plastically deformed cylinders and spheres", *Engineering Plasticity* (ed. J. Heyman and F. Leckie), Cambridge Univ. Press, 341–362.

Johnson, K.L. (1970) "The correlation of indentation experiments", *J. Mech. Phys. Solids* **18**, 115–126.

Johnson, K.L. (1983) "The bounce of 'Superball' ", *Int. J. Mech. Engng. Educ.* **111**, 57–63.

Johnson, K.L. (1985) *Contact Mechanics*, 361–366, Cambridge Univ. Press.

Johnson, S.H. and Liebermann, B.B. (1994) "An analytical model for ball barrier impact – II. Oblique impact", *Science and Golf II* (ed. A.J. Cochran and M.R. Farrally), E&FN Spon, London, 315–320.

Johnson, S.H. and Liebermann, B.B. (1996) "Normal impact models for golf balls", *The Engineering of Sport* (ed. S. Haake), Balkema, Rotterdam.

Johnson, W. (1972) *Impact Strength of Materials*, 303, Edward Arnold, London.

Johnson, W. (1976) " 'Simple' linear impact," *Int. J. Mech. Engng. Educ.* **4**, 167–181.

Kane, T. and Levinson, D.A. (1985) *Dynamics: Theory and Application*, McGraw-Hill, New York.

Keller, J.B. (1986) "Impact with friction", *ASME J. Appl. Mech.* **53**, 1–4.

Kelvin, W.T. and Tait, P.G. (1867) *Treatise on Natural Philosophy*, Vol. 1, Part 1, Clarendon, Oxford, UK.

Kerwin, J.D. (1972) "Velocity, momentum and energy transmissions in chain collisions", *Amer. J. Phys.* **40**, 1152–1157.

Ko, P.L. (1985) The significance of shear and normal force components on tube wear due to fretting and periodic impacting, *Wear* **106**, 261–281.

Kozlov, V.V. and Treshchev, D.V. (1991) *Billiards: a Genetic Introduction to the Dynamics of Systems with Impacts*, Transl. Math. Monographs, Amer. Math. Soc.

Lamba, H. and Budd, C.J. (1994) "Scaling Lyapunov exponents at non-smooth bifurcations", *Phys. Rev. E* **50**, 84–91.

Lankarani, H.M. and Nikravesh, P.E. (1990) "A contact force model with hysteresis damping for impact analysis of multibody systems", *ASME J. Mech. Design* **112**, 360–376.

Lee, E.H. (1940) "Impact of a mass striking a beam", *ASME J. Appl. Mech.* **62**, A129–A140.

Lee, Y., Hamilton, J.F. and Sullivan, J.W. (1983) "The lumped parameter method for elastic impact problems", *ASME J. Appl. Mech.* **50**, 823–827.

Leggett, A.J., Chakravarty, S., Dorsey, A.T., Fisher, M., Garg, A. and Zwerger, W. (1987) "Dynamics of the dissipative two-state system", *Rev. Mod. Phys.* **59**(1), 1–86.

Lennertz, L. (1937) "Beitrag zur Frage nach der Wirkung eines Querstosses auf einer Stab," *Ing.-Arch.* **8**, 37–45.

Lewis, A.D. and Rogers, R.J. (1988) "Experimental and numerical study of forces during oblique impact", *J. Sound and Vibration* **125**(3), 403–412.

Lewis, A.D. and Rogers, R.J. (1990) "Further numerical studies of oblique impact", *J. Sound and Vibration* **141**(3), 507–510.

Lifshitz, J.M. and Kolsky, H. (1964) "Some experiments of anelastic rebound", *J. Mech. Phys. Solids* **12**, 35–43.

Lim, C.T. (1996) "Energy losses in normal collision of 'rigid' bodies", *Proceedings of 2nd International Symposium on Impact Engineering*, Beijing, 129–136.

Lim, C.T. and Stronge, W.J. (1994) "Frictional torque and compliance in collinear elastic collisions", *Int. J. Mech. Sci.* **36**, 911–930.

Lim, C.T. and Stronge, W.J. (1998a) "Normal elastic–plastic impact in plane strain", *Math. Comp. Modelling* **28**, 323–340.

Lim, C.T. and Stronge, W.J. (1998b) "Oblique elastic–plastic impact between rough cylinders in plane strain", *Int. J. Engrg. Sci.* **37**, 97–122.

Lötstedt, P. (1981) "Coulomb friction in two-dimensional rigid body system", *Z. Angew. Math. Mech.* **61**, 605–616.

Love, A.E.H. (1906) *A Treatise on the Mathematical Theory of Elasticity*, 2nd ed. Cambridge Univ. Press.

Mac Sithigh, G.P. (1996) "Three-dimensional rigid-body impact with friction", *ASME J. Appl. Mech.* **58**, 754–758.

Marghitu, D.B. and Hurmuzlu, Y. (1996) "Nonlinear dynamics of elastic rod with frictional impact", *J. Nonlinear Dynamics* **10**, 187–201.

Maw, N., Barber, J.R. and Fawcett, J.N. (1976) "The oblique impact of elastic spheres", *Wear* **38**, 101–114.

Maw, N., Barber, J.R. and Fawcett, J.N. (1981) "The role of elastic tangential compliance in oblique impact", *ASME J. Lub. Tech.* **103**(74), 74–80.

May, R.M. (1986) "When 2 and 2 do not make 4 – nonlinear phenomena in ecology", The Croonian Lecture – 1985, *Proc. Roy. Soc. Lond. B* **228**, 241–266.

McLachlan, B.J., Beupre, G., Cox, A.B. and Gore, L. (1983) "Falling dominoes", *SIAM Rev.* **25**, 403–404.

Meyers, M.A. and Murr, L.E. (1980) *Shock Waves and High-Strain-Rate Phenomena in Metals*, Plenum, New York.

Miller, G.F. and Pursey, H. (1956) "On the partition of energy between elastic waves in a semi-infinite solid", *Proc. Roy. Soc. Lond. A* **225**, 55–69.

Mindlin, R.D. and Deresiewicz, H. (1953) "Elastic spheres in contact under varying oblique forces", *ASME J. Appl. Mech.* **75**, 327–344.

Mok, C.H. and Duffy, J. (1965) "Dynamic stress–strain relation of metals as determined from impact tests with a hard ball", *Int. J. Mech. Sci.* **7**, 355–371.

Morin, R.A. (1845) *Mécanique*, Libraire Scientifique et Industrielle de A. Leroux, Paris.

Mulhearn, T.O. (1959) "Deformation of metals by Vickers-type pyramidal indenters", *J. Mech. Phys. Solids* **7**, 85–96.

Newby, N.D. (1984) "Excitation of a composite structure by collisions", *Amer. J. Phys.* **52**, 745–748.

Nicholas, G. and Prigogine, I. (1977) *Self-organization in Nonequilibrium Systems*, Wiley, New York.

Painlevé, P. (1895) *Lecons sur le Frottement*, Gauthier-Villars, Paris.

Pereira, M.S. and Nikravesh, P. (1996) "Impact dynamics of multibody systems with frictional contact using joint coordinates and canonical equations of motion", *Nonlinear Dynamics* **9**, 53–71.

Poisson, S.D. (1811) *Traité de Mécanique,* Courier, Paris (Engl. trans. by Rev. H. H. Harte, Longman & Co., London 1817).

Regirir, S.A. (1989) "Active media with discrete sources and jumping waves", *Nonlinear Waves in Active Media* (ed. J. Engelbrecht) Springer-Verlag, Heidelberg, 176–184.

Routh, E.J. (1905) *Dynamics of a System of Rigid Bodies*, 7th ed. Part 1, McMillan, London.

Saint-Venant, B. (1867) "Memoire sur le choc lognitudinal de deux barres élastiques de grosseurs et de matières semblables ou différentes, et sur la proportion de leur force vive qui est perde pour la translation ultérieure; et généralement sur le mouvement longitudinal d'un système de deux ou plusieurs élastiques", *J. de Math.* (Liouville) **12**, 237–277.

Schwieger, H. (1970) "Central deflection of a transversely struck beam", *Exp. Mech.* **10**(4), 166–169.

Scott, A.C. (1975) "The electrophysics of a nerve fibre", *Rev. Mod. Phys.* **47**, 487–533.

Shaw, S.W. (1985) "The dynamics of a harmonically excited system having rigid amplitude constraints", *ASME J. Appl. Mech.* **52**, 453–458.

Shaw, S.W. and Holmes, P.J. (1983) "A periodically forced piecewise linear oscillator", *J. Sound and Vibration* **90**(1), 129–155.

Simon, R. (1967) "Development of a mathematical tool for evaluating golf club performance", presented at ASME Design Engineering Congress, New York.

Singh, R., Shukla, A. and Zervas, H. (1996) "Explosively generated pulse propagation through particles containing natural cracks", *Mech. Materials* **23**, 255–270.

Smith, C.E. (1991) "Predicting rebounds using rigid-body dynamics", *ASME J. Appl. Mech.* **58**, 754–758.

Sondergaard, R., Chaney, K. and Brennen, C.E. (1990) "Measurements of solid spheres bouncing off flat plates", *ASME J. Appl. Mech.* **57**, 694–699.

Spradley, J.L. (1987) "Velocity amplification in vertical collisions", *Amer. J. Phys.* **55**, 183–184.

Stoianovici, D. and Hurmuzlu, Y. (1996), "A critical study of the applicability of rigid-body collision theory", *ASME J. Appl. Mech.* **63**, 307–316.

Stronge, W.J. (1987) "The domino effect – a wave of destabilizing collisions in a periodic array", *Proc. Roy. Soc. Lond. A* **409**, 199–208.

Stronge, W.J. (1990) "Rigid body collisions with friction", *Proc. Roy. Soc. Lond. A* **431**, 169–181.

Stronge, W.J. (1991) "Friction in collisions: resolution of a paradox", *J. Appl. Phys.* **69**(2), 610–612.

Stronge, W.J. (1992) "Energy dissipated in planar collision", *ASME J. Appl. Mech.* **59**, 681–682.

Stronge, W.J. (1993) "Two-dimensional rigid-body collisions with friction–discussion", *ASME J. Appl. Mech.* **60**, 564–566.

Stronge, W.J. (1994a) "Swerve during three-dimensional impact of rough rigid bodies", *ASME J. Appl. Mech.* **61**, 605–611.

Stronge, W.J. (1994b) "Planar impact of rough compliant bodies", *Int. J. Impact Engng.* **15**, 435–450.

Stronge, W.J. (1995) "Theoretical coefficient of restitution for planar impact of rough elastoplastic bodies", *Impact, Waves and Fracture*, ASME AMD-205 (ed. R.C. Batra, A.K. Mal and G.P. MacSithigh), 351–362.

Stronge, W.J. (1999) "Mechanics of impact for compliant multi-body systems", *IUTAM Symposium on Unilateral Multibody Dynamics* (ed. C. Glockner and F. Pfeiffer), Munich, Aug. 1998.

Sundararajan, G. (1990) "The energy absorbed during the oblique impact of a hard ball against ductile target materials", *Int. J. Impact Engng.* **9**, 343–358.

Swanson, S.R. (1992) "Limits of quasi-static solutions in impact of composite structures", *Composites Engng.* **2**(4), 261–267.

Tabor, D. (1948) "A simple theory of static and dynamic hardness", *Proc. Roy. Soc. Lond. A* **192**, 247–274.

Tabor, D., 1951, *The Hardness of Metals*, 128–138, Oxford Univ. Press.

Taylor, G.I. (1946) "The testing of materials at high rates of loading", *J. Inst. Civ. Engng.* **26**, 486–518.

Thompson, J.M.T. and Stewart, H.B. (1986) *Nonlinear Dynamics and Chaos*, Wiley, Chichester, UK.

Thomson, W. and Tait, P.G. (1879) *Treatise on Natural Philosophy*, Vol. I, Part I. Cambridge Univ. Press.

Timoshenko, S. (1913) "Zur Frage der Wirkung eines Stosses auf einer Balken", *Z. Math. Phys.* **62**, 198–209.

Timoshenko, S. and Goodier, N. (1970) *Theory of Elasticity*, 3rd edition, McGraw-Hill, New York.

Tsai, Y.M. (1971) "Dynamic contact stresses produced by impact of an axisymmetrical projectile on an elastic half-space", *Int. J. Solids Structures* **7**, 543–558.

Ujihashi, S. (1994) "Measurement of dynamic characteristics of golf balls and identification of their mechanical models", *Science and Golf II* (ed. A.J. Cochran and M.R. Farrally), E&FN Spon, London.

Villaggio, P. (1996) "Rebound of an elastic sphere against a rigid wall", *ASME J. Appl. Mech.* **63**, 259–263.

Wagstaff, J.E.P. (1924) "Experiments on the duration of impacts, mainly of bars with rounded ends, in elucidation of elastic theory", *Proc. Roy. Soc. Lond. A* **21**, 544–570.

Walton, O.R. (1992) "Granular solids flow project", Rept. UCID-20297- 88-1, Lawrence Livermore National Laboratory.

Wang, Y. and Mason, M.T. (1992) "Two-dimensional rigid-body collisions with friction", *ASME J. Appl. Mech.* **59**, 635–642.

Wittenburg, J. (1977) *Dynamics of System of Rigid Bodies*, Teubner, Stuttgart.

Index

Made in the USA
Lexington, KY
12 November 2012